THE NEW NATURALIST

A SURVEY OF BRITISH NATURAL HISTORY

CLIMATE AND THE BRITISH SCENE

The aim of this series is to interest the general reader in the wild life of Britain by recapturing the inquiring spirit of the old naturalists. The Editors believe that the natural pride of the British public in the native fauna and flora, to which must be added concern for their conservation, is best fostered by maintaining a high standard of accuracy combined with clarity of exposition in presenting the results of modern scientific research. The plants and animals are described in relation to their homes and habitats and are portrayed in the full beauty of their natural colours, by the latest methods of colour photography and reproduction.

THE NEW NATURALIST

CLIMATE AND THE BRITISH SCENE

by

GORDON MANLEY

M.A. (Cantab.), D.Sc. (Manc.)

PROFESSOR OF GEOGRAPHY IN THE UNIVERSITY OF LONDON AT BEDFORD COLLEGE

WITH 41 COLOUR PHOTOGRAPHS
BY CYRIL NEWBERRY AND OTHERS
40 PHOTOGRAPHS IN BLACK AND WHITE
AND 75 MAPS AND DIAGRAMS

COLLINS
14 ST JAMES'S PLACE, LONDON

First impression 1952
Second impression 1953
Third impression 1955
Fourth impression 1962
Fifth impression 1971

Printed in Great Britain
Collins Clear-Type Press: London and Glasgow

"Our Trimmer is far from Idolatry in other things, in one thing only he cometh near it, his Country is in some degree his Idol; he doth not worship the Sun, because it is not peculiar to us, it rambles round the world, and is less kind to us than to others; but for the Earth of England, though perhaps inferior to that of many places abroad, to him there is Divinity in it, and he would rather die than see a spire of English Grass trampled down by a Foreign Trespasser."

GEORGE SAVILLE, MARQUESS OF HALIFAX in *The Character of a Trimmer*, from *Miscellanies*, London, 1704

"The southern artist has nothing to teach him of colour, everything to remind him of form . . . . . for us a single landscape varies every minute of the day in drifting colour; to the dwellers in the lands of fierce sunlight it is colourless, and its form is as immutable as the rocks of which it is made."

S. C. KAINES SMITH, in *Looking at Pictures*, Methuen, 1921 quoted by Cicely M. Botley in the Quarterly Journal Roy. Met. S., 1931

CONTENTS

COLOUR PLATES

FACING PAGE

COLOUR PLATES

It should be noted that throughout this book Plate numbers in arabic figures refer to Colour Plates, while roman numerals are used for Black-and-White Plates.

PLATES IN BLACK AND WHITE

EDITORS' PREFACE

If a census were taken of common topics of conversation amongst British people it is very probable that the weather would take first place. So much do its vagaries enter into everyday life of the people that the weekend activities of millions may be directly determined by the conditions of the moment. Twice daily at least, through the British Broadcasting Corporation, the Meteorological Office informs our fifty million people of their probable weather fate for the next 24 hours. Whilst the emphasis would thus seem to be on the variability of British weather and it has facetiously been said that Britain has no climate, only a succession of weather, in actual fact British climate is the most dependable in the world. November fogs, March winds, April showers may vary in intensity but those tremendous variations from year to year or fierce climatic features—the prolonged droughts, the delays in the 'coming of the rains', the destructive tornadoes, the violent cloudbursts, the destructive hailstorms, the abnormal snowfalls so common in other parts of the world—are scarcely to be understood by those accustomed to the even rhythm of British climate.

Realizing the fundamental influence of climate on every aspect of the natural history, ecology and scenery of Britain the Editors of the *New Naturalist* early planned for a volume in the Series which would deal with the almost untrodden no-man's-land between the field of the professional meteorologist and climatologist on the one hand and the naturalist on the other. Just as it is a matter of extreme difficulty to catch the fleeting moods of the weather in the camera, so is it difficult to convey the subtleties of the influence of those moods on the life of plants, animals and man. We believe Professor Gordon Manley has succeeded in his difficult task as no one else could have done. A Past President of the Royal Meteorological Society as well as Buchan Prizeman and Symons Lecturer there is no doubt as to his standing as a meteorologist. As Professor of Geography in the University of London at Bedford College he is concerned with the application of these specialist studies in wider fields just as he saw their application in other spheres as a wartime officer of the Cambridge University Air Squadron. Yet we can also picture him in his little observation hut on the highest Pennine ridge of Crossfell, personally experiencing the worst of the weather conditions of which he writes, just as we can see him deep in his study chair seeking with glee from long forgotten volumes those delightful references to climate which enrich this book. We are confident that this volume will give as much pleasure to many thousands of readers as it has to us.

THE EDITORS

AUTHOR'S PREFACE

FOR MOST OF US our knowledge of the climate of the island we inhabit derives from our experience. Our senses register its effects not merely directly, but also through the appearance, sound and even smell of the world around. For some of us knowledge gained through the senses is reinforced by instrumental observations. It is the purpose of this book to try and bring together the results of qualitative apprehension and quantitative recording, and to provide for those who wish to study any aspect of the natural history of these islands an introduction to its climatology, laying stress wherever one can upon the myriad ways in which the British scene in all its diversity is affected by the vicissitudes, regular and irregular, past and present, of our atmospheric environment.

This is not a meteorological text. Little or nothing will be said of the upper air. I have tried to confine attention to those phenomena which can be observed directly from the ground by any keen walker, or the many other travellers about this island whose lively interest takes them one day into the ripening wheatfield; next day while it rains they see the local art gallery or sample the local cheese; a month later their impressions may be totally different when they return from work on a blustery autumn evening. Hence the properties of the surface air masses and the characteristic results of their presence at the several seasons are reviewed. I have also tried to provide an abundance of references for further reading by those who wish to specialise, especially to the many papers in the *Quarterly Journal of the Royal Meteorological Society* which has recently celebrated its centenary as the senior society of its kind in the world. To that journal have contributed a great number of Fellows, amateur and professional alike, whose desire to make some contribution to the elucidation of the fascinating complexities of the British climate ultimately derives from their upbringing in an island of such delightful variety. It is indeed to be hoped that a lively body of meteorologists will continue to be one of the by-products of British weather and that at least a few of those who read this book will thereby be encouraged to embark on more serious studies. Physicists, biologists, geographers, geologists, mathematicians and many others can all find in the various aspects of meteorology a field of study of the utmost interest, breadth and complexity to which they may bring their particular training.

I have deliberately spread a net over many byways of climatological discussion. For those who wish to embark on the straightforward assessment of climatic data there are many better sources, some of which I have named; for those wishing to study underlying principles the fundamentals of meteorology are again presented in appropriate texts. Climate is however apprehended as a whole, and through several senses. Let the reader therefore try to recall not merely the meteorological situation, but all the feelings and associations of the landscapes at various seasons of which an illustration has been attempted.

For the use of many diagrams I have to thank the Council of the Royal Meteorological Society, also the authors of the papers concerned, notably Mr. E. Gold, Mr. C. K. M. Douglas, Dr. J. Glasspoole and others cited in the text, who have allowed me the privilege of reproduction from their publications; and the Controller of H.M. Stationery Office for permission to make use of officially-published daily and monthly weather reports and other meteorological data for purposes of illustration and in tables. Figures 3, 28, 29, 39, 40, 48, 49, 55, 65, 66 and 71 are based on material, or reproduced from the Society's "Quarterly Journal"; Figs. 14, 27, 72 from "Weather"; the charts and diagrams in Figs. 4, 5, 16 to 26 inclusive, 32, 33, 35 to 38, 41, 42, 46, 47, 58 to 61, 63 and 67 are based on H.M.S.O. publications. I have to thank one of the Honorary Secretaries of the Society, Professor P. A. Sheppard, of the Department of Meteorology at the Imperial College of Science, for his generous help and advice in regard to the first seven chapters. To Sir David Brunt and the Council of the Physical Society I am grateful for permission to include Fig. 74. I owe to Mr. Cyril Newberry a debt of gratitude for his indefatigable efforts to bring off one of the most difficult problems ever given to a colour photographer—that of catching the fugitive yet very real meteorological situations as they appear to the eye in typical landscapes. I am further indebted to Sir Nelson K. Johnson, Director of the Meteorological

Office, for certain anemograms, notably Fig. 50; to Mr. Paton of the University of Edinburgh for the reproduction of one of his remarkable auroral photographs; to Dr. Balchin of Kings College in the University of London for the illustrations (Plate III) of inversion fog in Somerset; to Dr. Julian Huxley for photographs of showery weather (Plate 20) in the same county; to Mr. E. L. Hawke for Plate XXVIb and for permission to include diagrams from his papers; to Dr. C. E. P. Brooks for Fig. 27; to Mr. D. S. Hancock for sunshine data at Bognor Regis; to Mr. B. R. Goodfellow for Plate XIIa; to Dr. P. R. Crowe for Fig. 71; to Mr. J. Wadsworth for Fig. 72. I have also to thank the Controller of H.M. Stationery Office for permission to use the snowfall and snow-cover maps (pp. 202, 204) from the Meteorological Magazine and Mr. E. F. Baxter and the University of Durham for permission to use published data from Durham Observatory. I have to thank Messrs. Constable for permission to quote from *A Reading of Earth* by George Meredith; Messrs. Chatto and Windus for permission to quote from R. L. Stevenson's *Poems*; Mr. Hilaire Belloc and his publishers, Messrs. Duckworth, for permission to quote from *The South Country* on p. 140; and to Dr. S. Petterssen and his publishers, Messrs. McGraw-Hill, for permission to base Figs. 6 and 11 on two diagrams from his *Introduction to Meteorology*.

Other acknowledgments and references have been made wherever possible in the text and bibliography. If I have unwittingly made references to the findings of recent writers without adequate acknowledgment I have in all probability done so on account of the many fruitful conversations and discussions which I have been privileged to join at the Royal Meteorological Society; and I should like again to commend the readers of this book to the splendid series of papers in its *Quarterly Journal* and to the articles in its more recent monthly, *Weather*, not forgetting the abundant aspects of climatological information comprised in the many publications of our Meteorological Office in which for a short time I had the honour to serve. Tables and data have been largely derived from the published data in the Monthly Weather Report; in this I have been greatly helped by Mr. D. S. Brock of Westminster School, a former member of the Department of Geography at Cambridge, not only for the labour of extraction and compilation but also for the drawing of many of the diagrams in the text, notably Figs. 34, 52, 69 and 70, and the necessary abridgment of many weather maps for small scale reproduction. I have again to thank Sir N. K. Johnson, Director of the Meteorological Office, Dr. J. Glasspoole of the Climatological Division and the many members of the staff who have helped me for the opportunities so generously provided; not least for the use of tabulations already made, in the task of bringing such things as rainfall averages up-to-date. In compiling such tables I have been at pains to choose a number of localities of representative interest to supplement those for which material is already available in the well-known standard text (E. G. Bilham, The Climate of the British Isles, London, Macmillan, 1938). Lastly no one who looks into the origins of our climatological information can fail to acknowledge an intimate debt to the long line of patient observers who have maintained a great tradition since the days of Robert Hooke; a part indeed of that observation of nature whose deep appeal we can ever share.

I have to thank Miss Alison Birch for much assistance with the reproduction of drawings and maps. I am grateful to Miss C. M. Botley for drawing my attention to the quotation from Mr. S. C. Kaines Smith's *Looking at Pictures*, 1921; and to Messrs. Methuen, the publishers, for allowing me to quote it here. Lastly, I have to thank Mrs. Audrey Hitchcock, a former student of the Department of Geography at Bedford College, for completing the index as well as for assistance in bringing the tabulations in the Appendix down to 1949. Tabulations are throughout based on Meteorological Office published data, wherever available; the reductions, calculations and averaging I have largely performed myself. References are in greater profusion than is usual in a book of this type; but the profusion of by-ways in meteorology is now so great that I hope it will thereby be found more useful.

GORDON MANLEY

NOTE TO THE FOURTH IMPRESSION

In this, the fourth impression, the opportunity has been taken to make minor additions, and to incorporate a number of corrections to which several correspondents have been kind enough to draw my attention; I am glad to be able to thank them. Certain tables and diagrams have been brought up to date (1961); I have to thank Miss Elizabeth M. Shaw for assistance with Fig. 65. With regard to the tables in the appendix it should be added that the Meteorological Office has now published averages for 1921-50.

G. M.

CHAPTER I

INTRODUCTION

Calm was the day, and through the trembling air
Sweet-breathing Zephyrus did softly play a gentle spirit
SPENSER

CLIMATE may be defined as an expression of our integrated experiences of "weather". As a word it comes to us from the Greek, from long before the days of instruments, so we may well look into the classical background. In the countries of the Mediterranean whence we derive many of the concepts implicit in our speech the overwhelming power of the summer sun has always impressed mankind. The short, sharp shadows of July, the white glare, the very rapid evaporation and the scorched remnants of grass are today a vivid recollection in the minds of many travellers. Throughout the Mediterranean coastlands there is moreover a very high proportion of sunny hours in the summer months. It is no wonder, therefore, that the Greek philosophers considered that climate was primarily dependent on the altitude of the sun, that is on latitude. Greek geographers were accustomed to recognise seven "climates" between the Equator and the Poles.

Around the Mediterranean and throughout much of South-Central Europe the seasonal rhythm is simple and relatively well-defined. The cloud, wind and intermittent rain of winter along the Mediterranean coasts stand in sharp contrast with the drought of summer. Further inland to the northward, there is a broad region where a definite season can be expected during which frost occurs, and there is a risk of snow; while through Burgundy and South-Central Europe the lowland summer is always warm enough for the ripening of crops. Occasional catastrophic thunderstorms and floods may occur causing local devastation but even the coolest and most cloudy summer is not calamitous over a wide area in the sense that is found towards the north-west margins of Europe.

Gradually as men move towards the north-west coastlands of Europe the wind plays an increasing part in their consciousness. For many months it can be said that the sun gives light, and at times an agreeable warmth. But it is the quality or "feel" of the air that we in the British Isles perceive first as we go out of doors. Ultimately this arises from the fact that the air arrives from a variety of different sources, and undergoes a varying degree of modification on its way towards us. The climate of the British Isles and the consequent appearance of the landscape owes much to latitude, notably in respect of the great seasonal variation in the amount and intensity of light; but still more is due to our position and maritime surroundings, especially with regard to wind.

Not that the Greeks were unmindful of the effects of wind. Winter in the Mediterranean is a season during which, as depressions pass, markedly different types of air pervade those coastal chaffering-places beloved by Mediterranean man. Boreas was the north wind, coming off the snowy continent of Europe to give cold clear air with good visibility and a nearly cloudless sky over the country round Athens; Notos, from the south, was recognised as warm, humid and enervating by comparison. In Rome blustery tramontana and languid sirocco play the like part. But the Atlantic shores of Europe offer the same contrasts with greater boisterousness throughout the year. It is a northern composer—Sibelius of Finland—who has given expression in music to the tremendous majesty, persistence and interminable energy of the northern winter storms. The northern mythology of the early Scandinavians characterised the sun as feminine. We can perceive the hint in the actions of those descendants who through the centuries have thrust, crept or clawed their way into the maternal bosom of Southern Europe, by contrast with those who faced the open Atlantic and even ventured across it.

The climate of the British Isles is such that the inhabitants enjoy, but are not subordinate to, the power of the sun. It has accordingly been stigmatised by Latin Europe. Tacitus left a renowned note on

---

PLATE I

NEWMARKET HEATH, Suffolk; in spring sunshine, May. North Sea, strato-cumulus in moderate N.E. wind, temperature 55°.

PLATE I

Cyril Newberry

PLATE 2

B. A. Crouch

the subject: "The climate is objectionable with frequent rains and mists, but there is no extreme cold". Dumas gave vent to the views of the romantic Sturm-and-Drang period, "L'Angleterre est un pays ou le soleil rassemble à la lune". By contrast, more discriminating observers have often found room for praise; and Englishmen themselves, especially those who have not dwelt for long elsewhere, or have resisted the seductions of lands nearer the equator, are evidently very proud of their climate. As a Venetian Ambassador said in 1497, "The English are great lovers of themselves and of everything belonging to them". Of Charles II, it is recorded that he never said a foolish thing and never did a wise one; a view which would not be entirely upheld by the Royal Society, formed with his encouragement in 1663. Charles was a keen observer; and his saying that "The English climate is the best in the world. A man can enjoy outdoor exercise on all but five days in a year" has been echoed by most energetic Englishmen.

In 1944 on a delightful May morning in Suffolk, an American enlisted man said, "I like this weather of yours. You can work all the year round without sweating". From America, too, the late Professor Ellsworth Huntington acknowledged the advantages of the climate of south-east England, no light tribute from such an energetic scion of Yale.

Appreciation of the British climate depends largely on temperament. That it has not been conducive to idleness has been reflected in the characteristics of the people; be it remembered that the urge to go and do something useful, to keep moving, to use one's intelligence, to protest against indolence, to stir up controversy and to censure the offcomer is also increasingly marked to the northward. Unreasonable activity and exertion are, however, gently damped down—the Englishman's own expression. Undue assertiveness in colour, music, architecture, opinion or sentiment is out of keeping; it is "not done"; gentle gradations of colour and fluctuations of mood are associated with the lack of the sharp shadows, the harder lines and fiercer contrasts of more southern lands. Some, be it noted would say "cruder contrasts"; others derive from the sharpness of contrast a stimulus to activity. Accordingly opinion as to the goodness or badness of the

---

PLATE 2
CRUMMOCK WATER, Cumberland: breezy afternoon, June. Fair weather cumulus at about 4,000 feet with fresh N.E. wind, temperature 60°.

British climate is likely to vary immensely from time to time between different leaders of opinion in this and other countries. Judgment can only be given by results, particularly in the realm of stockbreeding according to Mary Borden; and we may extend the argument to the results in the realm of originality of mind and accomplishment among the inhabitants for at least fifteen centuries past. To what degree this is the result of racial mixture, natural selection, or environmental influence it is not the purpose of this book to explore. We shall instead embark upon the discussion of climate, having especial regard to its significance as a factor in the moulding not merely of landscape but of the many other elements that go to make up the British scene.

Shakespeare's intuitive appreciation of the qualities of his ideal island led him to put into the mouth of Caliban—the unlettered but sentient native—

Be not afraid, the isle is full of noises,
Sounds, and sweet airs, that give delight and hurt not.

Yet there are but few days when this cannot equally be said of that English countryside which Shakespeare knew. That our airs give delight and hurt not goes far to explain the attractions as well as some disadvantages of this country of ours; a point of view to be developed in later chapters.

The many-faceted British scene indeed has the fascination of a well-cut diamond by contrast with the crude regularity of a simple crystal. Like a diamond it owes much to the artifice of man; the eighteenth-century landscape gardeners are responsible for the characteristics of much of Southern and Central England. Eighteenth-century landowners were likewise active in Scotland as the varied charms of Roxburgh and the Lothians still show. But the glitter of the facets owes much to the light; in regard to our varied British landscape this is primarily a result of our climate. We shall try to embark on a discussion of British weather particularly as its complex effects are perceived by our several senses when we go out-of-doors. Even indoors; our satisfaction with what many Americans are apt to cherish as the ruefully enjoyable ineffectiveness of our ancestral domestic heating arrangements also owes much to the curious qualities of our outdoor climate. In humid air approaching saturation when there is little or no wind vigorous exertion quickly leads to discomfort through overheating, although with the same amount of clothing it is too cold to sit still. Our normal winter indoor clothing allows for this compromise at somewhere between 50° and 55° in damp weather. With a bright source of radiant heat such as a fire, we sit and work quite comfortably in rooms between 55° and 60°, a temperature at which the gentle evaporation from the skin in the humid air does not add to that sense of dryness to which our skins are unaccustomed. Fires are generally dispensed with, in the daytime, as soon as the noon temperature begins to exceed 60°, that is from early May to early October in Southern England. But it is significant that Continental ideas of heating are beginning to spread in our greatest mart for fashion, propaganda and advertisement, and the urbanised monotony of the great shops and hotels of London might be exchanged without comment with that of like institutions in many great cities abroad. Is it any wonder that our city populace is showing signs of forgetting that the fundamental needs of life—food, water, shelter and transport—are still subject to one of the most erratic climates in the world?

Weather Maps

For those unfamiliar with the notation used in weather maps, of which a number appear in subsequent chapters, a short key will serve; Fig. 4 on p. 31 may be referred to. The winding lines are isobars; by joining places with equal barometer readings (after reduction to sea level) they illustrate the distribution and shape of regions of low and high atmospheric pressure at any given time. Pressures are in millibars (1000 mb. can be taken as 29.53 "inches of mercury"); mean annual pressure at Kew is 1015 mb. Figures beside stations are temperature in degrees F. Wind direction is shown by the shaft of the arrow, force by the number of flèches. One short and two long flèches means force 5 on the Beaufort scale, i.e. between 17-24 m.p.h. or, in sailors' language, a fresh breeze. Four long flèches means force 8 (38-45 m.p.h.) and is described as a gale. Lack of an arrow means a calm. State of the sky and prevailing weather are indicated by shading of the circle or by a symbol at the head of the arrow. Thus ◯ = blue sky, not more than one quarter clouded; ⦶ ⦶ ⦶ increasing proportions of cloud, ⦶ overcast; ● rain, ✱ snow, ✱ sleet; ≡ fog, ∞ haze. Fronts, that is lines on either side of which there is a marked difference in the qualities of the air, are shown: ⌓⌓⌓ a warm front, at which warmer air is advancing and rising over cooler air ahead; the symbol indicates direction of advance of the system.

▲▲▲ a cold front, i.e. cold air advancing and undercutting warmer air ahead; symbols again show the direction in which the front is advancing.

▲⌓▲⌓ an occlusion, across which there is in general little difference in temperature at the ground; the difference lies above and at some altitude warm air is being elevated, giving rise to extensive cloud and often rain or other precipitation.

CHAPTER 2

THE MAKERS OF THE OBSERVATIONS

Look how the gusty sea of mist is breaking
In crimson foam, even at our feet! it rises
As Ocean at the enchantment of the moon
Round foodless men wrecked on some oozy isle.
SHELLEY: *Prometheus Unbound*

APPRECIATION of the climate of the British Isles was perforce almost entirely qualitative until with the invention and diffusion of measuring instruments the accompanying spirit of scientific inquiry became widespread in the seventeenth century. For centuries before that, however, we can be sure that the sailors along the North Sea coasts had a very shrewd idea of the behaviour of the air, and that the farmers inland acquired a fund of sound knowledge regarding the merits of their fields under varying weather conditions. The sites of Anglian, Saxon and Danish villages, of Celtic croft and Norse farmstead betoken a keen eye for climatic advantage. Such knowledge, however, played little part in the minds of the literate; rarely there were signs of grace, notably in the Oxford clergyman who from 1337-44 kept a daily record of the direction of wind and the occurrence of rain. From Elizabeth's day onward a "wind book" appears to have been kept at the Admiralty; and the Elizabethan predecessors of to-day's geographers endeavoured to give some hint of the climatic features of the counties of England. Of Durham Speed (1610) wrote, "The air is subtle and piercing, and would be more, were it not that the vapours of the North Sea do much to dissolve her ice and snow." In such a way the Elizabethans endeavoured to explain the fact that the mean annual frequency of mornings with snow-cover is but eight at Tynemouth, eighteen at Durham, and eighty in those parts of upper Teesdale lying on what they knew as "Fiends Fell"—that bleak

Crossfell region which above 2,000 feet has been forsaken of man's habitations since at least 500 B.C. and perhaps since time began.

Instrumental recording began here and there soon after the invention of the thermometer and barometer. Early use of both instruments was made by Robert Hooke; his MS. for 1664 is preserved at the Royal Society. John Locke kept instrumental records as early as 1667. Instruments were, however, very imperfect, and the necessity of making adjustments and corrections was not fully understood, so that we can as yet make little use of our observations of temperature and pressure until after 1750.

Rainfall was first recorded over a considerable period in England by a most interesting character in a most appropriate place. A Lancashire squire, Richard Towneley, living at Towneley Hall near Burnley on the flanks of the Pennines, designed his own rain-gauge with much ingenuity and kept a record from 1677 to 1704. Let it not be forgotten that Towneley in this respect was a pioneer; far away from opportunities of discussion with the men of the nascent Royal Society, he showed that spirit of inquiry and desire to keep an accurate record which has appeared in one form or another in thousands of his countrymen. The same spirit of independent inquiry combined with a desire for accurate recording and transmission of knowledge flowered earlier in Wycliffe and Coverdale, and, some would add in Bede; it appears in Christopher Saxton, James Cook and William Scoresby; in John Dalton, Adam Sedgwick and John Phillips. The results of observation were intelligently applied by George Stephenson, Edward Pease, Joseph Whitworth, to whom many descended from the dales can be added today. Here we have named but a few of the men who were bred of the stocks which, like the sheep and cattle, have flourished for so long along the flanks of the windy northern moorlands; such men have also been part of the British scene. Elsewhere too our hilly districts appear to have been the original home of much of the observational science on which our modern technical advances rest: this seems to be as true of meteorology as of other sciences. Tyndale, Norden, "Strata Smith" and Huxley derive from Somerset. East Anglia, the home of so many early naturalists and landscape painters, has also provided its quota of early observers of weather; one of our earliest instrumental records (1673–4) comes from Wrentham near Lowestoft. From Norfolk the Marsham phenological records began in 1736.

Nevertheless even a moderate degree of instrumental accuracy was not attained in a day, and still there is room for improvement. The eighteenth century saw considerable progress in the realm of meteorology; scattered here and there about England and Scotland many men began to use the slowly improving types of instruments and to add daily readings to their "weather diaries" already in fashion. Thomas Barker, the brother-in-law of Gilbert White the naturalist, kept meteorological records at Lyndon in Rutland from 1736 to 1798. John Huxham, a physician at Plymouth, was another early observer of note; indeed during the 18th century the contribution of the medical men to our climatological knowledge was greater than that of any other group.

With his northern contemporary the Reverend Thomas Robinson of Ousby beneath Crossfell, William Derham, whose rainfall record at Upminster in Essex covers the years 1697 to 1716, was one of the first clergymen to keep instrumental records, and thus to set a fashion for many of his colleagues in later days. Derham also observed an early type of thermometer, but under conditions which are not easy to interpret. Robinson's readings have not come down to us; but he left us the first account of the helm wind (1696).

An unknown Edinburgh medical man kept the first Scottish temperature observations (1731–36) that have been reduced to modern standards. To satisfy the curiosity of the reader before going farther, it may be added that none of those early rainfall, temperature, or wind observations indicates a range of climatic variation greater than might be expected at the present day; and, of course, this is borne out by the seventeenth-century botanists who recorded the date of flowering of various plants, of leafing of trees, and the date of harvest.

Slowly it became possible to put these observations together: a considerable assemblage of rainfall data was made by John Dalton in 1799. Dalton rivals Barker as one of the most determined meteorological observers who ever lived; his daily records ran from 1787 to his death in 1844. He began recording rainfall at Kendal in 1787, with the result that Kendal soon took rank as the wettest place known. However, it was not until 1840 that a sufficient number of rainfall records became available to enable Joseph Atkinson, another amateur meteorologist from Carlisle, to compile the world's first tentative rainfall map.

The maintenance and compilation of meteorological observations demands considerable assiduity, as men were quick to note. Ralph Thoresby of Leeds visited Towneley in 1697 and saw the rain gauge; in his journal he notes that he thought much of keeping a similar record, but decided that he was not sufficiently patient. Dr. Short of Sheffield in 1750 was more downright; "for it being a dry subject, most gentlemen are soon weary of it," he wrote. In other countries support for meteorological recording was not uncommonly forthcoming long before 1800 at State observatories and the like. In Britain, almost all our earlier records were kept by independent amateurs. In this respect meteorology resembles all the other observational sciences. Not the least of the attributes of this island of ours is that if the normally cultivated countryside is left to itself, climatic factors lead to the establishment of a wild vegetation of great variety, on which order can only be imposed by prolonged co-operative effort. Without doubt the analogy holds with regard to our scientific accomplishment as a people.

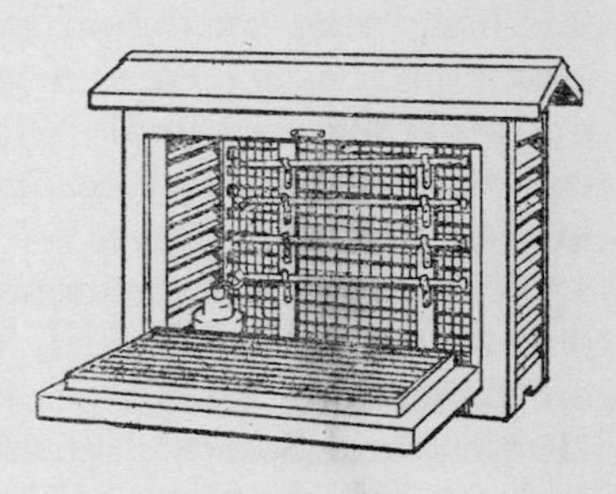

Fig. 1
Bilham's modification of the Stevenson screen

Co-operative effort was first essayed through the establishment of our many scientific societies. Among these the Royal Meteorological Society, founded in 1850, is the senior body of its kind in the world. It belongs to that Chartist decade in which the Chemical Society and the Rochdale Pioneers began their operations; a decade in which public health, water supplies, sanitation and engineering were also demanding co-operative effort, while at the same time leading many men to keep more precise records of weather. The Repeal of the Corn Laws followed the miserable summer of 1845.

Before the days of official organisations, temperature, pressure and other elements were observed in various ways For a long time, shade temperatures were thought to be sufficiently comparable if taken on a north-facing wall. Indeed, during the eighteenth century there was a long period during which many men kept their records by reading a thermometer *inside* a room without a fire. Such records are of course of little value unless very carefully used. This curious fashion probably arose among some of the doctors. In 1723, Jurin, then secretary of the Royal Society, recommended this method, perhaps because it suited his own arrangements in London, but more probably because as a medical man he knew that many men in cities pass most of their time indoors. Later in the century, however, the growing interest in gardening undoubtedly led to the keeping of better temperature records, here and there about the country. James Six, a gardener, invented the familiar "Six's maximum and minimum" in 1782. More reliable minimum thermometers began to come into use about 1810, but were not generally used until the forties. Before that date most observers recorded "fixed hour" readings twice or thrice daily. Gilbert White himself took observations of temperature, pressure and rainfall from 1768 to 1793; and scattered through the country were not only the doctors but men such as Holt of Liverpool with his keen observations on Lancashire agriculture (1794). Later we find that the Royal Horticultural Society's record in London was begun in 1825. Many early records were kept by the London instrument makers and we may conclude that they were not forgetful of the advertisement thus given. Among the best of these is John Cary's record at his shop in the Strand (1786–1846). The present-day record kept by Negretti and Zambra in Regent Street continues the honourable tradition begun by men such as Ayscough of Ludgate Hill in the 1750's; other examples include Casartelli of Manchester and Pain of Cambridge.

A long time elapsed before it was fully realised that the air temperature is not necessarily given by a thermometer on a north wall. In such circumstances, a thermometer, for example, on a warm day in March may very well be reading much too low, as it is recording the temperature of the wall rather than that of the air. Hutchinson, the remarkably lively Liverpool harbour master, tried to solve this problem by keeping his thermometer (1777–93) under a table on his roof.

In the nineteenth century increasing attention was given to these problems and from 1823 we learn of small groups founding "Meteorological Societies" for their discussion. From 1847 onward official encouragement was given to the collection of records kept under standard conditions. These were summarised by Glaisher, following

---

PLATE 3
LOCH ACHTRIOCHTAN, Glencoe, Argyll: September morning. Low stratus and fracto-stratus among mountains after drizzle, temperature 55°.

PLATE 3

B. A. Crouch

Cyril Newberry

Julian Huxley

his appointment (1840) as Superintendent of Meteorological Observations at Greenwich, in the Registrar-General's Quarterly Returns. After 1866 the Stevenson screen gradually came into use, and slowly began to supersede older methods of exposure; it was invented by Thomas Stevenson, the Scottish lighthouse engineer, who was also the father of Robert Louis Stevenson. Much encouragement was given to its use by the Royal Meteorological Society, already founded in 1850, which organised in the seventies a series of stations all over the country using this unexceptionable screen. The Stevenson screen is, of course, simply a box with ventilated sides in the form of louvres, and a ventilated floor and top; above the top is an air-space and a second roof. Such an arrangement ensures that the thermometers inside the box are recording the temperature of the air. The purpose of the top and sides is to prevent direct radiation from the sun reaching the thermometers; the slats forming the floor are a protection against radiation from the ground and surrounding objects, which may be very considerable, especially in hot dry weather. Anyone can perceive this who walks near a south-facing brick wall about sunset after a fine summer day.

Official recognition of the value of a network of standardised meteorological observations evolved, as we have seen, somewhat tardily in Britain, although advocates were not wanting. Even in the seventeenth century Hooke recognised that comparable data from a network of stations would be valuable. Another fillip to the exchange of records was given by Jurin. From 1773 onward observations of some kind were maintained at the King's private observatory at Kew, though not very regularly; and in 1774 the Royal Society initiated regular daily readings in London. Oddly enough, although Greenwich is the oldest surviving observatory in the world (1697) standard meteorological observations were only begun under Glaisher in 1841. With his encouragement the world's first synoptic weather maps were compiled with the aid of the newly-invented electric telegraph and for some time were shown daily at the Great Exhibition in 1851. The stage

---

PLATE 4

a. EAST MIDLANDS: warm afternoon verging on unsettled, July. Well-developed later afternoon cumulus. Light westerly wind, afternoon maximum 74°.

b. FOREST OF DEAN, Gloucestershire: August afternoon. Moderate S.W. winds, rather humid; 68°, heavy strato-cumulus and cumulus.

was set for an official organisation on a larger scale; and the Meteorological Office was founded as a department of the Board of Trade in 1854. For some years however, the issue of forecasts was frowned upon, while other countries went ahead. Daily synoptic charts and the issue of forecasts were finally adopted in the seventies, together with the collection of strictly comparable climatological data.

The decades 1840–1890 were the great days of the amateur. Around the Royal Meteorological Society there gathered a body of active observers. Among them were many clergymen; one of their best representatives was the Reverend Leonard Jenyns, vicar of Swaffham Bulbeck near Cambridge from 1823–53, who published his *Observations in Meteorology* in 1858; and a very pleasant work it is, not without profit to anyone who chooses to read it to-day as an epitome of what can be done by a keen amateur observer. As early as 1863 we can find the contributor of a paper to the *Quarterly Journal* surmising that the conflict of 'equatorial' and 'polar' streams of air was a necessary concomitant of what we now call 'depressions'. That this amateur tradition has continued to thrive is shown by the long series of observations kept by such men as Thomas Backhouse of Sunderland (from 1857–1915), John Hunter of Belper (from 1877–1931) and John Dover of Totland Bay (from 1886–1948), one of whose diagrams is reproduced on page 90. Among other observers the phenomenal energy of Clement Wragge who for many months climbed Ben Nevis daily (1880–81) can fitly be recalled. More recently the extensive kite-flying experiments of Mr. C. J. P. Cave of Stonor Hill did much to add to our knowledge of the behaviour of the atmosphere before aircraft came into use. One of Mr. Cave's superb cloud and landscape photographs will be found as Pl. Va, opposite p. 114. It is indeed a pleasure to be able to acknowledge permission to include the work of the late doyen of the amateur meteorologists of this country, some of whose photographs have been published in book form by the Cambridge University Press.

Our network of rainfall recording stations received a great fillip through the establishment, by George Symons in 1860, of the British Rainfall Organisation. Gradually the number of observers, largely voluntary, rose to over 5,000. In 1901 the direction was taken over by Dr. Hugh Robert Mill, one of the most eminent of British geographers and meteorologists; under his administration it became possible to map the distribution and rainfall with remarkable accuracy for almost any part of Great Britain. No better testimony to the great "amateur tradition" of British scientific inquiry can be found than this noticeably successful enterprise for the collection and dissemination of rainfall statistics. These, of course, have proved invaluable for numerous practical purposes, such as the design of reservoirs. Since Dr. Mill's retirement in 1919 the administration has been conducted from the Meteorological Office; and the carefully-compiled annual volumes of *British Rainfall* under Dr. Glasspoole's editorship are well known as a source of detailed information.

The story of the early investigations of rainfall is interesting and mention should be made of the efforts of the pioneer amateurs to record rainfall in more remote places. Especially is this true of our mountain districts. Nowadays it is common knowledge that places can be found here and there among our highest mountains with exceptionally heavy rainfalls. But in the eighteenth century the greater rainfall to be expected in hilly districts was scarcely even surmised.

In 1769, Heberden, a London physician and a member of a family of which several generations took an active interest in meteorology, showed that if a raingauge were placed at some height above the ground, *e.g.* on top of a tower, less rain was normally caught than by a gauge at ground level. For this purpose he used the towers of Westminster Abbey. This work drew the attention of John Gough, the famous blind "natural philosopher" of Kendal, who also encouraged the young John Dalton to make instrumental observations. Between 1787 and 1790 Gough set up a number of gauges at places near Kendal and it so happened that those placed on higher ground recorded in general less rain. There is little doubt that the reason for this lay in the imperfections of exposure. If a raingauge is placed in a very windy situation it is invariably found that less rain is caught by comparison with a more sheltered position nearby. This is due to the setting up of eddies round the gauge itself in windy weather, which have the effect of carrying a proportion of the falling raindrops outside the perimeter of the gauge. Obviously the same factors operate when the gauge is exposed on a high roof or tower, unless the roof is exceptionally broad and flat.

But many years passed during which the opinion prevailed that there should be less rain at a higher level. Theoretical reasons for this were adduced. Hutton, the Scottish geologist, had laid down a theory of rainfall in 1784, according to which rain resulted from the mixing of air currents of different temperatures. He based this argument on the known fact, recently determined by the eighteenth century physicists, that the capacity of air for water vapour increases more rapidly than the temperature. To illustrate: Saturated air at 48°F. contains 8·7 grams of water vapour per cubic metre. At 40° the weight will be 6·5 grams; at 56° 11·5. If, therefore, we mix one cubic metre of saturated air at the lower temperature (40°) with one at the higher temperature (56°), we shall have two cubic metres at the temperature intermediate (48°); but we shall also have rather more water vapour (18 grams) than the two cubic metres of air can themselves sustain (17·4 grams) at that intermediate temperature. Hence some of the moisture must be condensed whenever two saturated masses of air differing in temperature are mixed.

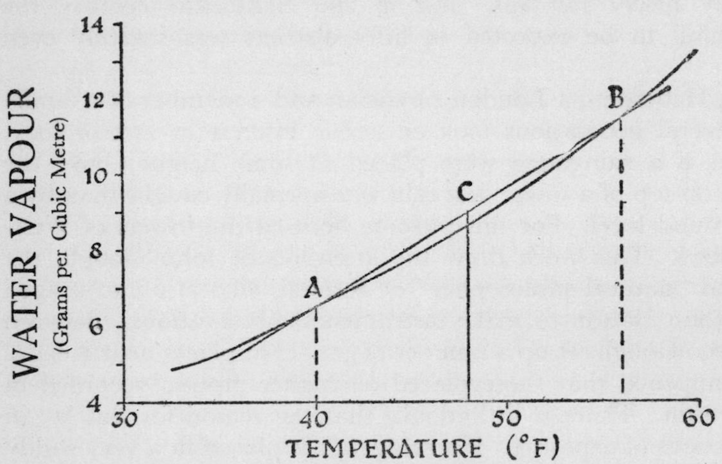

A, B : Air saturated C: Air super-saturated; some moisture must condense

Fig. 2

Now in nature this process does repeatedly occur and indeed it is extremely important; but the condensation so provoked is in general observed as a narrow belt of fog or low cloud along the boundary between the two air masses concerned. We may expect some mixing to take place along such a boundary; but the quantities of moisture thus condensed are very small and remain suspended in the air in the minute droplets of cloud or fog. Rarely if ever is there sufficient provocation in those circumstances for them to aggregate into the sizeable drops whose rate of fall justifies their description as drizzle, let alone rain. It may be mentioned here that a round figure for the diameter of cloud droplets is $\frac{1}{1,000}$ of an inch: drizzle droplets, which drift rather than fall, are of the order of $\frac{1}{50}$ of an inch; raindrops are of the order of $\frac{1}{25}$ to $\frac{1}{5}$ of an inch. Adjacent to the ground the stratification of the nearly saturated lowest layers of the atmosphere in the cooler months leads to the extensive development of fog along the boundary between them; as we shall see, this process must be invoked with regard to the thickening of nocturnal radiation fogs.

But as regards rain, it is evident that following such a doctrine, the cooler the air the less would be the resulting rainfall; hence early meteorologists were much puzzled by the fact that winter rainfall tends in the north to exceed that of summer. It was not for many years that it became widely recognised that such mixing processes might explain thin cloud and fog, and possibly a little drizzle; but were not adequate to explain rainfall. Moreover all normal cloud and fog is far too thick and contains far more condensed moisture than could be produced by mere mixing.

In the early nineteenth century, Dalton and others began to recognise that if a mass of air is for one reason or another compelled to rise and to expand, its temperature must be lowered; and it was Dalton who, in 1799, explained the behaviour of water vapour in the atmosphere. But considerable time elapsed before the necessary consequences of this cooling of air by expansion were worked out. Dalton himself, faithful as far as one can judge, to the memory of his early friend and teacher (Gough of Kendal) continued to skirt round the question whether rainfall did or did not increase on mountains. Another view, not uncommonly held, was that the moist air from the Atlantic impinged against the mountain sides and was there cooled by mixing with "the cool air adjacent to the mountain slopes". When men began to measure surface temperatures of the air on mountains this view again became untenable.

Conviction, however, of the existence of excessive mountain rainfalls again arose through the efforts of amateurs. In 1836 a certain Mr. Beck, a newly-arrived resident at Esthwaite near Windermere, recorded his rainfall. It was so much in excess of that at Kendal as to rouse the attention of others, notably John Fletcher Miller of Whitehaven. He set out on a very thorough investigation; first he supplemented the observation in his own garden by erecting a rain-gauge on the adjacent church tower. In the autumn of 1844 he went further and set up the first rain-gauge at Seathwaite, the group of farmsteads at the head of Borrowdale, which has since become known to thousands of Victorian and later schoolchildren as "the wettest place in England" (Pl. XVa, p. 202). So impressed was Miller by his first year's catch (of a thoroughly unexpected 152 inches) that in 1846 he set up additional gauges on the adjacent mountains; and for eight years his results were communicated to the Royal Society. While Miller's high level gauges suffered from over-exposure, they threw entirely new light on the problem of rainfall; and soon the inescapable conclusion was reached that the expansion and consequent cooling of air already near or at saturation point was the cause of the condensation necessary for formation of rain. Mountains provide, of course, one of several means by which moving air can be compelled to rise, expand and cool. The several other processes which give rise to our British rainfall will be described later.

We must not forget the great part played by little-known men such as Gough and Miller, Heberden and Hutchinson, Hoy and Barker in building up our knowledge of temperature and rainfall in days when the voluntary enthusiasm of individuals displayed the surplus energy arising from prosperous agriculture, nascent industry, and the splendid sense of freedom from external interference. Botanists and zoologists will recognise parallels in their own sciences; and geologists in particular can testify to the magnificent record of those Scotsmen, Welshmen and Englishmen who busied themselves with the unravelling of yet another aspect of the story of their own complex island.

The other elements of weather have all received their attention. Luke Howard, the London apothecary, has earned his fame through being, in 1802, the original inventor of the present universal nomenclature of clouds. He among other things strove persistently to establish a relation between the moon's phases and the weather, a notion which we now know to be erroneous. Admiral Beaufort's scale of wind force for use at sea (1805) has received world-wide acceptance. The measurement of wind force owes much to a clergyman of Armagh, the Reverend T. Robinson, who in 1846 devised the 'cup anemometer' still occasionally to be found at our older observatories. And that very beautiful instrument, the Dines anemograph now so widely adopted in one form or another at meteorological stations, is also a British invention by one of the most eminent of those later Victorian scientists who so often combined mathematical ability with the capacity to handle and design the necessary experimental apparatus. The Dines anemograph registers by means of a pen on a revolving drum, a continuous record of the fluctuations of pressure at the mouth of an open tube kept facing the wind. Momentary fluctuations of pressure in the open tube result from the variation in the speed of the wind flowing past the anemometer head. These are communicated to a pen which in a strong wind continually rises and falls with the momentary gusts and lulls. A Dines anemograph in full cry during a gale is indeed an elegant sight and from it a record becomes available, not merely of the average speed of the wind for every hour of the day, but of the extreme speeds attained in gusts and lulls, a matter of surpassing importance to engineers and all who are concerned with exposed structures (fig. 48, p. 152). Our knowledge of the behaviour of the lower atmosphere has also been vastly extended by its use. For purposes of comparison wind speed is in general measured at a height of 10 metres or 33 feet.

British meteorologists too have never been forgetful of the sun; there are probably more sunshine recorders in these islands than in any other country of similar size. 'Bright sunshine' is recorded very simply by concentrating the sun's rays, as the sun goes round, through a sphere of glass on to a card; hence the sun burns a trace on the card whenever it is bright enough. In general this means that the sun must be more than 3° above the horizon; on the most perfectly clear day the sun will not burn the card within about 20 minutes of its actual rising and setting. This type of recorder, also a British invention (Campbell-Stokes) has been in general use since 1880. Thermometers with blackened bulbs (absorbent of solar radiation) began to be used as an index of the intensity of 'sunshine' from 1860 onward, but the properties of one 'black-bulb' are wont to differ from the next and hence they cannot now be described as a satisfactory instrument for the purpose in view, even when enclosed, as they generally are, within an outer glass tube from which the air has been exhausted.

Much more elaborate instruments for the measurement of radiation have been devised in later years, some of which depend on the 'thermopile' principle; and the amount and qualities of radiation from clouds and from the earth as well as from the sun has been studied. At our major meteorological observatories (Kew and Eskdalemuir in particular) numerous other observations are made for special purposes which as far as we can see at present are not directly associated with those elements of weather which, when averaged, go to make up 'climate', so that they will not be mentioned here.

The formation and behaviour of dew was first explained by Dr. C. Wells of London (*Essay on Dew*, 1818); his writings and experiments have been described by Sir Napier Shaw as models of scientific method. Following him, it was soon observed, for example by Jenyns (already mentioned) in Cambridgeshire, that on a clear, calm evening the temperature of the dew-point is quite a fair index of the minimum that may be expected to occur before sunrise the next day. This remains a very useful rule for gardeners and fruit growers who from time to time are anxious how far the temperature will fall overnight. For, as soon as the process of condensation begins within the layer of the atmosphere near the surface, sufficient heat is liberated to offset the further fall of temperature of the ground. On the afternoon of 4 June 1939 at Durham the dry bulb read 81·0°, wet bulb 58·3°; relative humidity was 19%; dewpoint 36°. The previous night's minimum in the valley nearby was 30°, and that of the succeeding night 37°. Relative humidities below 20% are incidentally very rare in England; extreme values between 10 and 15% have once or twice been recorded.

Among other climatic elements, the frequency of occurrence of snow and snow-cover, of hail and thunder, and of ground surface minimum temperature below freezing point are all recorded. 'Earth temperatures' at various depths below the surface of the soil are also taken daily and are of value to agriculturists and others, notably in connection with public health.

Some mention should be made here of the instrumental recording of 'ground-frost', a figure subject to much misunderstanding and indeed abuse by journalists in search of sensational news of the weather. On a clear evening the earth's surface cools by radiation to outer space. The air adjacent to the surface follows suit; it cools partly as a result of contact with the ground and partly by radiation. The temperature of the surface of ground exposed to a clear sky falls farther than that of the adjacent air; and if a thermometer bulb is placed at ground level it too will radiate freely and register a lower minimum than a thermometer in the air, assuming that the latter is itself protected from radiation either to or from the bulb. Moreover the minute temperature differences set up owing to the varying rate of cooling of, for example, the different portions of a gravelly soil mean that even on the calmest night some slight movement and settling of the air adjacent to the ground must still proceed; this again tends to keep the air a shade warmer than the ground.

For purposes of comparison the minimum temperature of the air is generally registered by a thermometer in a Stevenson screen at the standard height, four feet above the ground. The effects of radiation from the bulb can be illustrated by putting a thermometer out on top of the screen; on a clear calm night such a thermometer will register a lower minimum than that in the screen which is representative of the surrounding air. If the thermometer is exposed on the ground it will commonly be found to register a minimum five, ten and occasionally even more degrees below that of the air; but a great deal depends on the nature of the ground and the vegetation cover, as well as the proximity of objects such as buildings or even fences. For purposes of comparison of 'ground frost', thermometers are placed with the bulb one inch above the ground over short grass.

If such an exposed thermometer has registered a given minimum overnight it is reasonable to assume that the surfaces of the outer exposed leaves of plants within an inch of the ground will have fallen to a similar temperature. Hence it is not unusual to find that, on a clear night with a 'screen minimum' of 35°, the tips of the leaves of young potato plants just appearing above the soil may have been browned and that an 'exposed' thermometer adjacent to the ground may register a minimum of say 28°.

But it is clear that the least degree of protection will check the fall of temperature due to radiation from objects at ground level. The bare branches of a tree provide such a check, as anyone can see at once following a night when hoar-frost has been deposited on exposed fields and roofs; little or no frost is visible under trees. Experiments have shown that even such things as thin muslin check the outward radiation, and many a gardener can save his young potatoes on a clear cold May night by draping them with sheets of newspaper. Moreover, as the plants grow, the upper leaves shade those below. Lastly, much depends on whether the soil is wet or not. A dry soil with plenty of air in it is a poorer conductor than a wet soil. Accordingly, heat is more readily conducted from the subsoil to the surface when the soil is wet, than when it is dry. Hence the biggest differences between 'minimum temperature in the air' and 'minima on ground' are apt to occur when the ground is dry, and when the air also is relatively dry.

Sensational ground-frost readings, however, bear little relation to those manifestations of greater significance to our daily lives, such as frozen pipes, ice thick enough for skating and the like. These depend on the fact that, whatever the ground readings may be, the air temperature in a stratum at least some hundreds of feet thick has fallen well below the freezing point. Moreover ground-frost readings can be shown to vary quite appreciably from point to point even in a small garden. Hence when a London evening paper declares that there were "thirteen degrees of frost on the ground" at Hampstead, while the minimum in the screen at Braemar was 18°, the Hampstead reader should not allow himself to believe that his morning nip on the way to the Tube was at all comparable with that associated with the tingling clarity of a Highland winter dawn.

No recollection is more vivid in the writer's mind than that of leaving London on a raw damp evening, then dismounting from the train at Blair Atholl to find a temperature of 7° at dawn on the clear Christmas morning of 1925. In London the sky was overcast with low stratus cloud and a light drift of air from the south-east prevailed. The air moving from the continent was rather cold, and evening temperatures in the town were about 35°. A depression lay in the Western Channel. Pressure was higher over Scotland, the night there was clear; there was a two-inch snow-cover in Perthshire over which radiation took place freely; lastly, Blair Atholl lies in a valley-bottom. Under such conditions temperature there fell very rapidly during the night and ski-ing was thoroughly enjoyable.

It will be evident that the slightest breeze begins to stir up the layers near the ground and to mix them with the air above. So long as there is a brisk wind capable of stirring up and removing the surface layers as fast as they form, valleys do not become so cold as the hill-tops, even on clear nights.

Some Present-day Sources of Climatological Data

The Meteorological Office compiles and publishes summaries of the observations of temperature, pressure, rainfall and other forms of precipitation, wind, cloud and sunshine, from upwards of 300 stations in Great Britain, at almost all of which the instruments are exposed and observed under comparable standard conditions. Some of the results are available in the *Daily Weather Report* (including those stations which telegraph their records daily for plotting purposes; data are plotted on the synoptic charts which provide the basic material for forecasting). But the results from many other 'climatological' stations are given in the *Weekly* and *Monthly Weather Reports*; these have been published since 1878 and 1884 respectively. Until 1911 many of the results were also summarised in the *Meteorological Record*, published by the Royal Meteorological Society from 1881 onward. We have accordingly quite a considerable number of British stations at which a strictly comparable series of observations has been maintained for a period of 50 to 65 years, using Stevenson screens, modern patterns of rain-gauge and the like. Readers are particularly referred to the *Monthly Weather Report* (which since 1911 has incorporated the *Meteorological Record*) for detailed information with regard to any district in which they are interested.

But from all that has been said in an earlier chapter it will be evident that comparable series of records covering longer periods are few. The strict reduction of the results given by old patterns of thermometer screen is a most troublesome proceeding, and the difficulties increase as we go farther back, so that even the best reductions are not altogether free from criticism. Several series of observations, however, are available for consultation by those who wish to study the extent and character of climatic fluctuations so far as they are revealed by instrumental data. The dates refer to the commencing year.

The longest officially established record, Greenwich, 1841, covers temperature, pressure and rainfall in great detail in the annual volumes of *Greenwich Observations*. Rainfall data are available from 1815 onwards.

OTHER TEMPERATURE RECORDS (MONTHLY MEANS):

Oxford (Radcliffe Observatory) 1815 onward	(Tables in *Radcliffe Observations* Vol. 55, Appendix 1932).
Durham (University Observatory) 1847 onward	(*Quart. Journ. Roy. Met. S.*, 1941).
'Edinburgh' (reduced from several stations) 1731–1736) 1764–1896)	(*Trans. R. S. Edin.*, 1897, 1900).
N.E. Scotland: Gordon Castle and other stations 1782–1892.	(*Journ. Scot. Met. S.*, 1893).
'Lancashire' (reduced from several stations) 1753 onward	(*Quart. Journ. Roy. Met. S.*, 1946).
'London' (reduced from several stations) 1763 onward	(*Philos. Trans. R. S.* 225A, 1925).
' Central England ' 1698 onward	(*Quart. Journ. Roy. Met. S.*, 1953).

Reduced with the aid of numerous scattered records.

Unreduced data are available from several other stations: Orkney 1827, Penzance 1807 (and district), Nottingham 1810; there are also Irish records.

RECORDS OF PRESSURE:

'Edinburgh', 1769–1896 (*Trans. Roy. Soc. Edin.*, 1900).

RECORDS OF RAINFALL:

Since 1860 these can be extracted from the volumes of British Rainfall. Starting before 1860, a table for Edinburgh (1785) is available in *Trans. Roy. Soc. Edin.*, 1900; for Oxford (1815) in the *Radcliffe Observations* named above; for York (1811) see *Q. J. Roy. Met. S.*, 1933. Data for Seathwaite (1845) and other stations can be found in *British Rainfall*, 1867 and 1895, also *Report of the British Association*, 1866. Rainfall variations over England as a whole are reviewed by Glasspoole in the *Meteorological Magazine*, 1928 ("*Two Centuries of Rain*, 1727–1926."). These are shown in the diagram on page 266.

Recorded extremes of temperature, both 'average' and 'absolute', are given, for a few stations, in the *Book of Normals*. Some of the sources cited above give data with regard to certain other elements such as sunshine duration. See also: *Climatological Atlas of the British Isles*. (H.M.S.O., 1952.)

Averages of temperature and sunshine duration are now published by the Meteorological Office (obtainable from H.M.S.O.). The latest available publication covers as far as possible the thirty-year period 1906–1935. For rainfall averages the later volumes of British Rainfall should be consulted. Other data are found summarised in many places, notably in many papers in the *Quarterly Journal of the Royal Meteorological Society*. For a number of coastal stations the volumes of the *Admiralty Pilot* are useful, especially with regard to frequency and strength of wind; also the more recent official publication *Weather in Home Waters* (M.O. 446) which gives detailed information for many coastal stations. Considerable caution is necessary before making comparisons between existing present-day stations. In Britain it is customary at the majority of stations to derive a mean temperature for each month from the average of the daily maxima and minima, the instruments being read and set once daily at 9 a.m. It will be evident that if in a cloudy mild winter month there is a single clear cold night, the temperature at 9 a.m. on the following morning will be little above the minimum entered to the previous 24 hours. Next day at 9 a.m. the minimum entered in the register will be that resulting from the setting of the thermometer the previous morning, and thus the effect of a single cold night is to appear twice in the register. If the thermometers are read and set in the evening this difficulty is largely avoided; but it will be clear that direct comparisons should only be made of the mean temperatures between stations using the same hours.

Elsewhere for the sake of continuity old-fashioned types of screen are still occasionally kept in use. That at Kew has been in continuous operation since 1868; but as it is attached to the wall of the building seventeen feet above the adjacent lawns the minima are often several degrees higher than those in the standard screen. The earlier Greenwich data are also to be viewed with care. With regard to rainfall much has been done to standardise the exposure of gauges in recent years.

The climatological observations of the frequency of snow, sleet and hail depend to some degree on the alertness of observers and must again be used with care. Remembering the inevitable difficulties attached to the assessment of eye observations a warning should be given. It is only too easy to make unguarded statements and draw plausible maps which will not stand up to analysis in respect of such material; moreover such ill-considered compilation does not do justice to the very real efforts made by the Meteorological Office to standardise our records on a firm basis.

SOME REFERENCES

Chapter 1.

BROOKS, C. E. P. (1929). *Climate, a handbook.* London, Benn.

HANDISYDE, C. C. (1947). The Climate of the Home. *Weather, 2:* 82–89 (gives further references).

KENDREW, W. G. (1949). *Climatology.* Oxford, University Press.
(1937). *Climates of the Continents.* Oxford, University Press.

MILL, H. R. (1928). *Climate of Great Britain,* in *Regional Essays.* Cambridge, University Press, ed. A. G. Ogilvie.

TANSLEY, A. G. (1939). *The British Isles and their Vegetation* (Chapters on Climate). Cambridge, University Press.

Chapter 2.

BROOKS, C. E. P. and GLASSPOOLE, J. (1928). *British Floods and Droughts.* London, Benn.

MARGARY, I. D. (1926). The Marsham Phenological Record in Norfolk 1736-1925. *Q. J. Roy. Met. S. 52:* 27–54.

MARGARY, I. D. (1927). Weather Observations at Wrentham, Suffolk, 1673–4. *Q. J. Roy. Met. S. 53:* 301–08.

MERLE, W. (1891). *Consideraciones temperiei pro 7 annis, 1337–1344.* London, reprod. and trans. under supervision of G. J. Symons. (The earliest known journal of the weather).

THORESBY, R. (1826–30). *Diary and Correspondence.* London, 4 vols.

TOWNELEY, R. (1694, 1700, 1705). Account of the quantity of rain . . . at Towneley, Lancashire. *Philos. Trans. Roy. Soc. London, 18,* p. 51; *21,* p. 47; *24,* p. 1877.

CHAPTER 3

SOME ELEMENTARY PROPERTIES OF OUR MOIST ATMOSPHERE

. . . not now, as ere man fell
Wholesome and cool, and mild, but with black Air
accompanied, with damps and dreadful gloom . . .
MILTON: *Paradise Lost*

IT IS our purpose to discuss British weather from within, and from the point of view of those who have to live in it, rather than to write a text-book of meteorology; knowledge of the physics of the atmosphere should be sought elsewhere, and in particular in Sir David Brunt's well-known works, of which the burden will be appreciated even by readers whose physics is rusty from disuse, or who have had little opportunity of study. In particular his recent book *Weather Study* is a compact and masterly introduction of the highest value to the student coming to the subject for the first time. It will suffice to remind the reader in this short chapter of a few elementary principles.

No one can appreciate the vicissitudes of our weather unless he is cognisant of the characteristics of Icelandic depressions and the Azores anticyclone; and no one can claim full knowledge of the effects of our weather unless he has some acquaintance with the relationships between types of cloud, precipitation and the behaviour of surface air.

The atmosphere which envelops us is in the main a mixture of gases; some of these are the permanent constituents, about one-fifth oxygen and nearly four-fifths nitrogen with small quantities of other gases. Water-vapour is the principal variable constituent; carbon dioxide is a minor variable constituent of potential importance. Water-vapour is found in the earth's surface atmosphere in quantities varying from a minute fraction to about 4% by weight; even in the driest deserts it is never wholly absent. Carbon dioxide comprises, as a rule, only about 0·03 %, though this figure is, of course, considerably higher in such places as enclosed crowded rooms. Lastly, there are minute amounts of solid matter as 'dust', and of liquid mostly in the form of extremely small water droplets; these for the most part float, or nearly so, in the atmosphere. Water-vapour is of overwhelming significance, inasmuch as its presence ultimately gives rise to the majority of the phenomena of weather. Most readers will be familiar with the facts; dry air at any given temperature can take up and retain a certain amount of water-vapour. Wet cloths, for example dry; puddles evaporate after rain. Water molecules escape from the surface of the liquid into the atmosphere, among whose molecules, the air being gaseous, they can move freely. This is the process of evaporation, which goes on from the surface of a liquid.

After the process of evaporation into a confined space has gone on for some time, however, just as many water molecules will be returning to the liquid as are escaping from it: the air, by this time relatively crowded with water molecules, is now described as saturated. The number of molecules (or alternatively the weight of water-vapour) required to saturate the air rises very rapidly with temperature. Expressed in ordinary units, a cubic metre of air at a temperature of 0°F. will, when saturated with water-vapour, hold almost 1 gram. But at 32°F. this figure rises to 4·87 grams; at 60°F., 12·91; at 80°F., nearly 25 grams; and 100°F., 50 grams. Frequently, instead of expressing the amount of water-vapour in grams, the vapour pressure is stated; that is, that part of the whole pressure of the atmosphere which arises from the water-vapour within it.

In general, however, air out-of-doors is not saturated; the atmosphere contains less water-vapour than the maximum permitted at that temperature, so that evaporation from damp objects or from a water surface can still take place. The ratio between what the air *does* contain and what it *can* contain at the same temperature is called the Relative Humidity and is customarily given as a percentage. Taken over the whole year for example, the average relative humidity near sea-level in England is about 80 %. At any given temperature, the lower the relative humidity, the more rapidly evaporation can take place.

Evidently if air containing water-vapour is cooled, a temperature will in time be reached at which the air is saturated; if the cooling is continued, some of the vapour will no longer be retained as such; it will condense into the liquid form, or directly into the solid form if the temperature at which condensation is initiated is below freezing-point. The surplus as liquid will either appear as very small drops in the atmosphere, or will be deposited on neighbouring objects as dew (liquid) or hoar-frost (solid); though it may be added that if the temperature of the ground on which dew has been deposited subsequently falls below the freezing point, the frozen dew-drops are also described as hoar-frost. If the very small drops suspended in the atmosphere are numerous, they will impede visibility and present the appearance of cloud or fog. The droplets in cloud (or fog) are commonly of the order of 0·02 to 0·05 mm. or $\frac{1}{1,000}$ to $\frac{1}{500}$ inch in diameter; the rate of fall of such small droplets is almost imperceptible and hence they are carried along in the stream of air. They may increase in size by further condensation upon them, or may diminish by evaporation. Either process however, is often rather slow; hence cloud formed over say, Devonshire is quite capable of being carried to Oxford and beyond given the right conditions.

The so-called dust, in the form of particles so small that many thousands are normally present in a cubic centimetre, is largely swept up from parts of the earth's surface such as deserts and areas of loose dry soil; a fact known to our Fenland farmers as well as those of Oklahoma. Some are composed of the particles of salt left after the evaporation of sea-spray; some may be of volcanic origin; some arise as smoke from imperfect combustion. The diffusion of such small particles through the surface layers of the atmosphere greatly affects visibility, or the degree to which objects can be clearly seen and recognised. The variations due to the drift of smoke from large industrial towns are only too conspicuous in countries such as Britain. Visibility at times is seriously limited at a distance of 60 miles down-wind from London, and instances have been known of visibility being affected as much as 200 miles from the source of the atmospheric pollution. In lands lying nearer desert areas, when in the afternoon there is plenty of surface turbulence, extensive dust-haze develops in a layer which may be as much as 8–10,000 feet deep, a fact well known to those R.A.F. men who were engaged in Libya.

Condensation as Fog or Mist

Having recognised the effects of water-vapour and also of dust, we may consider the various methods by which air can be cooled on a sufficiently large scale to provide widespread condensation.

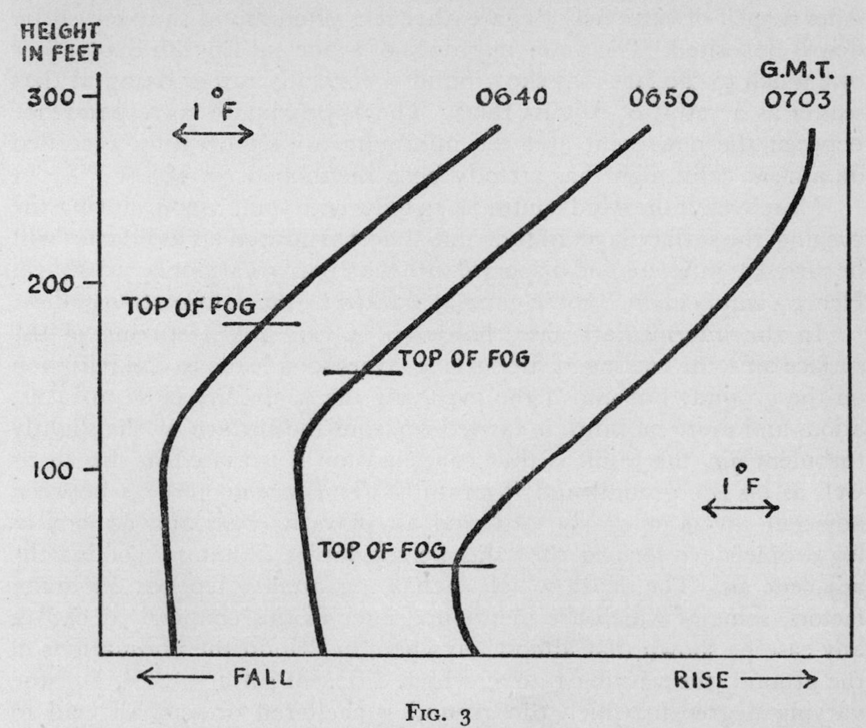

FIG. 3

Temperature Gradient at intervals during the dissipation of a September morning fog. Based on observations by Heywood, *Q. J. Roy. Met. S.*, 1931

With light stirring of the air after sunrise, the lowest temperature in each instance is above the ground but below the top surface of the fog

In the first place, this may arise from contact; when warm moist air rests on or flows over a cooler surface condensation of moisture at and near the surface commonly occurs, either as dew on the ground, or mist or fog in the air. For example, when damp ground is losing heat on a calm evening by radiation to a clear sky, some evaporation of moisture takes place at first into the thin surface layer of air immediately adjacent. But at the same time this layer is no longer being stirred up and removed, but is being further cooled in contact with the ground, and so the moisture from it later condenses on the cold ground as dew. There is no doubt that the formation of dew is considerably facilitated if the ground is already slightly damp. After a spell of extremely dry weather it is often found that very little dew is deposited. The dewy mornings of a normal English September owe much to the fact that the ground is normally rather damp at that season as a result of August rains. The approximate correspondence between the dew-point and the minimum air temperature recorded on a clear calm night has already been mentioned (p. 18).

Moreover, if the wind continues to blow with some vigour during the evening, the surface layer of cool and almost saturated air as it forms will be stirred at intervals and dispersed into the slightly warmer layers above. Hence a windy night is not in general marked by much formation of dew.

In the intermediate case, however—a very slight stirring of the surface air—the cooling of the surface layers soon leads to condensation on the ground; but also if the overlying air is already close to saturation, and more moisture is carried up from the surface by the slightly turbulent air, the result is that condensation is initiated in the air as well as on the ground and fog results. The mixing process between adjacent layers of nearly saturated air plays its part and as soon as fog droplets are formed they themselves act as radiators, cooling the adjacent air. The depth which such a fog attains depends on many factors, some of which are mentioned later in this chapter. It can in any case be shown that almost anywhere in Britain the undulations of the ground, the varying rate at which different surfaces cool, and the varying degree to which the ground is sheltered or not, all lead to slight local movements of air sufficient on many occasions to initiate the process of fog formation and accumulation, especially in the valleys and pockets of our varied countryside. Indeed such movements are necessary; air is a poor conductor, and were it to remain absolutely still it is considered that a night fog could not exceed about four feet in depth.

Bearing in mind that autumn months are often rather rainy, that with the waning power of the sun the temperature and the rate of evaporation are decreasing by comparison with summer, and that the nights are lengthening, we can readily see why morning mists and fogs are characteristic of quiet mornings towards the end of the year.

Further, on a clear night the air over long coarse grass tends to fall to a lower temperature than over short grass, or among trees.

The blades of grass are themselves radiators, and the tips are relatively well insulated against conduction from the roots. Hence the effective cooling surface is greater where the grass is coarse; this in turn cools the air among the grass blades. Hence we observe mists developing earlier, and lasting longer over damp marshy uncultivated hollows. Lack of drainage not only affects the moisture content of the adjacent air; it is often found that if the water cannot drain away, the movement of the dense cold air is also hindered. In a later chapter we shall see that the consequent ponding of cold air and development of frost-hollows and frost-pockets has very important effects.

The term 'fog' is used (by international agreement) whenever the visibility is less than one kilometre or 1,100 yards. With this criterion there is no doubt that Tacitus was right; by comparison with many other countries Britain is foggy, though not more so than Holland or Belgium or N.W. Germany. In this country we have adopted the term 'thick-fog' for visibility less than 220 yards, i.e. sufficient to provide an impediment to road and rail traffic and to impress the average London, Glasgow or Manchester citizen. Fog formed in the manner described above during a night of clear skies, is appropriately known as 'radiation fog'. In recent years careful studies have been made of the factors governing its development and the varying depth which may be attained overnight. It is very evident that the rate of variation of the moisture content of the air up to two thousand feet or thereabouts is important; in particular conditions favouring fog development occur when the moisture content of the atmosphere per unit volume slightly *increases* with height in the first two or three hundred feet. It is quite possible for this to happen in a moist air stream without its being quite saturated. But as soon as the layer at the ground is cooled below saturation point, the air a few feet higher is so near saturation that the very slight mixing due to turbulence is enough to cause the mixture to be saturated. Hence the fog, which begins to form as the familiar thin creeping layer over open damp spaces, quickly grows in thickness; and within an hour traffic is seriously impeded. In the East Midlands it is not uncommon for a night's radiation fog in November to attain a depth of five hundred feet. Moreover, a layer of fog is itself a radiator, and a partial reflector of any warmth it receives from above by day. Hence when the sun is low it is often unable to dispel a thick overnight fog, and if a second calm clear night follows the fog may well deepen further. Fortunately it is rarely that two or three days in succession remain sufficiently calm for this to occur. For the London fog, Dec. 1952, see p. 56.

It will now be evident that with a slightly more rapid decrease with height of the moisture content of the atmosphere in the lower layers, there will be many occasions when at any given level the mixing of the lower layer with that above will not quite result in saturation of the mixture. Fog will then not form, or will only begin to form after the temperature has fallen for some hours, say towards dawn; the balance is delicate, and careful measurements of the local humidity of several levels are recorded for forecasting. The importance of being able to forecast the times at which surface fog would begin to develop, and the depth it would attain, was very great during the war on account of the needs of returning R.A.F. pilots. Much of this recent investigation is due to Mr. W. C. Swinbank, one of the younger physicists who began his studies at the University of Durham.

In the light of what has been said regarding the approximate correspondence between the overnight minimum and the dew-point on a clear calm night it may be asked, how does the temperature continue to fall under conditions of dense fog; that it can do so, although rather slowly, is well known. The answer appears to reside in the fact, already mentioned, that outward radiation continues from the top of the layer of fog, and the cooled air from the top of the fog sinks to the ground level. The fall of temperature is however, comparatively slow. As the substance being cooled is mobile, the cooler elements at the upper surface as they begin to sink are continually being replaced by warmer elements from below.

If a dense fog occupies a valley with snowy uplands on either flank arising above the top of the fog, the air cooled by radiation over the uplands continues to flow down into the valley and extremely cold days result, as the fog is likely to prevent any penetration by the sun.

If the valley itself is also deeply snow-covered fog is not so likely to persist. Petterssen has shown that this arises from the fact that the saturation vapour pressure over ice is less than that over water. Expressed otherwise, if ice is evaporating into a dry atmosphere, saturation will be reached more quickly than if supercooled water were evaporating into air at the same temperature. Otherwise, if air is cooling towards saturation and there is a snow cover, the water

Graphic Photo Union

PLATE I: Kew Observatory, the principal meteorological observatory in Britain; with instruments. Dines anemometer head above the dome. Continuous temperature recording has been maintained on a large screen on the distant wall of the building, not visible here.

H. Rait Kerr

PLATE II: Alto-cumulus illuminated by the setting sun. London; July 1936.

vapour in the air will begin to condense on the snow-surface and so the moisture content of the air will remain just too low for condensation as liquid drops. Hence at low temperatures the presence of a snow cover often tends to prevent the formation of radiation fog in the air above. We thus see a reason why the Scottish Highland valleys are often deeply covered by snow, and very marked 'inversions' exist, that is, pools of exceptionally cold air occupying the valley-bottom; yet there is little or no fog.

But if the damp air overlying the snow is at a temperature *above* the freezing point, *i.e.* the snow is wet, the air is cooled by the snow and saturation may well be reached at a temperature above the freezing point. In this case condensation begins as liquid drops in the air, and a fog quickly forms which gradually falls to a temperature close to 32°, that of the thawing snow beneath.

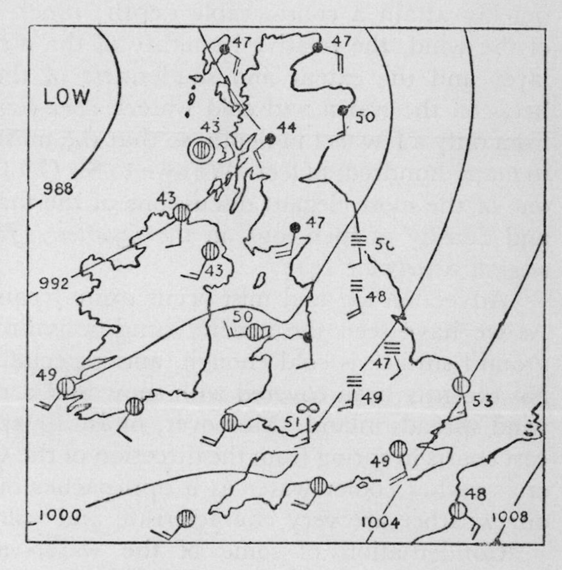

FIG. 4
Advection fog with smoke effects in Midlands and north east; 22 October 1937 (see p. 5)

Indeed, fogs over thawing snow in Britain are common, but although they can develop in the manner above, for the most part they are due to the advection of air; in this case the movement of a moist air current from a warmer source.

For we must recognise that moist air in motion, passing over a cooler surface, may also be sufficiently cooled to produce fog: and that this can, and does occur even if the sky is overcast. The best-known example in the world of persistent and frequent 'advection fog', as it is called from its mode or origin, is that of the Newfoundland Banks. Just to the east and south-east of Newfoundland the sea surface temperatures are those of the Labrador current, and are very low for the latitude. Yet the abnormally warm waters of the Gulf-Stream Drift lie only a short distance farther south. Hence almost any movement of air from points between south-west, south and east brings moist warm surface air on to a much cooler surface, and fog results.

It is to be noted that movement of air implies some degree of turbulence over the surface; packets of saturated air in which some condensed droplets are already to be found are therefore carried upward from, and downward to, the surface. Fog, therefore, may quickly attain a considerable depth; much depends on the strength of the wind, the relative humidity of the air above the chilly surface layer and the extent and suddenness of the temperature difference between the warm and cold waters. Sea-fogs of this type may range from only a few feet in depth, so that the masts of ships are intervisible, to many hundreds of feet. We owe to Sir Geoffrey Taylor of Cambridge one of the most elegant discussions of the factors governing the depth and density of such fogs, in the *Quarterly Journal of the Royal Meteorological Society* for 1917.

Advection fog and mist occur quite frequently in the British Isles. As we have seen the needful conditions are provided whenever the ground surface is cold enough, and especially when the country is or has recently been covered with snow and a mild damp south-westerly wind spreads inland. Moreover, moist air approaching our south and east coasts in spring from the direction of the Continent must frequently cross rather cooler water as it approaches our shores. The effects on our weather are very characteristic and will later be discussed.

Condensation of some of the water-vapour is likely to occur in a belt along the boundary between two nearly saturated currents, as we have already seen in the last chapter. On a small scale this goes far to explain the development of radiation fog, rather than mere dew. On a large scale, some might argue that a belt some miles wide of 'horizontal mixing fog' would appear to be formed when for example, a slow-moving moist but cool current from the continent, and a warmer but also very moist current from the ocean converge. But it must be remembered that the amount of moisture condensed due to this process is very small and unless it is confined to a very narrow layer near the ground it does not suffice to make a fog, largely because pure horizontal mixing is unlikely. The warmer current tends to override the cooler; we get low cloud rather than surface fog.

With a light southerly wind in the English Midlands such a state of affairs is not uncommon in autumn and early winter. Moreover, smoke from the towns drifts along in the colder surface layer and cuts down the visibility. If the converging air streams are more lively, forming a minor warm front, drizzle or rain from the cloud above falls into the chilly moist air below and brings it to saturation point, so that a belt of 'frontal fog' is then found. Over the damp, retentive clays of the Midlands it is quite rare to find good visibility in autumn and winter unless there is a strong wind. Advection fog due to a warm southerly airstream creeping inland over the cooler ground; valley radiation fogs deep enough to be farther spread by a light wind over the neighbouring country; and the occasional frontal fog are all reinforced by smoke. Hence, as a whole, the area most subject to mist and fog lies in a long belt from London to South Lancashire and Yorkshire, with patches of greatest frequency over our great inland cities, and local upland districts such as the High Chilterns which may stand above some of the radiation fogs. Allowing for such local exceptions, throughout this belt fog, that is visibility less than 1,100 yards, is observed on upwards of 50 days yearly. (Cf. Durst, p. 55).

Some local condensation also takes place at the base of the atmosphere if a very cold current of air blows over a much warmer water surface. This produces the same effect that we observe in the 'steaming' of the surface of a hot bath; locally on the surface the cool air is first saturated and condensation begins. But the saturated air has not to move very far before it is again mixed with the unsaturated air above; hence the 'steam' continuously forms along the surface, but as quickly dissipates. We may often observe the steaming of unfrozen lakes, rivers, and ponds in frosty weather. Apart from such slight effects, however, 'steam-fog' has no significance in Britain; but on the Canadian Great Lakes an outburst of Arctic air early in winter may sometimes give trouble through this cause to the local shipping, at the season when the waters are still ice-free.

The differing behaviour of our atmosphere with regard to fog formation plays its part in scenic effects. Under English conditions the combination of a deep snow cover with a temperature well below freezing point and clear skies is rare. In the Scottish Highlands it is more common and Dr. Fraser Darling reminds us in his *Natural History in the Highlands* in this series how men and animals alike enjoy the brilliant frosty days following a heavy snowfall. Under such conditions a rarely-seen amethyst colouration is perceptible in the landscape and especially in the shadows on the snow, given the really pure atmosphere of those fortunate areas far removed from towns. It is probably attributable to the fact that the air is so pure and free from dust particles that even the blue wave lengths are not so much scattered as usual. (Cf. chapter 8, p. 146.)

Normally when the land is snow-covered and the sun shines from a cloudless sky, shadows are blue in colour because they receive all their illumination from the brilliant blue sky, largely reflected by the snow. Exceptionally brilliant cold days early in March 1947 will long remain in the memory of many country dwellers, and Plate 29, p. 206 will remind us of the unforgettable splendour of the Derbyshire moors on that occasion. But in England at least it is much more usual to find that our occasional bright winter day with the more normal snow-cover is at the same time decidedly hazy (Plate 11, p. 94). This is largely on account of the fact that there is generally a surface inversion some hundreds of feet deep in which smoke-haze is penned, seriously limiting the visibility in combination with the light mist that so often drifts from neighbouring districts in which less snow has fallen, or where it has already melted. Hence the brilliance of mid-February in the upper Gudbrandsdal, let alone Pontresina, is very rarely matched in England, though it sometimes develops for brief spells in the clearer air of the Scottish Highlands. Can we ascribe to such influences the marked regard of the Norseman for brighter colours? Although his descendants in our north-western dales show in the style of their barns—as in many other respects—the traditions of their forefathers, those who observe the fell farms of Lunds from the northbound Scottish express as it crosses Aisgill beyond Garsdale will see no sign of that cheerful Scandinavian red. From Reyniháls the valley of Litli Langadalur (Wrynose and Little Langdale in the Lake District) is indeed beautiful; fell and dale alike bear witness to climatic action of the kind that has moulded Norway; but barn and wall remind us of the fundamental difference in our winter climate, which nowadays but rarely gives us a sample of the dazzling snowy brightness of the North.

Condensation as Cloud

The visible results of condensation of water-vapour in the form of cloud are evident to us all. The droplets which go to form cloud are of the same order of magnitude as those in fog, *i.e.* they 'float' rather than fall. With sufficiently low temperatures the liquid droplets are replaced by minute ice crystals; in general, however, as we shall see, clouds in British latitudes are not wholly composed of ice crystals until we attain levels above 20,000 feet.

Air, as a mixture of gases, obeys the laws of gases according to which the pressure, volume and temperature of a gas are closely related. For example, if we take an enclosed volume of air and allow it to leak through a small hole into an adjacent enclosure from which the air has been partially extracted, we shall find that the pressure of the air on the side of the first vessel is lowered, but also that its temperature will be lower. Air consists of large numbers of molecules in constant motion; with expansion of the air into the adjacent vessel, the decrease of molecular activity is manifest to us as a fall of temperature.

A mass of air rising from the earth's surface expands owing to the decreased pressure upon it of the air above; and provided that it is not saturated with moisture, it falls off in temperature by 5·4° F. for each 1,000 feet that it rises. Normally air contains a good deal of water-vapour, though not enough to saturate it; but if it continues to rise it will sooner or later cool to a temperature at which it is saturated, and condensation will begin; the small drops forming what we call cloud. We shall see that there are occasions when moist air rising from the surface is checked in its ascent; if the check takes place before the rising air has cooled to saturation point, cloud will not form.

Masses of air can be caused to rise from the earth's surface in three principal ways. First, by heating of the surface layers to such a temperature that the rate of fall with height or 'lapse-rate of temperature' exceeds the figure given above. Secondly, by the movement of the air over obstacles such as hills or mountain ranges, or movement of lighter air over denser air in its path, or by convergence of air currents differing in density. Thirdly, by turbulence; vigorous movement of air over the earth's surface is accompanied by considerable friction, and on a windy day packets of air from the surface are continually being carried upward and downward in a layer many hundreds of feet thick, if not more, as a result of the disturbed flow.

Clouds form with little difficulty over the British Isles for several reasons. Fundamentally of course the surface air has only too frequently moved within the previous few hours from over the sea and is generally moist, that is, it has a high relative humidity. Little cooling is therefore needed to initiate condensation. The cooling of the air may be brought about, as we have already seen, by contact with a cooler surface beneath, in which event we get fog. Or it may be brought about by ascent and consequent expansion; and it follows that the more humid the air the lower the cloud base is likely to be when the air ascends.

Ascent may be the result of the heating of the surface layers adjacent to the ground; this is the normal process on a summer morning, forming cumulus clouds. If the process is sufficiently developed so that the ascending moist currents boil up to a level at which further condensation takes the form of ice crystals, the top of the cloud acquires a "fibrous" appearance (Pl. 20a, p.127) and is called 'cumulo-nimbus'; this is the cloud in which thunder-showers so often occur.

We may add a little more on the heating of the surface layers. This may be effected by the heating of the ground, which heats the air in contact with it; but also if cold air flows over a warmer surface, similar effects are produced. These are very important over our seas in winter.

Within an area of low pressure there is necessarily some convergence of air-stream towards the centre and hence ascent; so that areas of low pressure frequently tend to be cloudy quite apart from the additional factors leading to production of cloud discussed in Chapter 4, p. 63.

Just as fog once formed itself loses further heat by radiation, so a cloud sheet towards nightfall may thicken as a result of cooling due to radiation from its upper surface. Thin sheets or layers of cloud ('stratus' types) can sometimes be attributed to this cause, especially when they appear rather quickly in a clear sky. We should distinguish the types of cloud associated with the rapidly rising currents due to surface heating as 'cumulus' and 'cumulo-nimbus'.

Dissipation of Cloud

It follows that descent of air leads to compression and warming; and the descent or subsidence of saturated air with the consequential rise in temperature means that the air will no longer be saturated. If cloud was present, it disappears. Evidence of the slight descent of air after crossing quite small ranges of hills is frequently given by the existence of a gap in the clouds to leeward. For example, when low stratus cloud arrives on a south-west wind on the Welsh coast, it is not uncommon to find extensive breaks in the cloud sheet over Shropshire. We shall see that much depends on the height of the hills and the height of the clouds. Hills do not, in general, affect the flow of the air to more than about three times their own height. We shall see, too, that some descent of air is a necessary consequence of the existence of a region of higher atmospheric pressure, or an anticyclone, so that there is an *a priori* tendency for widespread clear skies; in a later chapter we shall discuss the extent to which in Britain this remark needs amplification.

Supercooling

It is extremely important to bear in mind certain additional properties of small water droplets once they are formed. In the first place, provided they remain very small, they can be cooled far below the freezing-point without turning into ice; but when drops are thus super-cooled, a very small disturbance such as the impact of the wing of an aircraft or even the movement of a cyclist or walker causes them to freeze immediately. The finest cloud droplets which are normally present can be cooled to about zero Fahrenheit before freezing. We therefore find that the great majority of clouds seen in Britain up to an altitude of 20,000 feet are still principally composed of water droplets; but above this level we can assume that with rare exceptions, high clouds are composed of ice crystals in the form of minute hexagonal crystals or plates.

Confirmation of the composition of the thinner cloud-sheets is often forthcoming. If the sun or moon shines through a thin veil of 'cirro-stratus', *i.e.* a sheet of ice-crystal cloud generally above 20,000 feet, some of the rays are refracted by their passage through the crystals. To an observer on the earth these rays appear to arrive from a position in the sky 22° in angular distance from sun or moon. Hence a narrow ring of light is seen to surround the luminary, often forming a complete circle when the veil of cloud covers a large area between the sun or moon and the observer. Not uncommonly the ring is seen with the colours of the spectrum more or less developed, the ice crystals behaving as small prisms with regard to those rays passing through them. Occasionally a variety of additional phenomena appear including mock suns (or moons) and haloes of larger radius; for a full discussion the reader should refer to writers on atmospheric optics.

When shining through a thin layer of cloud at a lower level, however, the sun or moon generally appears surrounded by faintly coloured rings (coronae) close to the luminary; these arise from diffraction or the scattering of light by the 'grating' of small water droplets through which the rays are passing. It will be observed that these appear even through a thin layer of alto-stratus at heights of the order of 10,000 feet in winter, confirming that although the temperature at that height is very far below the freezing point very small liquid droplets still predominate in the composition of the cloud.

But if water-vapour condenses at a temperature below 32°F. it is liable to take the form of ice crystals. We can see, therefore, that if we have an ascending mass of saturated air in which condensation began at say 45°, as the air rises and cools degree by degree its vapour content for saturation must decrease. Condensation will continue; and the liquid drops so formed may be carried up by the rising air to levels at which the temperature is far below freezing. But that part of the condensation which takes place below 32° will be directly as ice; and so we find that in the upper part of a large rapidly ascending current, such as is manifest in a big cumulo-nimbus cloud, ice crystals and water drops are found together; finally, at very high levels, ice-crystals predominate. This can often be observed; cumulus cloud retains the rounded edge characteristic of water-drop cloud, until it grows beyond a certain level, after which the top assumes a characteristic fibrous appearance. The ice crystals are swept out by the wind, just as they are in the well-known high cirrus 'mares' tails' we so often see; and this fibrous structure at the top of the high-piled towers of cloud is the distinguishing mark of 'cumulo-nimbus'. The predominance of ice crystals in the upper part of the cloud arises on account of the fact that the vapour pressure required for saturation over ice is less than that over water. Hence in an atmosphere containing ice crystals and supercooled water drops, further condensation takes place on the ice crystals which thus tend to grow in numbers and size. If the growth is rapid they will begin to fall; and it has been argued that the greater part of the precipitation as rain falling from clouds in temperate latitudes arises in this manner, becoming rain at lower levels. For if a cloud is merely composed of minute water drops, it is by no means clear why they should begin to aggregate. Nevertheless, it is known that raindrops can form in clouds in which no ice whatever can possibly exist; hence the problem of the actual formation of raindrops within a cloud, simple though it may appear, is not yet fully explained. It is known that other processes may have to be invoked at temperatures above 50°F. In a turbulent cloud the droplets formed as a result of condensation at different levels are carried upward and in the same part of the cloud droplets differing considerably in size may be found. If this is granted, then momentarily the vapour pressure over the bigger drops will be less than that over the smaller drops; hence further condensation is likely to take place on the bigger drops, which consequently increase in size. Moreover clouds in which there is considerable turbulence appear to be essential if rain is to fall. Hence it was also thought that coalescence of the droplets by collision was an adequate explanation; but this will not do, because the small droplets themselves 'float' in the turbulent air and there are many very turbulent cumulus clouds from which rain does not fall. Thus we need further explanation. Quite recently some physicists have expressed their belief that even at lower temperatures there are frequently not enough ice crystals present to initiate extensive growth of raindrops, and yet it rains. There is much room for further study of the properties of water in the atmosphere at low temperatures, as the results already obtained by the research group under Professor G. M. B. Dobson at Oxford have already shown. Reference should also be made to recent papers in the *Quarterly Journal* by F. H. Ludlam and B. J. Mason.

TABLE OF CLOUDS

Height	*Form*	*Description*
High Clouds. Mean Lower level 20,000 ft.	Cirrus (Ci)	Detached clouds of delicate and fibrous appearance, without shading, generally white in colour, often of a silky appearance.
	Cirrocumulus (CC.)	A Cirriform layer or patch composed of small white flakes or of very small globular masses, without shadows, which are arranged in groups or lines, or more often in ripples resembling those of the sand on the seashore.
	Cirrostratus (Cs.)	A thin whitish veil, which does not blur the outlines of the sun or moon, but gives rise to halos.

TABLE OF CLOUDS—*contd.*

Height	*Form*	*Description*
Middle Clouds. Mean Upper level 20,000 ft. Mean Lower level 6,500 ft.	Altocumulus (Ac.)	A layer (or patches) composed of laminae or rather flattened globular masses, the smallest elements of the regularly arranged layer being fairly small and thin, with or without shading.
	Altostratus (As.)	Striated or fibrous veil, more or less grey or bluish in colour.
Low Clouds Mean Upper level 6,500 ft. Mean Lower level close to the ground.	Stratocumulus (Sc.)	A layer (or patches) composed of globular masses or rolls; the smallest of the regularly arranged elements are fairly large; they are soft and grey, with darker parts.
	Stratus (St.)	A uniform layer of cloud, resembling fog, but not resting on low ground.
	Nimbostratus (Ns.)	A low, amorphous and rainy layer, of a dark grey colour and nearly uniform.
Clouds with Vertical development. Mean Upper level 20,000 ft. Mean Lower level 1,600 ft.	Cumulus (Cn.)	Thick clouds with vertical development; the upper surface is dome-shaped and exhibits rounded protuberances, while the base is nearly horizontal.
	Cumulonimbus (Cb.)	Heavy masses of cloud, with great vertical development, whose cumuliform summits rise in the form of mountains or towers, the upper parts having a fibrous texture and often spreading out in the shape of an anvil.

Classification of Clouds

Luke Howard, the London apothecary, writing in 1802 was first to name the characteristic types of cloud; he distinguished cumulus (heap cloud), stratus (or layer cloud), cirrus (the high wisps or streaks) and nimbus (from which rain falls). These terms have gradually been combined and enlarged and the present-day international classification of the main types of cloud is universally adopted. For the reader's convenience they are tabulated here with some details of their occurrence and characteristics; a fuller discussion will be found if desired in most meteorological textbooks. The highest clouds in the British Isles are very rarely above 40,000 feet. The normal limit, formed by the base of the stratosphere, is about 33,000 feet.

Precipitation

Moisture may be precipitated from the air in the form of drizzle, rain, snow, sleet or hail, which fall; or dew, hoar-frost and rime which are directly deposited on exposed surfaces.

Drizzle is nowadays distinguished from rain inasmuch as drizzle droplets, being small, drift down on the wind rather than fall. In general, a drizzle drop is of the order of $\frac{1}{50}''$ to $\frac{1}{100}''$ in diameter, *i.e.* of the order ten times the diameter of a cloud-drop. Raindrops range from about $\frac{1}{25}''$ to $\frac{1}{5}''$ diameter. Any drop of water of larger size, if it forms, is quickly torn apart again due to its own limited surface tension and to frictional effects as it falls through the air. Further, the resistance of the air imposes a maximum velocity on falling raindrops depending on their size; for a drop of $\frac{1}{5}''$ diameter this is about 25 feet per second, or 18 m.p.h. It follows that no rain can fall if the 'vertically' rising currents exceed this figure. Not uncommonly with vigorous convection on a humid summer morning when vertical currents of the order of 30–40 feet per second prevail, cumulus and cumulo-nimbus continue to grow for some time until at length the first large drops begin to fall. They immediately cool the ground and the lower air by evaporation; the vertical currents are then further diminished and the rain is suddenly released in a way known to all Englishmen who watch cricket-matches.

Snow occurs when the minute ice crystals resulting from condensation below the freezing-point fall; at very low temperature (below zero Fahrenheit) single ice crystals fall in the form of a very fine dust. At higher temperatures complex aggregations of crystals develop in the infinite variety of patterns we know as snowflakes and when the temperature is close to freezing-point, those may in turn stick together so that the snowflakes become quite large. Snowflakes will only reach the ground if the temperature of the air through which they fall remains cold enough. Upon the ground there may sometimes be appreciable accumulation of wet snow with the air temperature at 34°, but it soon melts unless the temperature falls. The persistence of a snow-cover is greatly favoured if the ground surface is already below freezing point.

Snow which has partly melted in its fall, so that some of the drops fall as rain, is known as sleet. As English winter temperatures frequently lie in the neighbourhood of 40° at sea level, precipitation there is most commonly observed as rain. But a climb of 1,000 feet on a rainy and windy day with saturated surface air means that temperature will fall to about 36·5°, at which some flakes of melting snow are likely to be observed mixed with rain. A further climb of 1,000 feet is likely to bring the air temperature down to 33°, at which heavy wet snow is most probably falling; and if the temperature falls below 32°, dry snow flakes will be swept before the wind and drifting will occur. Clearly one of the most characteristic features of the British winter climate is the rapid increase with altitude in the frequency with which snow or sleet is observed either to fall or to lie.

There are, however, occasions when although the surface air remains below freezing-point, raindrops fall from a warmer layer above. Such raindrops are slightly supercooled and freeze immediately on impact, and if the surfaces of objects at ground level are already chilled below the freezing-point the accumulation of 'glaze' as it is called may occasionally be serious. In January 1940 enormous damage was done to trees, telegraph wires and transmission lines by a phenomenal ice-storm of this kind throughout the West Midlands; rain fell for several hours with an air temperature in the region of 28° (fig. 5). Similar events occurred in Sussex in March 1947 (Plate 13, p. 98).

It appears that a small raindrop carried through a stratum of air below freezing-point may freeze to form a small hail pellet and around it supercooled cloud droplets freeze on impact. As some air is included the resultant aggregation appears opaque. It is probable that in cool, rather stormy winter weather raindrops at moderate altitudes are often carried just sufficiently upward in the turbulent air to freeze, and the small opaque pellets of 'soft hail' result. Soft hail is common enough on our coasts in winter, especially farther north, and can often be attributed to the slight uplift and disturbance caused when a cool stream of air, usually from a westerly point, passes over the land.

Such air streams tend to be more frequent and to blow more strongly in more northern latitudes, as we might expect. Hence at those observatories at which an alert watch is kept, soft hail is observed to fall on about 10 days yearly in the south, 15 days yearly in Lancashire, upwards of 20 in Western Scotland and at Aberdeen and over 30 in Lewis and the Shetlands. No precise comparison of the incidence of soft hail can however be made as its recording depends so largely on the alertness of observation.

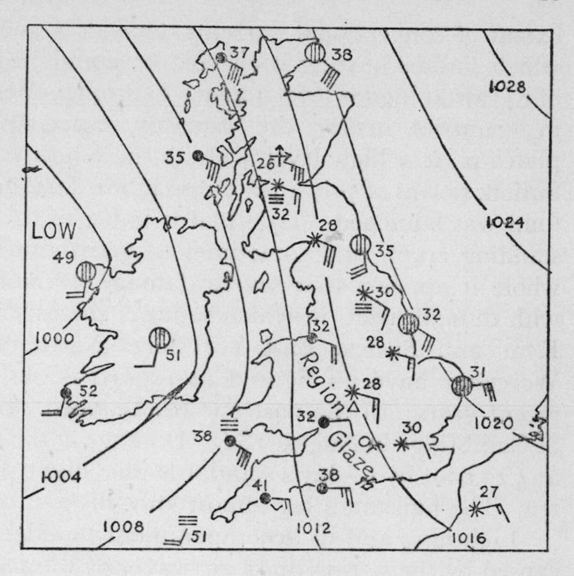

FIG. 5
Glazed frost, 28 January 1940; note blizzard at Dalwhinnie. Glaze, from rain falling at temperatures below freezing point, from Wiltshire to the North Wales border; very heavy snowfall followed in Lancashire and the north-west

True hail, as it is often called, is much more frequent in summer and falls from thunderclouds. The violent vertical currents developed in a big cumulo-nimbus cloud carry the raindrops formed near the base of the cloud to great heights when they freeze; at the same level ice crystals are found and some of these accumulate on the outside of the pellet of ice. The pellet may then fall and partly melt; but only to be swept up again and re-frozen. Sometimes the repeated up-and-down tossing results in the formation of hailstones of considerable size; when these are cut, the 'concentric' structure often is revealed. In England hailstones the size of marbles are regarded as large; in this country the limit for single stones so far reported is about the size of a golf ball. Bearing in mind that the region in which such violent upward and downward currents prevail in a cloud is often quite small, we can at once see why the occurrence of really large hailstones is somewhat capricious and fortunately limited to very small areas. The extent of convectional currents depends a good deal on the extent to which surface heating is effected by comparison with the temperature of the air at higher levels; hence hailstones often attain greater violence in countries nearer the Equator, especially when they comprise plateaux at a high level such as the South African veldt. The West Suffolk storms of July 1946 did so much damage to crops that an appeal fund was launched. True hail is liable to do considerable damage to standing crops, and sometimes to greenhouses and the like. On the whole it appears that severe damage by hail tends to be associated with those districts in which summer thunder is most frequent. Inland Kent and Sussex, East Yorkshire, Northamptonshire, Suffolk and Worcester have all figured in reports of serious damage by hail in recent years. In the past one of the worst storms on record occurred at Cambridge in August 1843. Damage in the town was then estimated at £25,000; by to-day's standards this might perhaps be multiplied by ten. The hailstones lay almost knee-deep.

Lightning and its accompaniment thunder, which is a sound effect caused by the setting up of air waves by the passage of a spark, are also associated with violent convection in cumulo-nimbus cloud. As a result of electro-static effects arising from the breaking-up of larger raindrops, perhaps also from their passage through the air, and from friction among the ice crystals, different parts of the cloud become highly charged and spark-discharges occur between them, or from cloud to earth. The sheet lightning low down towards the horizon, so often seen on a warm and cloudy summer night, is mainly attributable to the reflection of distant flashes from the under surface of the clouds.

Little has been said with regard to the processes by which the small cloud-drops become the much larger drizzle—or rain-drops—and begins to fall. It was at one time considered that aggregation by collision in the turbulent rising currents might be a sufficient explanation, but this view is not wholly acceptable, as we saw on p. 39. We may, however, observe that many of the drizzle- or rain-drops in a cloud are considerably larger than those which on the average go to form the cloud itself; and, however the process of aggregation is effected, the formation of drizzle and rain only occurs when there is considerable turbulence within the cloud. Coagulation of droplets by collision may be possible if some of the falling drops are already of larger size. In particular, a great deal of the formless layer cloud with which we are provided gives no precipitation. It can be shown that within such cloud, turbulence in general is slight, and in particular vertical ascent of the condensed moisture is prevented above a certain level so that the cloud is generally limited in thickness; and that such a cloud is probably homogeneous, *i.e.* all the droplets are of the same size.

With regard to the formation of the deposits of dew, hoar-frost and rime, dew and hoar-frost have already been discussed (p. 18).

Rime is frozen fog; we have already seen that fog below freezing-point is in general still composed of liquid drops. Cycling through such a fog, and even walking, impacts the supercooled drops so that they at once turn to ice on one's coat, while motorists quickly note the deposit on the windscreen of a car. The slightest wind deposits the fog as minute frozen ice crystals against the sides of fences and the like, and gives us a small-scale reminder of the processes which lead to icing on aircraft. Rime-deposits associated with persistent low cloud and strong wind on our mountain summits are often very conspicuous. On Ben-Nevis 'frost-feathers' sometimes grew outward to a length of five feet on the exposed masts carrying the instruments in the days of the old Observatory. The scenic effects associated with widespread rime-deposit on trees are often extremely beautiful.

Before we end these reminders of the part played by our vaporous atmosphere in producing 'weather', we may also recall that water-vapour has important effects with regard to the transmission of radiation. Large quantities of vapour in the atmosphere act to some extent as a screen especially with regard to the long-wave terrestrial radiation, that is outward radiation at night; hence the greatest daily ranges of temperature occur when the air for many thousands of feet above the surface is decidedly dry (p. 172).

Stability and Instability of Surface Air

Having reviewed the part played by water-vapour in regard to the formation of cloud and precipitation we must now discuss its significance with regard to 'stability' as we call it, in the atmosphere. The term is used with reference to the equilibrium of a small mass of air. Normally in air which is not saturated the rate of fall of pressure from the earth's surface upward, consequent on the decreased weight of the air above, is such that a mass of air rising from the earth's surface, expanding and cooling as it does so, would fall in temperature by 5·4°F. for each 1,000 feet of ascent. This is known as the 'dry adiabatic lapse rate', the term adiabatic implying that no further heat is transferred to, or from the rising mass of air on its way; while 'lapse-rate' is a convenient term for the lapse or decrease of temperature with height. On a clear and breezy afternoon in summer the rate of fall of temperature in the disturbed air near the surface approximates very closely to the dry adiabatic for two or three thousand feet above the ground. Typical daily variations are shown in Fig. 6.

Let us suppose that on a given occasion the rate of fall of temperature with height for several thousand feet above the ground is 4°F. per 1,000 feet. Such a prevailing lapse rate is shown in simple form in Fig. 7, p. 48, plotting temperature against height. Consider now what happens if a small mass of air adjacent to ground level is warmed by five degrees. It will then be less dense than the surrounding air at the same level, and will accordingly rise. When it rises its temperature must fall off by 5·4° for each 1,000 feet of ascent, assuming for the moment that there is no further heating of the rising mass due to some external cause. By the time it has risen 3,000 feet (approximately) its temperature will have fallen to the same value as that of the surrounding air at the same level, and as it will then have the same density as the surrounding air it will not tend to rise any farther. Due to its momentum it may rise a little farther, but its rate of ascent will quickly diminish and come to a stop.

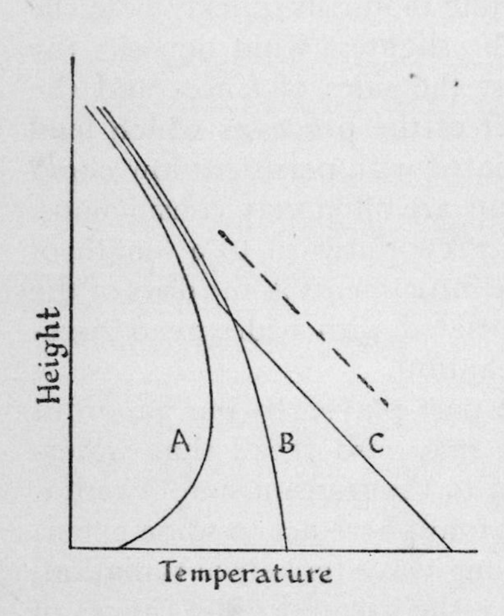

FIG. 6

Diurnal variation of stability over land. A, early morning; B, midday; C, evening; broken line, dry adiabatic (based on Petterssen, by courtesy of Messrs. McGraw-Hill)

On the other hand if the prevailing lapse-rate in the air as a whole were say, 6° per 1,000 feet, a rising "bubble" from the surface (and 5° warmer, as above) would find itself only 5·4° cooler after 1,000 feet whereas the environment would be 6° cooler. That is, the air which was warmed to begin with now finds itself on attaining a level of 1,000 ft. relatively warmer, lighter, and more buoyant than the air

PLATE III: *a.* Anticyclonic weather; inversion fog in a Somerset valley, mid-April

b. From the same view-point, free from fog.
W. G. V. Balchin and N. Pye (from the Quarterly Journal of the Royal Meteorological Society).

PLATE IV*a*. *The Times*

b. *R.A.F. Crown Copyright Reserved*

surrounding it at the same level, and so, not only will it continue to rise but it will tend to do so more quickly than at the surface.

Suppose now that instead of heating calm surface air as above, the surface air is in vigorous motion; a strong wind is blowing. As the air moves rapidly over the irregular surface of the ground, slight obstacles are continually causing small masses to rise. If in the air as a whole the lapse-rate is greater than 5·4° per 1,000 feet it is evident that a small mass of air deflected upward by an obstacle will continue to rise (as in *B* above). But if the lapse-rate in the air as a whole is *less* than the critical value of 5·4° per 1,000 feet a small mass of air deflected upward will find itself cooler than its environment and will tend to fall back to the surface from which it rose. In the first case the air is in unstable equilibrium; a small displacement is followed by further displacement. In the second case the air is in stable equilibrium; a small displacement is at once followed by a return to the original position.

It will thus be evident that the meteorologist must take into account the prevailing lapse-rate throughout the lower atmosphere before he can decide whether rising air currents due to surface heating will come to a stop near some particular level as (A) above, or whether they will continue to rise. In the event of turbulent flow over the surface he must decide whether a sufficient degree of heating or cooling will also take place as the air moves to cause the air to become more stable or more unstable (cf. the paper by R. M. Poulter cited at the end of this chapter).

Further, the stability or otherwise of an air current is complicated by the existence of water-vapour. Rising air cools and may reach a level at which the water-vapour begins to condense.

Just as the process of evaporation, that is the conversion of a substance from the liquid into the gaseous state at the same temperature, requires heat, so the process of condensation liberates heat. Suppose therefore that the temperature of the air at the surface is

---

PLATE IV*a*: Summer afternoon clouds over the English Channel; an infra-red photograph from 20,000 feet. The line of cumulus clouds formed a little way inland where the sea breeze rises is a frequent development when the general direction of the wind lies along the coasts. Distant alto-stratus with some cirro-stratus above.

b: View over N.E. France looking westward from 26,000 feet, June 1944. Lower cloud sheet breaking into normal summer cumulus below. Distant frontal cloud spreading over southern England.

50° and the moisture content is such that the relative humidity is 80%. If air at a temperature of 50° is saturated it contains 9·33 grammes per cubic metre of water-vapour; hence the sample in question must contain 80% of this, *i.e.* 7·46 grammes. We find from the tables that this is the amount appropriate to saturated air at 43·7°F. Therefore if a packet of this air rises from the surface its temperature will begin to fall at the expected dry adiabatic lapse-rate

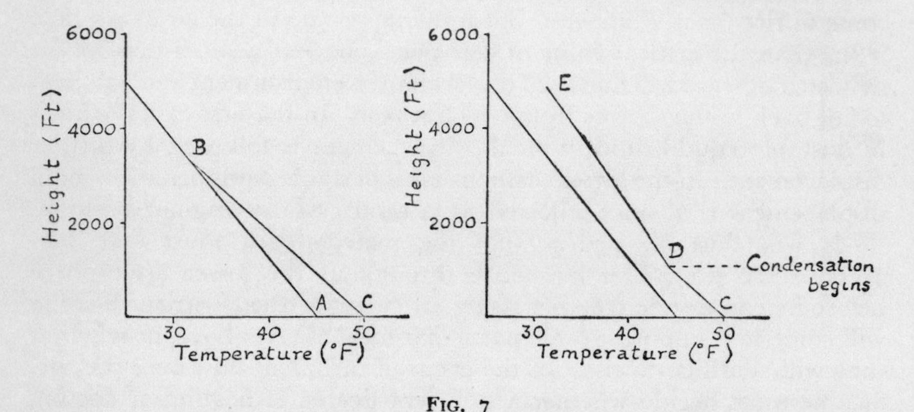

FIG. 7

Diagrams illustrating the behaviour of rising air masses up to 5,000 feet *left.* Warmer unsaturated air at C is 5° warmer than its environment A; it can rise to B. *right.* Rising warm air from C reaches saturation at D; as it is still warmer than its environment it will continue to rise, and its temperature will fall as for saturated air along D–E

of 5·4°F. per 1,000 feet, and a temperature of 43·7°F. will be attained at about 1,200 feet; at this level, therefore, the air will be saturated. Should the air continue to rise, expand, and cool, some of its moisture must continually condense. But as the process of condensation liberates heat, this will be communicated to the rising air mass, and the resultant rate of fall of temperatures will be considerably less than 5·4° per 1,000 feet; at the temperatures given above the lapse-rate for saturated air is about 2·8° per 1,000 feet rise. On account of the fact that very warm air holds much more moisture, when condensation begins the amount of heat liberated is greater; hence at temperatures round 90° the saturated adiabatic lapse-rate has decreased to about 2° per 1,000 feet, while at zero it is nearly 4° per 1,000 feet.

It will therefore be evident from the adjacent diagram (fig. 7) that if a warm bubble of air rises in an atmosphere in which the lapse-rate is represented by AB, and the warm bubble has initially a temperature C, much depends on whether saturation is reached. If the rising air is initially moist and saturation is reached at a low level (D) the temperature of the rising packet of air will decrease along a line DE so that it will remain warmer than its surroundings and continue to rise. This

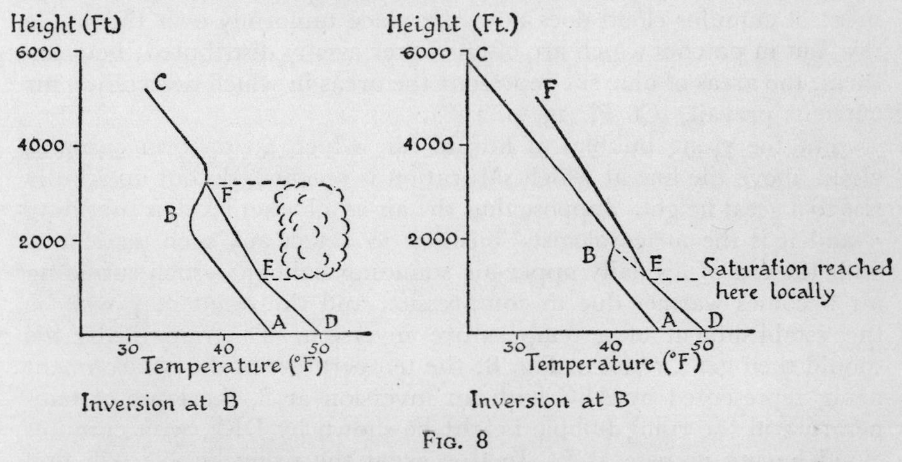

FIG. 8

Diagrams illustrating typical behaviour of rising air masses up to 5,000 feet *left.* Cloud, if formed, does not ascend above F. *right.* Cloud, if formed, ascends rapidly to great heights. Saturated air at F is still considerably warmer than its environment given by B–C

is the normal result of convection over heated ground when the air supply is moist. Indeed on occasions the rise becomes so rapid that the vertical currents assume the dimensions of a gale. Glider pilots and others, caught in the centre of large cumulus clouds, have at times reported rising currents of 50–60 feet per second even in Britain. Abroad, where humid air masses at high temperatures are most common, the very gradual fall of temperatures represented by the saturated lapse-rate means that on many occasions the difference of temperature between the rising air mass and its surroundings tends to increase with height and the uprush of air becomes even more violent. This fact helps to explain the greater height reached by cumulonimbus clouds in regions such as the Southern United States and the greater intensity of all the associated phenomena—lightning, hail, violent up-currents and the like.

In practice our notions of the behaviour of rising air 'bubbles' should be modified to some extent to allow for some mixture along the boundary as they rise. Further, once convection is established on a fine day the rising bubbles, breaking away from the surface every few minutes, are replaced by descending air which is in turn heated from below. Hence a circulation is set up which explains why the development of cumulus cloud does not take place uniformly over the whole sky, but in patches which are often rather evenly distributed; between them, the areas of blue sky represent the areas in which descending air currents prevail. (Cf. Pl. 23, p. 138).

But the rising bubbles of humid air, which go to form cumulus cloud above the line at which saturation is reached, do not invariably rise to a great height. Suppose that the air at a higher level, is subsiding —and it is the meteorologists' business to detect any such tendencies with the aid of his daily upper-air sounding balloons—such subsiding air becomes warmer due to compression and the result may well be the establishment of a temperature inversion. Diagramatically we should then get a result as Fig. 8: the temperature of the environment being represented by ABC with an inversion at B, the lapse of temperature in the rising bubble height be shown by DEF, with cumulus cloud having its base at E. In this event the rising bubble will find itself at the same temperature as the environment at the point F. If it were to climb farther, it would immediately be colder than its surroundings and accordingly must sink back.

We very frequently find that the upward growth of cumulus clouds is checked in this way, especially when Britain lies in the westerly air stream on the northern flank of a summer anticyclone. In such circumstances the surface air stream is fairly moist and is warmed up in the daytime over the land. But as subsiding air is characteristic of anticyclones we frequently find that a slight inversion, sufficient to check the upward growth of cumulus, is present at a height of the

---

PLATE 5

a. LAKE DISTRICT: broken strato-cumulus in turbulent air over mountains at sunset, October. Westerly wind, maritime polar air.

b. HERTFORDSHIRE: tailed and plumed cirrus, September evening. Wisps probably indicative of strong wind at high levels with eddies forming the "plumes"

PLATE 5

B. A. Crouch

Cyril Newberry

B. A. Crouch

Cyril Newberry

order of 3,000–4,000 feet. Hence the frequent sight of the fleets of 'fair-weather cumulus'—so-called because we can be sure that if the cumulus clouds show no great vertical development there will be no chance of showers. On the other hand, if once they reach the ice-crystal level—shown by the tell-tale fibrous appearance of the upper edge—as towering cumulo-nimbus, showers may at once be expected. The part played by the ice crystals is mentioned earlier (p. 38).

FIG. 9
Flow of air over hills
left. Stable atmosphere, cloud rising to no great height over hills and generally smooth outline of upper surface. *right.* Unstable atmosphere, cumulus cloud rising to a much greater height. Compare Fig. 8, p. 49

It will be evident that if the warm air at a higher level is replaced by colder air, the likelihood of an unstable atmosphere in which vertical currents, once initiated, can ascend to great heights unchecked is greatly increased. The diagrams above are simplified, and the actual state of the atmosphere as shown by upper air soundings is often much more complicated. It will be evident that there will be many occasions

---

PLATE 6
a. ESSEX: late summer sunset, September. Daytime strato-cumulus decreasing and settling down to patchy stratus.

b. ALLESTREE, Derbyshire: summer sunrise, August. Underside of alto-cumulus sheet illuminated by the rising sun, still below the horizon; red rays predominate through surface haze. Higher and more distant cirro-stratus already fully lit in clear sunshine.

when the atmosphere will remain stable if the rising air at any given level remains unsaturated; but if, locally, saturation is just reached the resultant instability may be marked.

Moreover, with a moist wind blowing the effect of a mountain range is frequently to cause much more cumulus development over the summits than over the plains below. Nothing is more characteristic, with a light westerly wind on the northern margin of a High, than the clusters of cumulus among and above the hills of Snowdonia or the Lake District while over the sea and the coastal plains the sky remains clear (Plate 32, p. 223, of the Leven-Kent estuary).

> "Over the smooth sands
> Of Leven's ample estuary lay
> My journey, and beneath a genial sun
> With distant prospect among gleams of sky
> And clouds, and intermingling mountain tops . . ."
> (Wordsworth, *The Prelude*).

This can be shown to be partly due to the lifting of the whole block of air, and partly to the localised heating of the ground surface either in constricted valleys, or at higher levels on the uplands. Instability is indeed more likely to develop over heated upland, for in sunny weather the ground heating is just as great as it would be in the plains, while at the higher level the prevailing atmospheric environment is cooler. For this reason some of the Western Pennine valleys in which unstable humid air from the south-west finds itself heated and constricted show rather a high frequency of summer thunderstorms (cf. p. 264; also the paper by Sir Charles Normand cited at the end of this chapter).

It is not intended that this chapter should swell into a meteorological text, but a brief mention should be made of another characteristic climatic feature so frequently superimposed on any British prospect, namely, the tendency to grouping and alignment of the fair-weather cumulus cloud which is so common. Cumulus growth in calm weather can be regarded as the result of a series of 'cells' in each of which ascending air in the centre is complemented by descending currents on the exterior. If a breeze is blowing the clouds drift gently over the country. But if at the level of the top of the clouds the wind direction is appreciably different in speed and direction from that at the surface, the clouds dispose themselves in different patterns; sometimes these take the form of 'cloud-streets' or longitudinal rolls aligned roughly in the direction of the surface wind, but at others they lie transverse to the wind. On other occasions a sheet of cloud breaks up as a result of convection set up within it into a series of rounded masses indicating that a vertical circulation has been set up in the cloud sheet, leading to regions of ascent (where the masses of cloud are thick) and descent where the cloud is thin, or absent. Assemblages of evenly-disposed cloudlets whether at low or at high levels, in the form of more or less globular masses are very familiar to all who observe our British skies. Other familiar cloud formations arise, as we shall see, in association with 'waves' forming along the boundary between upper and lower streams of air. For a detailed account of the recent theories of formation of some of our familiar cloud-patterns reference should be made to papers by Sir Gilbert Walker and Sir David Brunt, in the *Quarterly Journal of the Royal Meteorological Society*, 1932 and 1937.

The development of longitudinal cloud streets under convection is of especial importance to the glider pilot. Providing as they do continuous lift over a long stretch of country, with this aid flights of many miles can successfully be made.

At this point the reader may well ask whether in extreme instances any limit is ultimately imposed on the vertical ascent of moist air from the surface and the consequent upward growth of cloud. There is indeed an upward limit imposed at the base of the stratosphere. Disregarding small inversions, which in general are not more than a few hundred feet thick, it is universally found that temperature decreases with height up to a level which averages ten miles above the surface at the equator, nearly seven in Britain (33,000 feet) and about four miles over the Poles. Above that height balloon ascents, and now aeroplane ascents, reveal that the temperature remains stationary or begins slowly to rise. The region in which the temperature as a whole falls with height is called the troposphere; that above is the stratosphere.

Rising air masses cannot climb through an inversion (p. 49) and as far as the troposphere is concerned the base of the stratosphere fulfils the same function. The maximum height to which moist air from the surface can rise is thus determined; and for practical purposes no cloud whatever is observed in the stratosphere. Indeed recent investigations by Professor Dobson and his team at Oxford go to show that practically no water-vapour is present in the stratosphere at all. The height of the base of the stratosphere is known to vary; it is generally higher over anticyclones and may attain 43,000 feet or so. This may be regarded as the greatest height at which any cloud is ever to be seen over Britain. The tops of the highest cumulo-nimbus may occasionally tower to nearly 30,000 feet. Hence, on a clear winter day with unstable polar air, the tops of high-piled cumulo-nimbus are characteristically visible around the horizon at great distances, upwards of a hundred miles being common. They are thus a common ingredient of the stormy winter sunsets over the sea.

For it will now be evident that convectional activity is not confined to the land on warm summer days. In winter it is the sea that is the source of warmth, and instability is liable to develop as we shall see whenever a colder air stream blows across it. Convectional activity too is not confined to the daytime under such conditions, hence towering shower clouds are equally to be seen at night when in winter the hard north-wester from Greenland roars over the Hebrides behind a vigorous depression.

Those whose interest has been aroused and who wish to pursue these problems of the physics of the atmosphere may be commended to a variety of reading matter. In addition to Sir David Brunt's delightfully clear introduction to meteorology (*Weather Study*, London, Nelson 1941), in which more advanced texts are named, readers of different tastes may appreciate the following recently published works, all of which are generally obtainable.

(i) KENDREW, W. G. (1949). *Climatology*. Oxford, University Press. A clear and comprehensive text book of wide appeal.

(ii) MILLER, A. A. (1950). *Climatology*. London, Methuen. A comprehensive introductory text especially addressed to the university student; the latest edition contains valuable additional matter.

(iii) BOTLEY, Cicely M. (1948). *The Air and its Mysteries*. London, Bell, latest edition. A charmingly concise and most informative smaller book which will be found very acceptable for the armchair.

(iv) BILHAM, E. G. (1938). *The Climate of the British Isles*. London, Macmillan. The well-known standard résumé of the material collected under the auspices of the Meteorological Office, with informative tables and diagrams; an indispensable reference. To this add the *Climatological Atlas of the British Isles* (M.O. 488: H.M.S.O. 1952).

(v) BROOKS, C. E. P. (1949) *Climate through the Ages*. (second ed.). London, Benn. The standard work of its kind in English; an invaluable digest of an enormously wide subject.

(vi) BROOKS, C. E. P. (1950). *Climate in Everyday Life.* London, Benn.
(vii) BRUNT, D. (1939). *Physical and Dynamical Meteorology,* Cambridge, University Press. For more advanced students with a sound mathematical and physical background. Cited repeatedly in the meteorological literature of every country in the world.
(viii) KIMBLE, G. T. (1951). *The Weather* London, Pelican Books. An exemplary introduction.
(ix) HARE, F. K. (1953). *The Restless Atmosphere.* London, Hutchinson. Recommended; compact and stimulating, especially for students.

To these may be added two well-known comprehensive Service text-books published by the Stationery Office:—*Admiralty Weather Manual* (A. G. Forsdyke); *Meteorology for Aviators* (R. C. Sutcliffe), together with the invaluable *Meteorological Glossary* and *The Weather Map* obtainable from the same source; these last are still astonishingly good value at five shillings each. Coast-dwellers will appreciate *Meteorology for Seamen* (C. R. Burgess, 1950. Brown and Ferguson, Glasgow).

SOME REFERENCES

BREWER, A. M. (1946). Condensation Trails. *Weather, 1:* 34–40.
BRUNT, Sir David (1937). Natural and Artificial Clouds. *Q. J. Roy. Met. S. 63:* 277–88.
DARLING, F. Fraser (1947). *Natural History in the Highlands and Islands.* London, Collins' *New Naturalist.*
DOBSON, G. M. B. (1946). Temperature of the Upper Atmosphere. *Weather, 1:* 58–65, 73–77, 115–22.
(1949). Ice in the Atmosphere. *Q. J. Roy. Met. S. 75:* 117–30.
DURST, C. S. (1940). Winter Fog and Mist Investigation in the British Isles, 1936–7. Meteorological Office (M.O.M. 302).
(1949). Meteorology of Airfields (M.O. 507). H.M.S.O.
HEYWOOD, G. S. P. (1931). Wind structure near the ground and its relation to temperature gradient. *Q. J. Roy. Met. S. 51:* 433–55.
HOWARD, Luke (1803). *On the Modifications of Clouds.* London.
LUDLAM, F. H. (1951). The production of showers by the coalescence of cloud droplets. *Q. J. Roy. Met. S. 77:* 402–417.
NORMAND, Sir C. (1938). On instability from water vapour. *Q. J. Roy. Met. S. 64:* 47–68.
PETTERSSEN, S. (1941). *Introduction to Meteorology.* New York and London, McGraw-Hill.

POULTER, R. M. (1938). Cloud Forecasting: the Daily Use of the Tephigram. *Q. J. Roy. Met. S. 64:* 277–92.

(1946). The Depth of Forecasting. *Weather, 1:* 137–40.

SCHUMANN, T. E. W. (1938). The Theory of Hailstone Formation. *Q. J. Roy. Met. S. 64:* 5–20.

SHEPPARD, P. A. (1947). The constitution of clouds and formation of rain. *Science Progress, 35:* 185–92.

SIMPSON, G. C. (1941). On the Formation of Cloud and Rain. *Q. J. Roy. Met. S. 67:* 99–133.

SMITH, K. M. (1946). Forecaster's Progress: the Tephigram. *Weather 1:* 233–38.

SWINBANK, W. C. (1943). Synoptic Division Technical Memorandum 52. Meteorological Office, London.

TAYLOR, G. I. (1917). The Formation of Fog and Mist. *Q. J. Roy. Met. S. 43:* 241–68.

WALKER, Sir Gilbert (1933). Clouds and cells. *Q. J. Roy. Met. S. 59:* 389–96.

NOTE ON THE LONDON FOG, DECEMBER, 5-9, 1952

This four-day fog, during which the temperature in the Thames Valley remained persistently near freezing point, gave rise to peculiarly unpleasant consequences. Much throat irritation, bronchitis and pneumonia developed, with a sharp rise in the death-rate, especially of the elderly. Discussions focussed on the probable accumulation of sulphur compounds, among others, in addition to the coarser products of combustion. (Cf. Q. J. Roy. Met. S. 80. 261-278).

CHAPTER 4

THE ATMOSPHERIC CIRCULATION OVER THE BRITISH ISLES AND NEIGHBOURING SEAS

Be patient, swains; these cruel-seeming winds
Blow not in vain.

THOMSON : *The Seasons*

OUR ISLANDS occupy a northerly latitudinal position in which there is a great difference between the power of the summer sunshine and that of winter, arising largely from the much lower angle of elevation of the sun's rays above the horizon in winter and partly from the length of the day. At that season the rays must traverse the atmosphere by a much longer path with the result that a greater proportion is absorbed or scattered on the way. To this we may add that much of the remainder is reflected from the upper surface of the extensive low cloud of winter. Even with clear skies we must also at times allow for reflection by a snow surface of much of the incident radiation during the day, and consequently little gain of temperature by the adjacent air compared with the nocturnal loss. Over the earth as a whole about half the total solar radiation incident upon the atmosphere reaches the earth's surface. Over open country that part which has penetrated the atmosphere is very largely absorbed in a thin layer adjacent to the ground surface. Because of this absorption of the warmth in a thin layer the air in contact with the ground is also warmed; whereas at night the loss of heat by outward radiation from the surface is accompanied by cooling of the adjacent air. Some of the outgoing and incoming radiation is absorbed by the air; on balance under normal conditions the temperature of the surface air reaches a minimum just before sunrise, and a maximum about two hours after noon. This lag is due to the finite time necessary for the heat to be transferred upward from ground to screen level. On the Eiffel Tower in summer the maximum is normally recorded about 4 p.m.

Associated with the daily rise of temperature and the consequent stirring-up of the surface air the average wind speed at ground level also tends to attain a maximum early in the afternoon. This daily stirring-up goes far to explain why, in higher latitudes where there is little winter sunshine, the daily range of temperature at ground level is still noticeable. The diagram below (fig. 10, p. 60), showing the normal diurnal range of temperature at Parc St. Maur in Paris and at the top of the Eiffel Tower (984 ft.) shows how great a part is played by surface heating and cooling. Similar effects are found on isolated mountain summits, for example on Ben Nevis.

With us the observed daily range between maximum and minimum depends not only on the march of the sun; it is also affected by changes in the source of the air, and by the cloudiness or otherwise of the sky and the extent to which the surface layers continue to be stirred up at night. In winter the cumulative effects of such changes are sometimes large enough to make the maximum for the 24 hours occur during the night. The range of temperature is also much lower by the sea, where the air is much more likely to be in motion as a result of the temperature difference between sea and land. The day-to-night variation of sea surface temperature is almost negligible compared with that of the land, largely because water is semi-transparent to much of the solar radiation. Hence the heating effect must be spread over a considerable depth rather than a mere few inches. Statistics show that at small islands in the Hebrides where cloud, wind and maritime location combine their effects the average range between the daily maxima and minima is about half that at inland stations.

In Britain, as the tables show, the average daily range varies between about 6° in December and 12° in June at island and coastal stations, and about 10° in December to 20° in June at inland stations.

PLATE 7

a. TANERA BEG, Summer Isles, N.W. coast of Ross-shire: May. Fine clear sky above, patchy strato and stratus-cumulus developed over the sea.

b. EXMOOR, Somerset: March. Cumulus nearly becoming cumulo-nimbus. Unsettled showery weather in unstable maritime polar air.

PLATE 7

James Fisher

B. A. Crouch

Cyril Newberry

Cyril Newberry

On the great majority of days in these islands the air to a depth of many thousands of feet is moving more or less rapidly. To us, therefore, it is in general more important to know where the air has come from; and afterwards to consider what the sun, for example, will be capable of doing to it when it arrives. The temperatures we experience on any given day of summer or winter depend very largely on the origin of the air and its life-history, that is, on what has happened in that air on its way to our shores. So far as our perceptions are concerned the effects of solar heating or nocturnal cooling are generally secondary in importance to those arising from the qualities of the air masses we receive.

We must, therefore, consider why these islands are relatively breezy, and why they are subject to such marked and frequent changes in the qualities of the air. This involves a summary of the average distribution of surface barometric pressure. On our rotating globe with its surface composed of land and sea and subject to isolation through a partly absorbent atmosphere, our atlases remind us that an average disposition of surface pressure arises such that pressure is generally high in the central Atlantic and low from South Greenland to North Norway.

It will at once be evident that we in the British Isles lie in winter on the south-eastern side of the region in which low pressure predominates near Iceland. To the eastward, pressure rises steadily in the direction of the 'great Siberian High'; this is one of the most prominent features of any pressure map of the Northern Hemisphere between November and March. In the central Atlantic between the Azores and Bermuda the average pressure is also higher than it is to the north and south.

In summer this 'Azores High' occupies a more northerly latitude and becomes more emphasized by comparison with the average

---

PLATE 8

a. DUNSTABLE DOWNS, Bedfordshire: April. Gliding site of the London Gliding Club. Fine April afternoon with detached cumulus bounded by an inversion above. Moderate N.E. surface wind blowing from the cool North Sea. Temperature 52°.

b. ALLESTREE, Derbyshire: August afternoon. Cirro-stratus above strato-cumulus, heralding the approach of a depression. Note how the decreased convection, due to the oncoming cirro-stratus, has led to the degrading and flattening of afternoon cumulus.

pressure over the continents in the same latitudes; and while the 'Icelandic Low' is still present, it is much less marked than in January. To the eastward, pressure is low over south-west Asia throughout the summer.

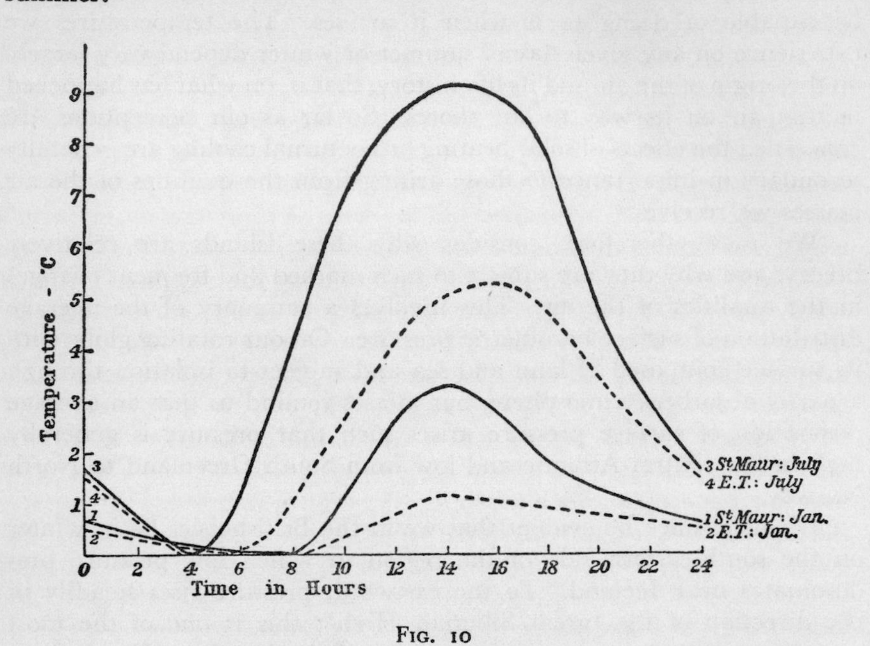

FIG. 10

Diurnal range of temperature at Parc St. Maur Paris, and at the top of the Eiffel Tower (from Lake's *Physical Geography*, by courtesy of the Cambridge University Press)

Owing to the earth's rotation air in motion in the northern hemisphere is subject to a force, the so-called 'deflecting force', proportionate to its velocity and acting at right angles to the right of its path. If then the force on the air arising from a difference of pressure, and acting from high to low pressure is exactly balanced by the deflecting force, these two forces must act in opposite directions so that the air will move along the isobars with low pressure to the left. The greater the pressure force, *i.e.* the closer the isobars, the greater the deflecting force required for balance, and hence the greater the wind speed. It is in fact found that the winds of the free atmosphere at about 2,000 feet around Britain conform quite closely to this balance.

At the surface however, the movement of the air is less rapid owing to friction and the deflecting force and pressure force are not so well balanced, hence while we observe that the direction of the surface wind lies somewhat across the run of isobars, the deviation as a rule is of the order of 10° over the sea, and 30° over the land.

At 2,000 feet over the sea or an open low-lying plain, however, the frictional effects are less, and both the speed and direction of movement of the air closely approach those which would prevail in theory. Hence if we observe the surface wind direction in the open country on a breezy day, we shall find that the motion of low clouds at 2–3,000 feet level is steadier and more rapid than that of the surface wind, and from a direction a few degrees farther round the compass.

The effects of friction can be illustrated by the rough rule that in windy weather, the surface wind at sea blows at about two-thirds the speed of that at 2,000 feet; inland among trees and buildings the average speed may be less than one-third of this. Occasional gusts, however, approach the velocity recorded in the free air above. With fairly closely spaced isobars and a west wind of Beaufort Force 8 at 2,000 feet (40 m.p.h. *i.e.* gale force) we may find a mean wind speed over an hour of about 25 m.p.h. (between force 5 and 6) at, say, Blackpool; and 15 m.p.h. (force 4) at typical inland stations. But occasional gusts even inland may attain 40 m.p.h. and it will readily be understood that the fluctuations in wind speed, between gusts and lulls, are greater and more frequent at the inland station.

The considerable variations in the mean wind speed at British stations arising from exposure produce marked results with regard to vegetation, and hence greatly affect the appearance of the landscape. From the point of view of our own perceptions these variations are extremely noticeable, and go far to explain the varying climatic reputation of many places of resort. Examples abound; the traveller on a northbound express on the old North-Western route cannot but notice the gradual change in the attitude of exposed trees, particularly as the line quits the relative shelter of South Cheshire in the lee of Wales for the windier Lancashire plain north of the Mersey. If from Leeds he takes the old Midland route through the Pennines, the roar of the south-wester as the train crosses the exposed viaduct of Ribblehead goes far to explain the struggling ash beside the fellside farm. We can be sure that in the eighteenth century the beauty of Wetheral Woods flanking the deeply incised Eden was the more appreciated by north-bound travellers after the miles of battering rain on the top of a coach over Shap. And to-day no Cambridge man will deny the significance of wind across the Fenland by contrast with the shelter of a great city, when on a February night the searching north-easter scours the station platform as he dismounts from the London train. Neither will Edinburgh men forget the bleakness of Carstairs Junction when on such a night they too are required to leave the Glasgow-bound express.

Variations in the strength of the surface wind arise from many other features. The trend of water inlets such as the Solent (fig. 44, p. 147), the Bristol Channel, or the Firth of Clyde leads to some 'canalising' of the flow of air which must be allowed for by all whose business takes them into the neighbourhood of the water. Steep slopes and summits experience additional strength in the wind from particular directions. There is for example, a peculiar bleakness even in the northern Chilterns near Whipsnade, on days when a strong north wind crossing the snow-covered Midlands besets those exposed slopes. Much may depend on the detailed local contour of the hills; most lee slopes are sheltered, but some are surprisingly harried by wind. A noteworthy example is the south-west slope of the Crossfell escarpment; given the right conditions the north-easter blows down this treeless scarp with exceptional local strength, as the widely known 'helm wind' (see p. 148).

All these effects arise as a result of the varying amount of distortion and frictional disturbance of the great streams of air crossing the country; more must be said about the origin and characteristics of these air-streams.

The average distribution of pressure outlined above indicates at once that for a great part of the year air will flow over these islands more or less in accordance with the average isobars, namely from south-south-west in mid-winter, west-south-west in summer. Moreover, the average pressure gradient, with pressure falling from S.E. to N.W., is much greater in winter; hence we might correctly deduce that stronger winds tend to prevail then. But the frequency with which the wind comes from all other points of the compass besides those mentioned, leads us to recognise at once the extent to which the systems of pressure indicated above are only the 'average' of a very large number of daily readings of pressure taken over many years. From day to day the distribution of pressure and the direction and strength of wind can vary very widely from the 'average state of affairs' shown above, so that while far more often than not our air supply has recently crossed a wide stretch of sea it does not necessarily arrive from south-westerly points.

Barograph records soon reveal that pressure during a whole month fluctuates considerably; and instead of merely considering average pressure for the month, we should instead give more attention to the results of the movement of those regions of high and low barometer which give rise to the barometric fluctuations we observe and which we call anticyclones and depressions. Hence we can first embark on a short review of the characteristics of the majority of depressions in our part of the world.

If, for a moment we consider the average distribution of pressure shown in the diagrams as the result of planetary conditions it is clear that in a region like the North Atlantic very different types of air will frequently be juxtaposed. Air which has followed the isobars round the Azores High has moved slowly (in accordance with the gentle pressure gradient) over a long stretch of warm ocean. This air spreads northward as a wind from the south-west over a sea whose surface gradually becomes cooler. Hence the surface layers of air tend to fall in temperature, and at the same time their relative humidity rises nearer the saturation point. Very little lifting of the surface air is now required to form cloud and hence the northern parts of the middle Atlantic are a region in which extensive low cloud is frequently found.

But, if pressure is low in the neighbourhood of Iceland, the air starting as a north-wind and following the isobars from the direction of Greenland is found spreading from westerly points across the middle North Atlantic. This air is moving from a cool source; possibly some part can be attributed to cooling over the Greenland plateau, but probably much more is due to the neighbouring ice-covered coastal waters. It then moves across an open water surface whose temperature increases to the southward. By the time it comes up against the air that has come round the Azores High it differs markedly in character; it is still considerably colder than the sea over which it is moving, whereas the air from the flanks of the High is warmer. Two such air streams of different temperature, and, therefore different density, do not quickly mix; the lighter warmer air flows over the colder and denser air.

We may now adopt the meteorologists' language and call such streams respectively 'tropical' and 'polar'. In our latitudes the term tropical is conveniently used for air whose temperature in the lower layers as a whole remains above that of the sea surface; while in polar air the surface layers are cooler than the sea surface over which they are moving. Considering for convenience of description, the North Atlantic, it will be observed that throughout the year surface pressures tend to rise a little in high latitudes towards the pole. On the edge of this region of slightly higher pressure air more or less follows the isobars but with an outward component of motion; expressed otherwise, north-easterly winds are dominant in high latitudes.

At first it might appear that the north-easterly stream of polar origin, and the south-westerly stream from the Azores High might flow parallel to each other in opposite directions on either side of a well-defined boundary. In practice, however, such a system never persists. If we plot very carefully the observations of pressure, temperature, cloud and weather over such a boundary we soon find that conditions at the boundary are evidently unstable. Small masses of air may rise on very slight provocation and evidence of the process will be seen on the map as a very small area of slightly lower pressure at the boundary of the currents. But as soon as this occurs the air on either side of the boundary which we may now call the 'polar front' deviates slightly to follow the isobars round the incipient low. Such a movement leads at once to the stage shown in the third diagram below. Warm moist air rides forward and over the colder surface air north of the boundary.

West of the centre, a tongue of cold air pushes southward, and in the form of a nose undercuts some of the warmer air at the surface. Near the tip of this advancing nose of cold air the moist warm air from the south is rapidly elevated; a belt of clouds and showers is characteristic of this area. If observations are plotted for some hours the air is soon seen to be roughly following the isobars round the centre of a well-defined area of lower pressure which may after a day or two be upwards of a hundred miles in diameter.

East or north-east of the centre moist 'tropical' air from the south is ascending steadily over the colder surface air beneath; accordingly a wide area of heavy low cloud develops in which the droplets are continually being formed and enlarged. Hence continuous precipitation occurs in what we frequently describe as 'the rainy sector of a depression'.

Once initiated and developed in this manner, the whole system tends to move more or less eastward along the boundary formed by the two types of air. Invariably, however, the protruding tongue of cold air on the south-western flank advances relatively rapidly and after some days careful plotting of the observations reveals that it has overtaken the line which we call the warm front; bounding the warm

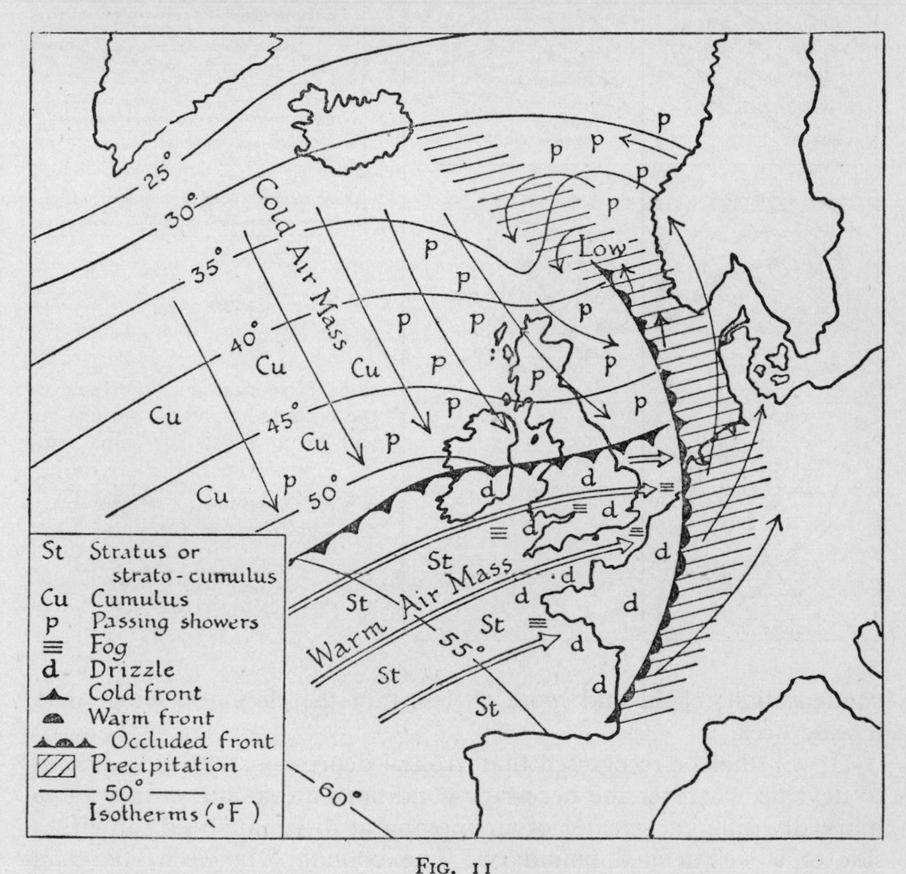

FIG. 11
Meeting of Polar and Tropical air masses (based on a diagram by Petterssen, by permission of Messrs. McGraw-Hill)

air and the rainy sector. The whole of the advancing warm air supply has thus been undercut. When this process has occurred the depression is said to have been 'occluded', and a sharp difference of temperature is no longer to be found at the surface. The warm air above, however, is still as a rule undergoing uplift, and hence cooling with further

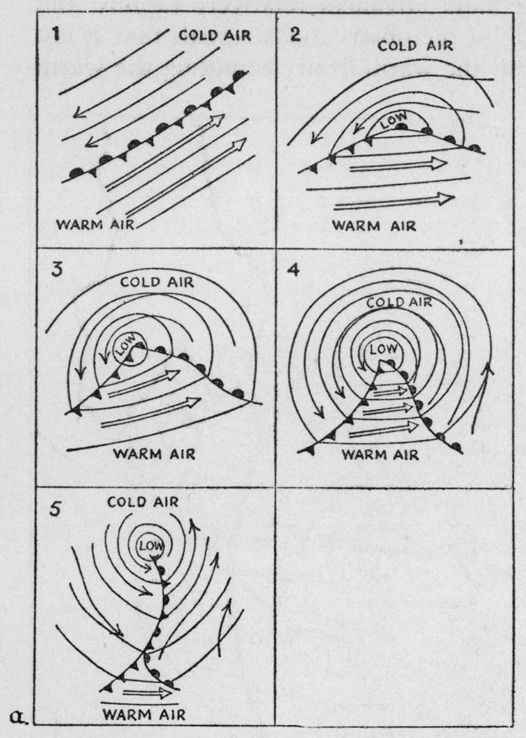

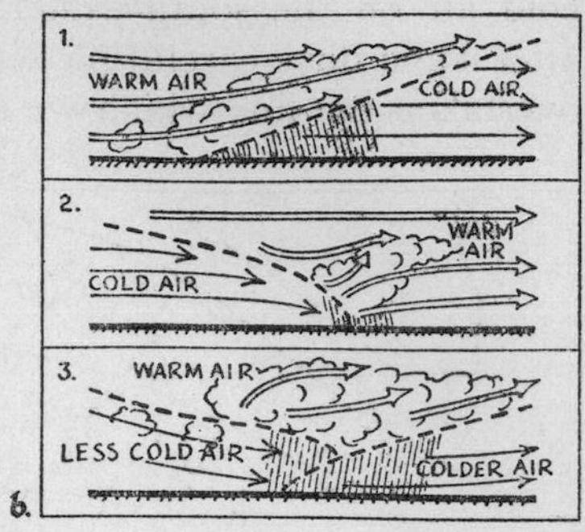

Fig. 12

a. 1–4. Stages in the growth of a depression. 5. Occluded depression. Note that on either side of the occluded front the surface air will be at much the same temperature.

b. Vertical section: 1. Warm front. 2. Cold front. 3. Occluded front

condensation; cloud and more or less rain therefore still occur over a wide area.

It will thus be recognised that 'frontal depressions' can be expected to develop wherever the necessary conditions are found, namely, two markedly different streams of air moving at different speeds on either side of a well-defined boundary. The boundary between cold and warm air masses in the North Atlantic is found repeatedly in the neighbourhood of Newfoundland—Iceland—N. Norway, as we might well surmise from the geographical conditions in that latitude. The great majority of the depressions which affect us originate in this region and their centres travel from S.W.–N.E. along tracks to the north of Scotland. As each low passes the barometer falls and rises so that throughout the year there are frequent and considerable fluctuations; these in general are much greater in winter. Winter is the season of deeper depressions, steeper pressure gradients from margin to centre, and stronger winds. The energy of the depressions is no doubt associated with the extent of the contrast between the two air masses. In winter the seas to the south of Iceland are still open, and relatively warm; in January the average surface temperature is still 50° off N.W. Ireland, and 40° on the S. coast of Iceland; not more than 10° cooler than in July. But a little to the northward, for example in East Greenland in 70° N., coastal stations show that the air in January is on the average as much as 40° colder than July. This gives an adequate reminder of the greater intensity of winter contrasts across the polar front.

It will now be evident to the reader that the Icelandic Low shown on the atlas maps of average pressure for January is not an indication of permanent low pressure throughout the month. Such a map is rather to be interpreted as a demonstration that the centres of the deepest depressions are most often to be found passing over that region; on any given day, however, it is quite possible that for a brief interval after the passage of a low, pressure over Iceland may be quite high until the approach of a new low is heralded by a new fall of pressure. Further, the tracks followed by our Atlantic lows vary very much; their speed of movement varies too. It is broadly true that the majority of lows by the time they have reached the British coasts are occluded, and that occluded lows move more slowly and rather erratically. This incidentally provides additional difficulties for all who would forecast British weather. Depressions, once occluded, begin to fill up more or less rapidly.

The average life of a well-developed North Atlantic Low travelling from near south Greenland to the north coast of Norway is of the order of 5–7 days. The speed of movement, however, varies considerably, and while the great majority move more or less eastward, great variations in detail are observed. The diameter of the area affected by the circulation in a fully-developed depression is commonly of the order of 500–800 miles, but may range from less than 100 to as much as 2,000 miles on occasion. For a well-developed anticyclone 1,000 to 1,500 miles is normal in our latitudes.

Many depressions also approach us on more southerly tracks; for example, their centres may move along the south coast of Ireland and up the English Channel. Frequently such lows develop as secondaries to a primary centred north of Scotland; for just as the primary developed on the boundary between tropical and polar air south of Iceland, when the nose of polar air advances southward, the cold front at its tip is equally a line across which a marked contrast of temperature exists and it is very common for additional lows to be initiated upon it, and to approach us from a more southerly part of the Atlantic. It is also very important to recognise that contrasting types of air sufficient to engender active fronts may arrive from the neighbouring continent, especially in winter when the oceanic air is commonly decidedly warmer than that lying over Central Europe. Even on our Atlantic coasts a distinction can frequently be observed between polar air which has had a long fetch over the ocean, and that which has arrived by a shorter route. For this reason it is commonly found that several minor fronts, each accompanied by a belt of cloud and more or less precipitation, are to be found over the eastern Atlantic, marking the boundary between polar air-streams with a longer or shorter track over the ocean waters.

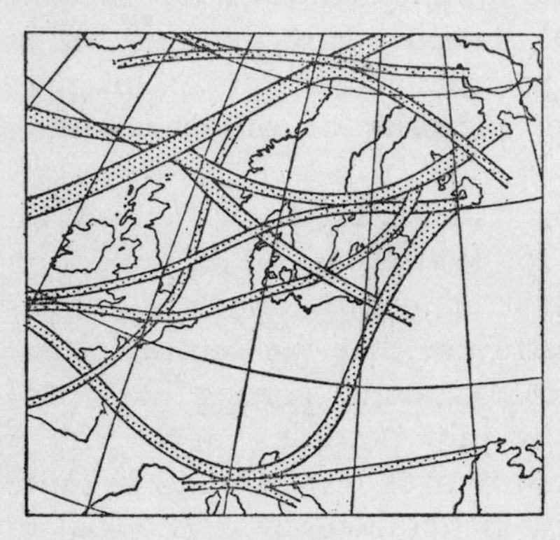

FIG. 13
Characteristic tracks of centres of depressions

Much of our rain in Britain falls in association with the passage of occlusions. As we have seen the passage of an occlusion is not marked by any sharp change of temperature. In theory, as the diagrams show, the air mass present at the surface is the same throughout.

But in practice it is generally found that the polar air which has come all the way round the low does differ in temperature a little from that which it is now overtaking. In winter for example such cool air may have travelled from Scotland across a wide stretch of the Atlantic before it swings back across England and the North Sea; when that has happened it is often a degree or two warmer than at the start. Similarly, in early summer there are times when the surface layers of the air crossing a wide stretch of the cool sea may be cooled a little compared with the temperature at the start.

Hence we find that there is often a *small* rise or fall of temperature as an occlusion passes, and hence 'warm' and 'cold' occlusions may be distinguished. This is noteworthy because the belt of rain associated with a cold occlusion is generally narrower than with the warm type. But when we consider what complicated tracks the air can follow among our islands, for example the effects of the varying breadth of the Irish Sea on the temperature of the air carrying it, it is clear that occlusions cannot always be easily labelled. The prediction of the precise length of time during which the rain is likely to continue is therefore not always practicable as yet.

It is important to remember that as the passage of occlusions is often an indication that a depression is coming to a stop or filling up, winds are in general nothing like so strong as with an active cold or warm front. Slow-moving occlusions in summer, in which a great deal of very moist warm air has entered above, are often productive of several hours of nearly windless rain. The early days of August 1948 were marked by the passage of several such occluded fronts over Southern and Central England, and many places accordingly received more than the normal expectation of rain for the whole month within the first week.

Persistent heavy rain with little wind is associated in the minds of many British travellers with summer holidays on the Continent. From what has been said it will be clear that the slow-moving occlusions associated with the shallow lows of summer are equally likely to develop where humid air supplies have spread into Europe. The present writer remembers too well the horrid puddles of Bonn and the dreadfully persistent rain at Salzburg, in an August when more Mediterranean air than usual had spread into the Tyrol, to be under any illusions that the weather is always better abroad. And the discomfort of twelve hours of remorseless August downpour at Boston, Massachusetts, with the temperature around seventy degrees remains as an even more disagreeable recollection.

Lastly, even our cold fronts in Britain have rarely that sharpness of definition that we might in theory expect. One must imagine the cold air mass advancing over the land not in the manner of an evenly extended line of troops, but rather as a series of platoons in echelon. As such a front advances on a winter afternoon, two or three well-developed squalls with their characteristic clouds and showers may pass over at half-hourly intervals. Between the onset of the first and the tail of the last temperatures may fall by as much as 8° or 10°, in three steps of about 3°. The 'echelon' arrangement can be attributed to the varying frictional hold-up of the advancing cold air over hilly areas and plains, and also to such things as the local variations in sea temperatures round our coasts.

For these reasons, too, the occurrence of lightning and thunder in association with cold fronts tends to be somewhat sporadic. Cold fronts are best seen advancing over the sea or over open plains, for example that of West Lancashire in westerly weather, or over the Fenland if the wind is more northerly. Elsewhere, as the meteorologist would say, they are apt to be somewhat diffuse, and rather rarely do they give the dramatic violence of those of the Great Plains of North America.

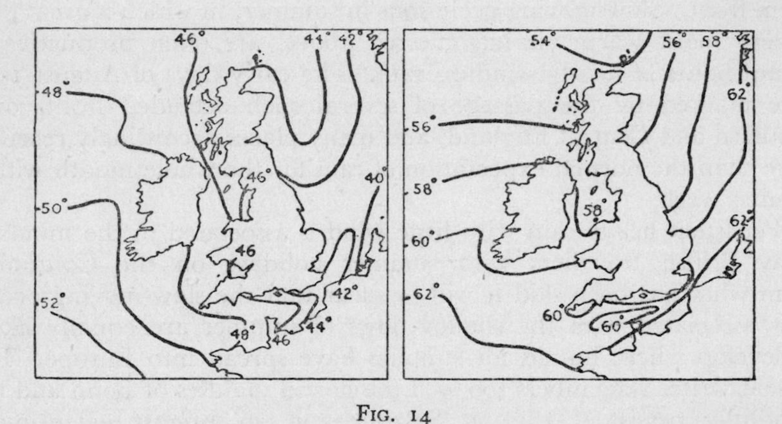

FIG. 14

Sea temperatures round Britain

left. Average isotherms (F°) for February. *right.* Average isotherms for August

CHARACTERISTICS OF THE AIR MASSES REACHING THE BRITISH ISLES

Now that we have reviewed the workings of our great Atlantic depressions it will be clear that, so long as their eastward route of travel lies to the north of our islands, we shall experience winds from between southerly and westerly points; occasionally a few hours of northerly wind may be recorded, for example if the centre of a Low moves southward from the Norwegian coast into the North Sea. If a depression is not occluded we ought then in theory to experience during the day or two days of its passage, a spell of relatively warm air of tropical origin, making its way towards us from the region of the Azores; then a well-marked shower as the cold front passes, and a veer of wind from S.W. to W. with a sharp fall of temperature as the polar air arrives.

But in practice, we find that the centres of some depressions moving on more southerly tracks pass south of England, giving us easterly or north-easterly winds which may originate over the continent, or may be derived, for example, from the northern Norwegian Sea. Sometimes, instead of receiving our polar air by a rather long sea route over the Atlantic, we may get it directly from the Arctic. Air coming from the edge of the Arctic pack-ice near Jan Mayen crosses less than 900 miles of open sea to Scotland: whereas if it reaches us from the region of South Greenland it may have crossed anything from 2,000 to 3,000 miles of ocean.

According to the source-region and the route of travel, therefore, we may classify the six principal types of air reaching Britain. Most of the air reaching Britain can be traced directly from one of three source-regions; regions, that is, in which the air has been relatively quiet for some days and has had time to acquire considerable homogeneity of temperature and humidity throughout its mass over a wide area. Sufficiently quiet regions for this purpose are found in the permanent Azores high-pressure region, and over the high Arctic latitudes of the Polar Basin; in winter another very well-marked source, this time of

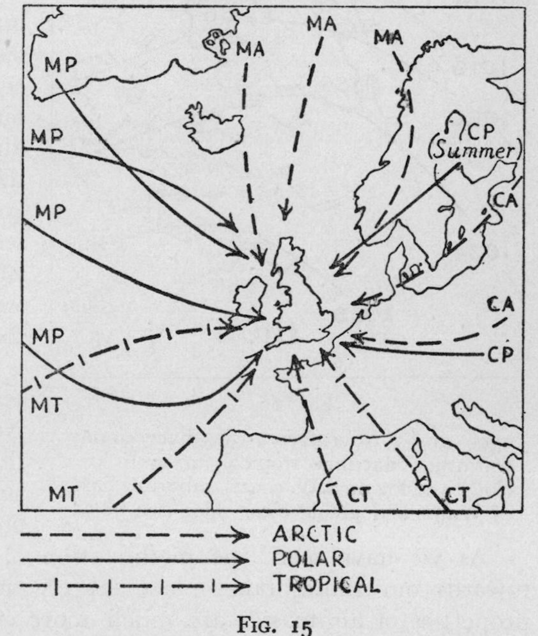

FIG. 15

Generalised directions of movement of the characteristic air masses reaching Britain

extremely cold surface air, is found from time to time in anticyclones centred over Northern Russia, representing a north-westward displacement of the continental anticyclone mentioned earlier in this chapter. In summer too, conditions are often quiet enough on the continent for the air to acquire during two or three days' travel well-marked characteristics of warmth, and hence dryness.

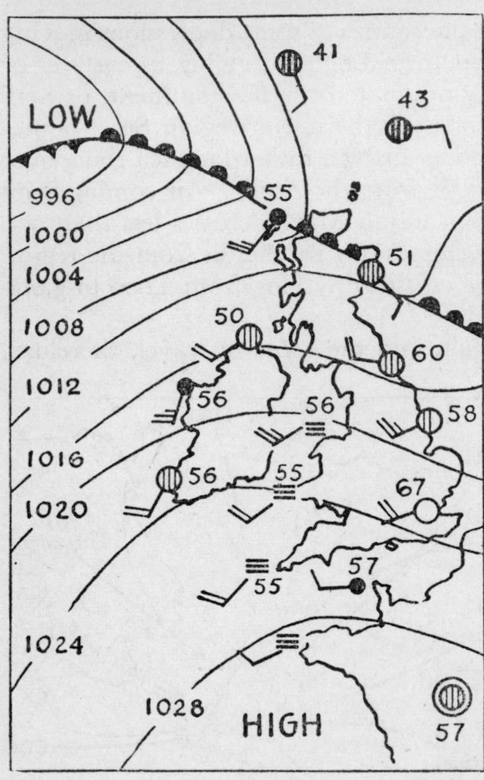

FIG. 16

1300 hrs. 5 November 1938. Exceptional warmth. Maritime tropical air with low cloud and fog on S.W. coasts, subsiding east of Wales and giving clear skies (see p. 5)

The six types of air are known as Maritime and Continental Arctic; Maritime and Continental Polar; Maritime and Continental Tropical. These titles are often abbreviated as mA, cA; mP, cP; mT and cT. It may be added that in low latitudes 'equatorial' air is distinguished from 'tropical' air, but that equatorial air never reaches our latitudes. Our tropical air can be traced as a rule to the region between 25° and 40° N. As we have already seen, its principal characteristic is that over the North Atlantic it is warmer than the sea surface over which it is moving. Several phenomenally warm nights late in November 1947 were due to the arrival of a tropical air-stream from as far south as 20° N. in the central Atlantic.

As we have seen, the modification of air masses on their way towards our shores takes place largely in the surface layers. The properties of air masses are much more conservative at high levels. For example, maritime polar air leaving the Greenland seas may have a surface temperature of 10° and at ten thousand feet −20°F. Arriving off the south-west coast of Ireland in January when the sea temperature is still about 48°, it blows over the land with a temperature of 40° or so in its surface layers; but at ten thousand feet the temperature may still be −10°. We thus see how necessary it is for the forecaster to have his upper air data at hand. Moreover, the properties of the air masses reaching us play an integral part in the impression we obtain of the country; visibility, the intensity of light, the type of cloud and its extent are all at once affected.

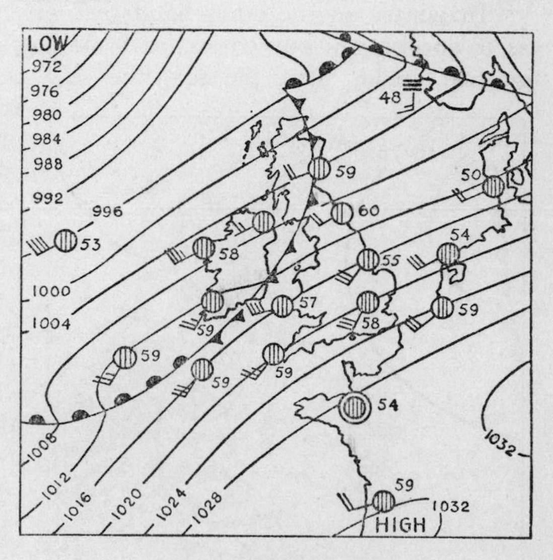

FIG. 17

Synoptic chart of maritime tropical air over England. 1800 hrs., 21 November 1947 (see p. 5)

Maritime Tropical Air fulfils the conditions described earlier; flowing as a humid current over a colder surface, its relative humidity by the time it reaches Britain is high. If its motion is exceptionally slight it may give surface fog over our south-western approaches and on coasts such as those of S. Cornwall, S. Pembroke, and Anglesey, especially if on its course it has been warmed slightly over N. France at the season when the sea is relatively cold, namely in late spring or early summer. In general, however, it is moving fairly briskly with appreciable turbulence over the sea; hence the saturated air is found a few hundred feet above the surface in the form of very extensive low stratus or strato-cumulus cloud. This is particularly noticeable in winter when stronger winds prevail and the air, being cooler, requires less water-vapour for saturation. Hence, the cloud-base of the characteristic St. and Sc. tends to be lower in winter than summer. Above the low cloud over the sea skies are often clear; and the surface air is stable, being slightly cooler than that above the cloud. *mT* air moving inland in summer is moving on to a land surface which in the daytime is receiving some radiation, even through the cloud. Hence the surface layers become warmer; cloudbase, therefore, tends to rise to a higher level. As we have seen, however, the skies above the cloud are generally clear and the air relatively dry. It is, therefore, not uncommon to find that in summer the cloud in *mT* air moving inland becomes thinner, and in the daytime breaks up into detached cumulus; for as the sky clears the ground warms sufficiently to give the needful rising currents in moist air.

In winter on the other hand, *mT* air has already become cloudy as it approaches our coasts, and moving over the land which is in general colder than the sea, the low cloud persists and thickens. Indeed mist or even fog may prevail at ground level especially if the ground has been previously chilled by overnight radiation, or for example, by recently melted snow. English readers will recall the many milder days of winter when a light south-west wind, a dull grey sky, and humid surface air prevail all over the Midlands: visibility is rarely more than two miles or so. In the nearly saturated air under the grey lid above, smoke from the great cities drifts slowly over the country, so that large areas for many miles to leeward of the cities record even poorer visibility and greater dullness. Temperature on such days on a December afternoon hovers about 50° on the Cornish coast, while 46°–47° inland is characteristic. In July on the other hand, afternoon temperatures may lie

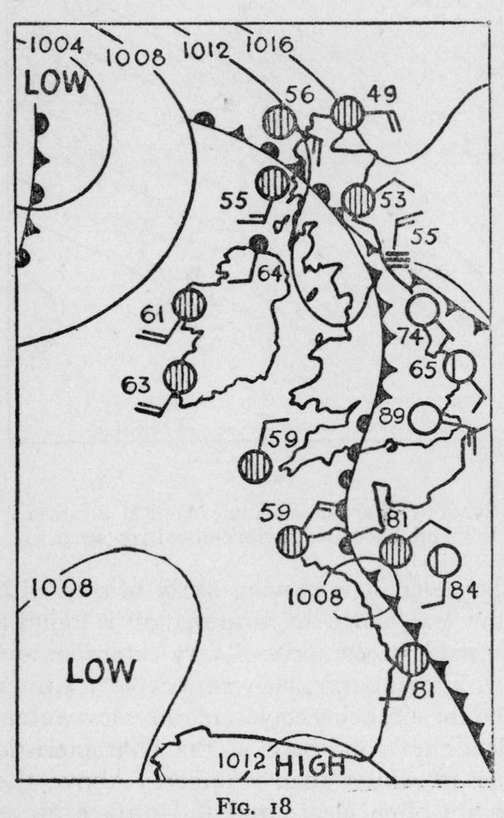

FIG. 18
Synoptic chart of Continental Tropical air over S.E. England. 1200 hrs., 3 June 1947

in the region of 66°–68° on the Cornish coast, but with clearing skies the Midlands have a humid relaxing afternoon, only too well known at Oxford and Cambridge, with a maximum of the order of 74° and intermittent sunny intervals between widespread lumpy cumulus with summits of varying height.

Continental Tropical Air. By contrast this is warm and dry; but even in the south-east of England it arrives rather infrequently in summer, practically never in winter. Sometimes in summer pressure is relatively high over the continent, while a shallow depression approaches slowly from the Atlantic. South to south-easterly winds bring air which may have been warmed for some days over S.E. Europe; the dry clear air travels across the well-warmed continent and is capable of giving on rare occasions very high temperatures indeed in N. France and S.E. England, the Channel not being wide enough here to effect more than a narrow surface stratum of the air. If the stream of air maintains its course northward over the land, the temperatures may attain high levels in Central Scotland, but more rarely and for an even briefer period than in the south as the adjacent map suggests.

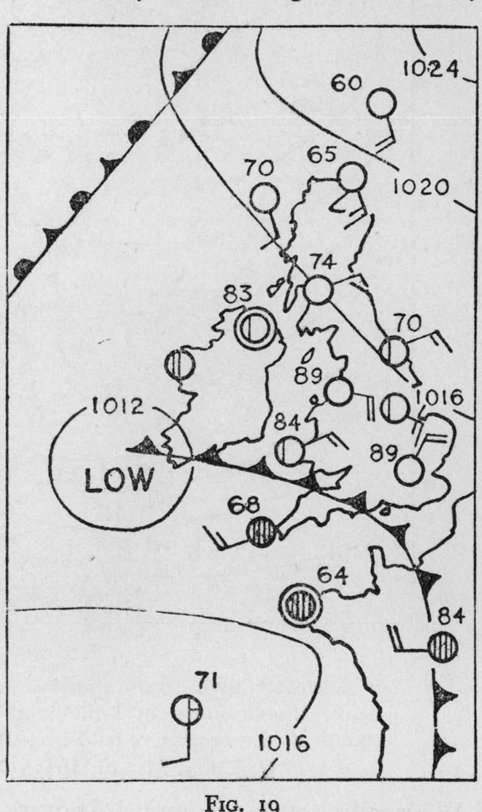

FIG. 19

1200 hrs., 29 July 1948. Continental Tropical air; great heat in Manchester

In winter it is very rare indeed to receive air which is both warm and dry from the south. Even North Africa at that season is relatively cold. Recently the phenomenal records of 70° on November 5, 1938 and 75° on

March 9, 1948 give rare hints of the possibilities of winter warmth but it is doubtful whether the origin of the air on these occasions really lay so far south as N. Africa. (Cf. the charts shown on pp. 72 and 88).

It will be evident that in summer the approach of the above-mentioned Atlantic depression with its cooler maritime air from the

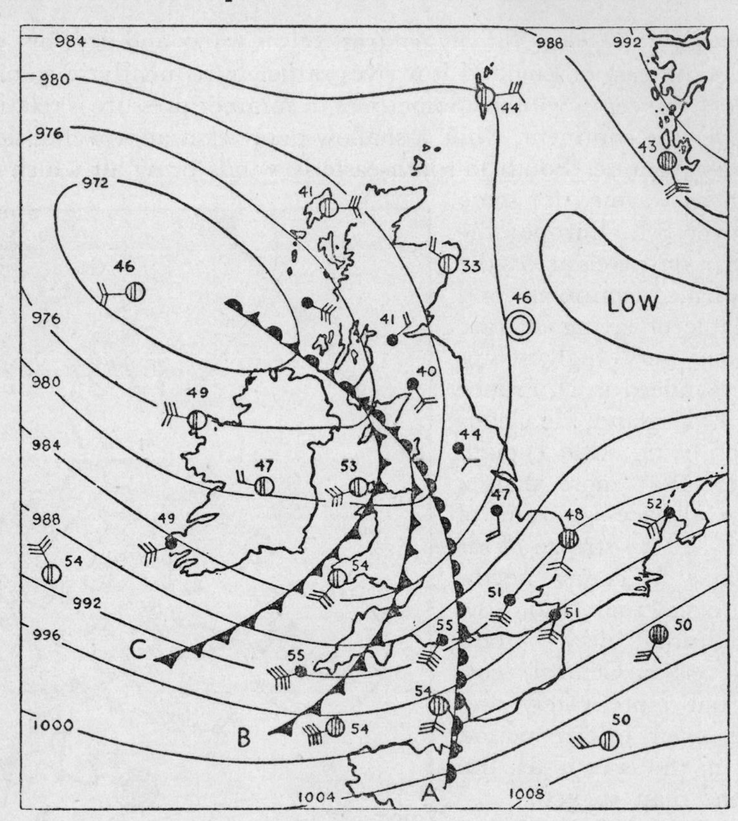

FIG. 20

2 November 1930. Rain ahead of advancing warm front A; heavy thunderstorm in London at 10.30 a.m. with severe squalls at the passage of cold front B (noted by F. H. Dight, *Q.J. Roy. Met. S.*, 1931) (Notation, p. 5)

S.W. usually breaks up any brief hot spell; this will be discussed again later when we consider the great thunderstorms which usually terminate hot weather.

Maritime Polar Air. In its many varieties this is without doubt the most common type of air we receive. Let us look into its properties.

In winter cold air is approaching over a warmer sea. The surface layers therefore become warmer, and at the same time pick up moisture; while above, the air is still relatively cold and dry. At the surface, air coming from S. Greenland can frequently be expected to increase by 30° in temperature in January; but at 10,000 feet the increase is only about 10°. Hence as the air travels towards us the lapse-rate increases to a point where the surface air is generally unstable; and the intermittent ascent of moist air from the surface gives extensive belts of cumulus over the sea. These are well seen in winter on our westward coasts almost whenever there is a strong westerly wind, and the hard-edged cumulus gives characteristic stormy but vivid sunsets.

Sometimes when the air is exceptionally cold above, the towering clouds become cumulo-nimbus and thunder and lightning occur, especially when the wind blows strongly against coasts which give a slight extra impetus to the rising currents, for example, the West Highlands, and even the cliffs of Cornwall and Sussex. In all these regions winter thunderstorms are reported with some frequency, compared with the Midlands. Occasionally a well-marked cold front causes thunderstorms to travel right across England even in a winter month. The weather map above shows a depression followed by two cold fronts, the first of which gave a prolonged thunderstorm lasting nearly an hour in London from 10.30 onward on the November day in question; it was discussed by Mr. F. H. Dight (*Q. J. Roy. Met. S.*, 1931).

The degree of instability in *mP* air depends a great deal on the length and route of its travel over the warmer ocean. Given a very large low west of Ireland the polar air coming round it may have travelled far to the south, returning towards us as a wind from the south-west. Under these circumstances it has travelled some distance over a sea warmer than our own waters, and thus as the surface layers approach us they become more stable and, like tropical air, may give considerable low stratus cloud and very similar temperatures, both in summer and winter. In such returning polar air in summer, however, the upper air is still quite cold. The low strato-cumulus breaks up over the warmer land (in a similar fashion to *mT* air above) and the rising currents give cumulus. In *mT*, however, the upper air was warm and dry; but with returning *mP* it is considerably colder. Hence once convectional cumulus has been started it may ascend to high levels through the cold environment above, and such rapid growth may result in inland showers, sometimes accompanied by thunder.

Maritime Arctic Air may conveniently be considered here; it is merely an exaggerated form of polar air with a relatively short sea travel. In winter it feels bitterly cold, as normally it reaches us as a strong north wind, often attaining gale force on exposed coasts. Such cold air flowing over a warm sea is extremely unstable, and gives characteristic sharp showers generally of snow and hail in the colder months. These developing over the sea, become especially noticeable in many coastal regions of Britain, such as those of East Kent, North Norfolk and Lindsey, Holderness, Cleveland and Durham, Banff, Moray, Nairn, Caithness and Sutherland; also North Wales. In every case the unstable air from the north impinges on a sharply-rising coast; the additional ascent of the surface air gives vigorous showers. On such days of Arctic air the towering cumulo-nimbus clouds over North Norfolk can often be seen from Cambridge; those over Northumberland and Durham can be seen from the sunlit summits of the Lake District in the lee of Scotland. For only the surface layers of the air are moist; hence when they have descended on the lee side of a mountainous region such as the Scottish Highlands they are drier and much less likely to give cloud than where they have just

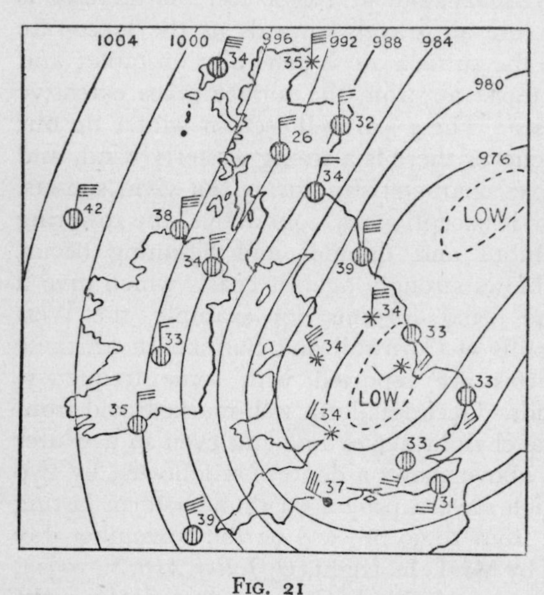

FIG. 21
28 February 1937, 7h. Maritime Arctic air

PLATE 9
DUDDINGTON, Northamptonshire: on a fine mild January afternoon. Temperature 44°. Characteristic patches of strato-cumulus in mild south-westerly air stream round a High over N. France. (Cf. fig. 32, p. 98.)

PLATE 9

Cyril Newberry

PLATE 10

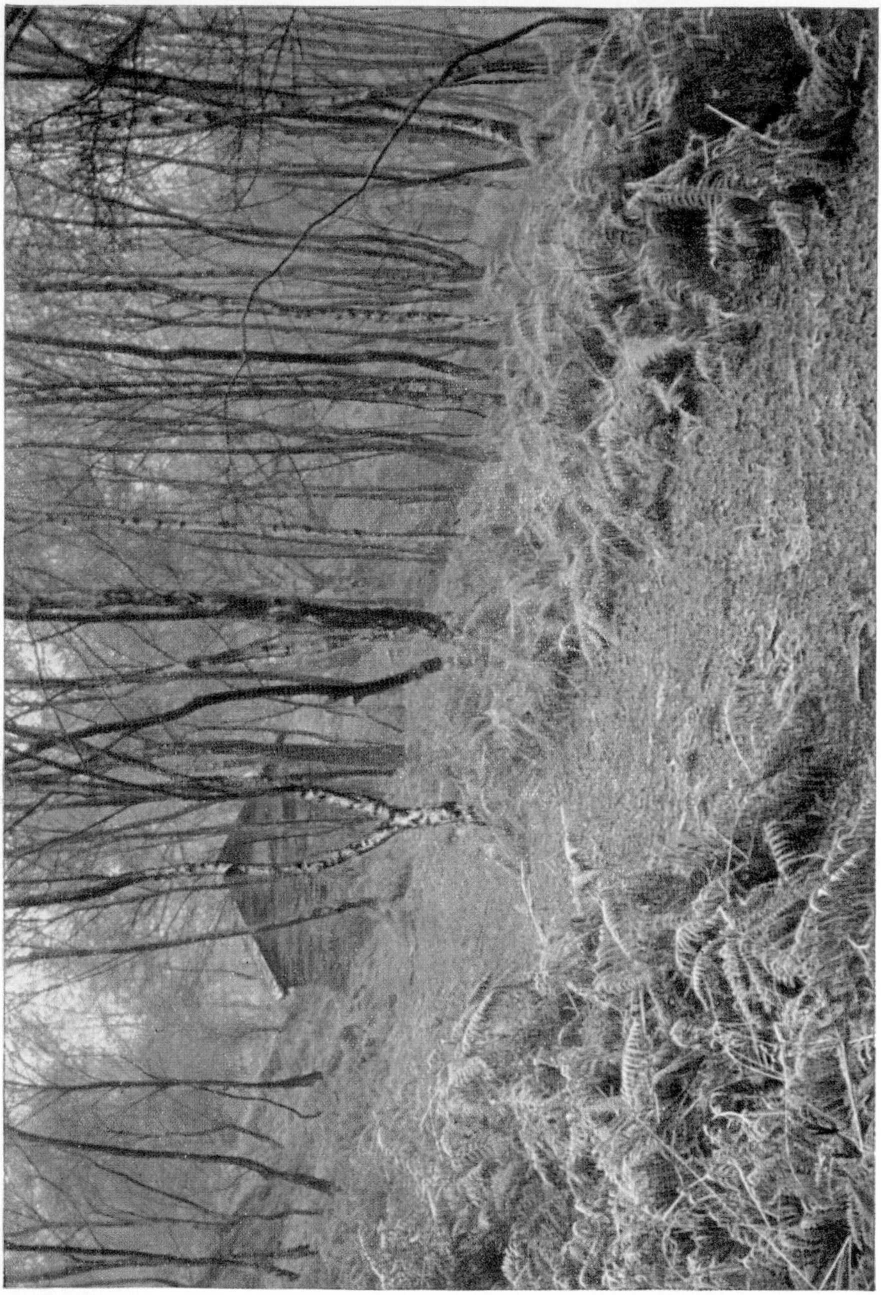

Cyril Newberry

traversed a wide stretch of open sea. Hence with Arctic air the Lake District enjoys long spells of superbly clear sunshine, and the like is true of South Wales.

Even in May, *mA* if it arrives is likely to give snow showers in all the districts named above. In a winter month such as February, on the north coasts of Scotland it may arrive with a surface temperature of 30° or even a little lower, and if it is blowing hard day-time maxima are scarcely likely to exceed 35° even in the south. Maritime Arctic air sometimes reaches us from the N.E. across Scandinavia, having come originally across the open sea from the north.

Continental Polar and Arctic Air may similarly be considered together. Both reach us from the south-east, east or north-east, in association with high pressure over Scandinavia and the Baltic region. In summer, *cP* in its surface layers becomes quite warm as it travels across Central Europe, and it reaches us as a rather hazy easterly wind cool on the North Sea coasts but warm inland. The origin of the air can be traced to Central Russia. In winter, cold, dry and frequently cloudy weather prevails; the cloud being largely the result of the passage of the air across the North Sea. While the North Sea is still fairly warm, as in late autumn, the instability resulting when cold air flows over the warmer sea may give rise to showers near the east coast.

Continental Arctic Air derived from Northern Russia or even N.W. Siberia gives us our most severe winter weather. If the continental high of winter extends, generally in the form of a separate cell with a centre lying across North Russia to Scandinavia, a large area of intensely cold stagnant air is still further cooled to a considerable depth as a result of radiation over the snow, and makes its way outward across Germany and N. France to England; if it reaches Scotland, it generally does so from the south or south-east rather than from due east. Air of this type crosses the Russian frontiers with a temperature which is sometimes below zero; crossing Germany its temperature slowly rises and it becomes drier, largely as a result of mixing with warmer air ahead and above. But even in Holland and Eastern

---

PLATE 10
LITTLE EATON, Derbyshire: birch wood with hoarfrost. February. Very characteristic of quiet winter days when temperature has just fallen to the freezing point at night.

France the thermometer may still read below 10°F. (see chart below) with a strong east wind to add to the discomfort; for towards the margin of the anticyclone the pressure gradient is steeper, and under such conditions a gale is often experienced down Channel. The effect of the sea, however, is very apparent on the English coasts; temperature and humidity rise considerably and with the addition of moisture in the turbulent surface air a good deal of low stratus cloud forms frequently covering almost the whole sky; cloud base lies between 1,500 and 3,000 feet. In December 1938 a severe outbreak of this type of air gave a maximum temperature, on the 21st, of only 22° at Lympne in Kent, 25° at Brighton, 26° in London. But on the same day maxima on the adjacent continent were far lower, 10° at Calais for example. Perhaps the most appalling outburst of such conditions we know of occurred at the end of December (O.S.) 1739. There is reason to believe that with an easterly gale blowing the temperature in London was below 15° on that occasion; in Holland the gale was accompanied by temperatures from −2° to +2°F. for many hours.

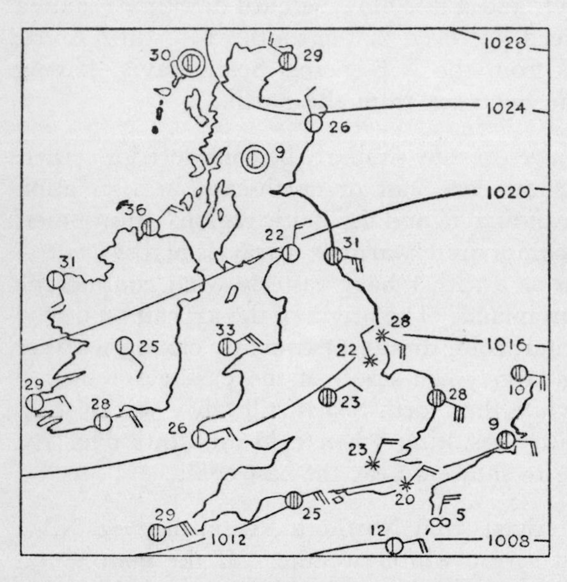

FIG. 22

20 December 1938. Continental Arctic air. An exceptionally cold outburst; warming over sea gives instability snow showers in E. England, frost at Scilly Isles and S.W. Irish coast

The warming effect of the sea is very evident. Even the coldest east wind down-channel rises to temperatures near the freezing-point, and on the coasts of North-East England and E. Scotland the same slight rise is evident. At the same time however the surface layers become more unstable and a severe outbreak of air of this type frequently gives rise to snow showers on and near these coasts, although they are more conspicuous as a rule in the maritime-Arctic air previously described.

Continental polar air in winter differs merely in degree from that derived from more northerly sources. The normal keen south-east to east wind with a temperature just above the freezing-point at inland stations and a grey sky is more often than not derived from the great plains of S. Russia, associated with high pressure over Germany. The course followed by the air depends largely on the position of the region of highest pressure. (Cf. fig. 15, p. 71).

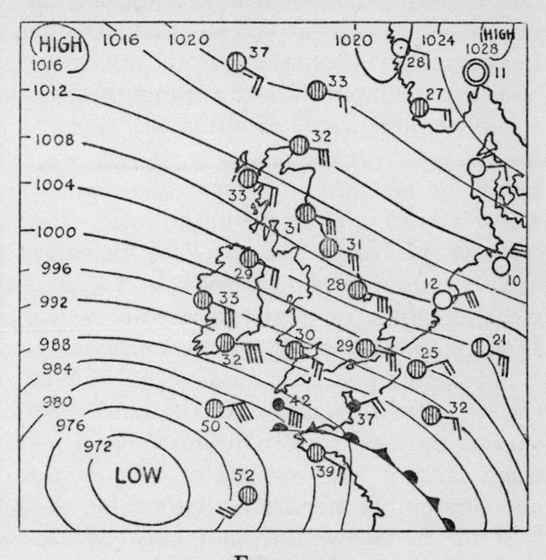

FIG. 23

Synoptic chart of Continental Arctic air; note increased cloud after crossing North Sea. 0600 hrs., 8 February 1947 (Notation p. 5)

In describing the characteristics of the six principal types of air reaching this country, we have considered their behaviour in high summer and the depth of winter. Little has been said with regard to spring and autumn. At these seasons there is still a well-developed high-pressure area in the Central North Atlantic providing a source region for our maritime tropical air. Towards the continent, however, 'the great Siberian High' is no longer found. In spring it is more common to find relatively high pressure over the region of the Baltic and Scandinavia, and sometimes over the northern Norwegian Sea.

It will readily be seen that with persistent high pressure in these regions we may expect many days of winds from the north-east quadrant; and one of the most characteristic features of the British climate is the relatively high frequency of winds from that direction from March to May. Moreover, with regard to continental polar air in a month such as May, the surface layers become quite warm and dry as they cross Central Europe as the skies are generally clear; but the North Sea at that season is still cool. We find then that relatively warm air is approaching us over a cooler sea, producing as a rule an 'inversion-layer' at an altitude of the order of 2,500–3,000 feet. Below this layer, the turbulent surface flow often carries up enough moisture from the sea surface to form extensive cloud. Hence the frequent formation of the 'North Sea Cloud'—stratus or strato-cumulus—in dry weather in spring, and early summer, extending for many miles inland from our east coasts, sometimes right across Southern England. In late autumn on the other hand, we have seen that the same pressure distribution still tends to give cloud in the air crossing the sea. But the sea is relatively warm while the land is cooling; hence the air coming from land to sea is not as a rule so stable as in May. Cloud is more disturbed, and showers are more frequent. The effect of the North Sea on cloud formation has been carefully studied, and a summary has been given by Mr. E. Gold, formerly of the Meteorological Office in a lecture to the Royal Meteorological Society on *Weather Forecasting*. It will be found in the Society's *Journal* for 1947.

Maritime polar air tends to become unstable due to warming over the sea in winter, and over the land in summer. At the intermediate seasons the chance of cumulus development and of showers is if anything slightly less, and the relative dryness (on the average) of April and September is partly attributable to this.

Our air-masses may not only be somewhat modified, depending on the breadth of the adjacent sea which must be traversed before they reach us; they also can be expected to behave slightly differently in the several months. At all times of year, however, the march of the Atlantic depressions towards Europe continues to play a major part in our weather. They and their secondaries draw in the air-masses we have described; these in turn with their characteristic cloud and weather, give rise to the changeful effects of light and colour in our landscape which play so large a part in the beauty of these islands. The principal departures from the normal changefulness arise whenever the march of the depressions is either temporarily interrupted or diverted. We may now discuss why some years are characterised by considerable spells of settled weather.

'Settled Weather': The Movement of Anticyclones

An anticyclone is a region of higher barometric pressure surrounded by closed isobars, according to definition; and we have already used the term for the large high-pressure region which generally covers the central Atlantic between the Azores and Bermuda, and for the Siberian High of winter.

At the surface it is observed that the actual flow of air is slightly outward across the isobars; this means that the great cushion of dense air which we describe as an anticyclone leaks. To compensate for this there must be some subsidence in the centre; as we have seen, this means compression and warming of the descending air, hence a decreased relative humidity and cloudless skies.

Smaller transient anticyclones occur from time to time, as detached masses of relatively quiet air; in general they are larger in area than depressions and move more sluggishly.

Detached areas of subsiding warm air up to fifteen hundred miles in diameter frequently spread north-eastward from the Azores High. In winter, surface masses of cold dense air from the Arctic or from Siberia sometimes spread and rest for several days over Scandinavia and the Baltic. A large, slowly-moving High tends to hold up the circulation *i.e.* the movement of depressions. Yet such Highs have been described as themselves a necessary part of the circulation, inasmuch as considerable forced ascent of surface air occurs in depressions and this is compensated by the large scale subsidence in anticyclones. Highs indeed must exist; but exactly why in some years they remain faithful for weeks to some particular part of the North Atlantic or N. Europe, while in other years they scarcely develop there at all, is one of the biggest puzzles our atmosphere has to offer.

Occasionally a persistent anticyclone in an unwonted part of the Atlantic produces very marked departures from average conditions over several weeks. Examples can be cited in February 1932 and April 1938. In both months pressure remained persistently high to the west of the Hebrides. The normal Atlantic depressions repeatedly followed tracks far to the north beyond Iceland, or skirted the anticyclone to the southward; Britain in general was only affected by the movement of air on the margin of such lows.

Both months were extremely dry. In February 1932 light northerly winds prevailed, but the month was not cold; for it will be evident that the surface air had either moved gently round the High from the region of the North Atlantic, or had subsided from higher levels; the north wind was not of Arctic or even Continental origin. With clear skies at night there was a little frost, but the days were relatively warm;

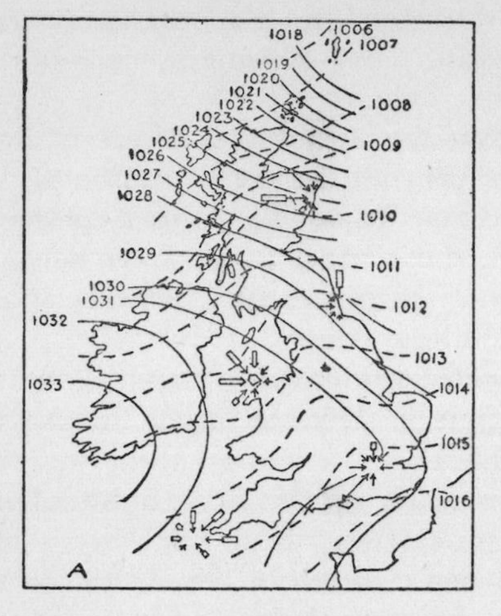

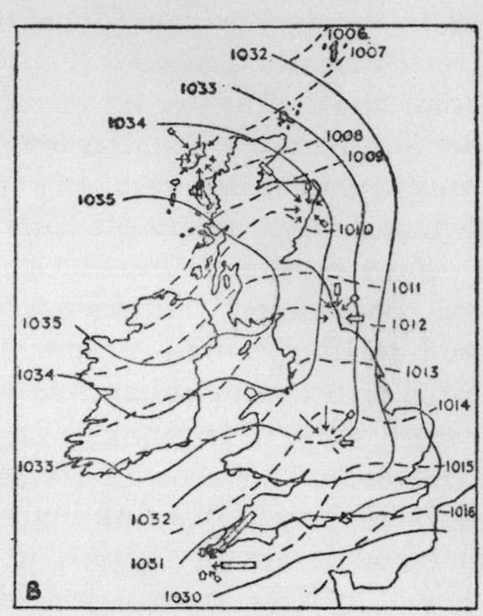

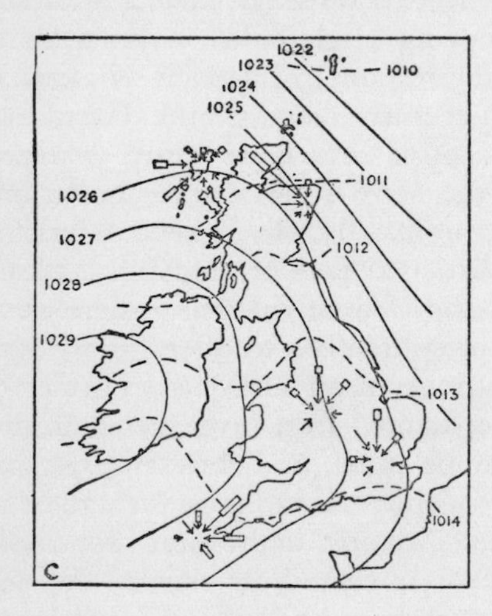

FIG. 24

Three exceptional anti-cyclonic months; showing wind roses and mean pressure in mb.
a. February 1934, *b.* February 1932, *c.* April 1938.

Normal pressure distribution shown by dashed lines
(For notation, see p. 5)

rather cloudy with low stratus towards the east, but frequently sunny on the west side of the country, and so dry that at places in the wettest part of the Lake District no rain whatever was measured during the month. For the first three weeks of April 1938 conditions were very similar.

February 1947 was characterised by an extremely obstinate anti-cyclone towards the Norwegian Sea and North Greenland, that is, centred farther away from these islands. On its southern margin, very persistent easterly wind prevailed, completely reversing the normal circulation as the map shows (Fig. 25, p. 86).

Among the most remarkable summer months dominated by a large and persistent anticyclone were June 1925, August 1947 (Fig. 25, p. 86), August 1955 and September 1959. In June 1925 no depression-centre approached within 250 miles of our coasts; much of Southern England was rainless. The skies were so cloudless, especially where the gentle flow of air had come over a long stretch of land, that places on the Cornish coast had an average of 12½ hours of bright sunshine daily, nearly 80% of the maximum that could be recorded. In August 1947, the distribution of rain was even more remarkable; no rain whatever fell in Glasgow and over much of the W. Highlands, in exceptional contrast to the normal experience.

Winter anticyclones in our maritime climate, however, frequently give extremely persistent cloudy and overcast skies. Even in a summer anticyclone the increased movement of air towards the outer margin may be enough to build up a good deal of low stratus or strato-cumulus where the air crosses a cool sea. In autumn and winter the establishment of a High following a spell of typical rainy weather means that the ground is damp, the skies clear, and the air quiet; all these conditions are extremely conducive to the formation of extensive radiation fog. After a calm November or December night this is often sufficiently thick to last through the next day, at the season when the sun's rays have little power to penetrate the fog and warm up the ground beneath. Meanwhile the air above the fog is commonly subsiding and becoming warm; hence the inversion-layer in which the fog exists becomes more marked. Any movement of air within the fog is not likely to lift it through the inversion; instead, the fog tends to drift along the surface. In this manner we frequently find that radiation fog formed over the land in quiet weather drifts gently over the adjacent estuaries or narrow seas. The mouth of the Thames and nearby Channel is thus liable to be foggy in cold winter weather; the same applies at the mouth of the Mersey.

An extensive anticyclone in which fog has formed generally tends to move eastward in winter and to merge with the high pressure area towards Central Europe. Hence the winds on its western flank gradually become southerly and increase in force. Fog is then lifted off the ground, exactly as with the advection fog over the sea which was described in the last chapter; but such fog only becomes low stratus. It cannot ascend into the warmer air above, and so it is liable to persist for as long as the anticyclone remains in our neighbourhood. Very persistent cloud of this type prevailed during the cold February of 1942 over Southern England, and even more during the severe December of 1890. This still ranks as the most sunless month on record for most of England, and gave rise to the phrase 'anticyclonic gloom' frequently used by British meteorological writers during the succeeding decades.

From March onward to October anticyclones may give considerable low cloud on their flanks while the air has crossed the cooler seas as we have already seen; but large areas are free from cloud, local radiation fog is easily dissipated early in the day by the more powerful sun, and high sunshine durations prevail especially in the lee of mountains. April 1938 and June 1925 have already been cited.

Nevertheless, the extent to which we are affected by transient highs varies very greatly from year to year, and meteorologists have

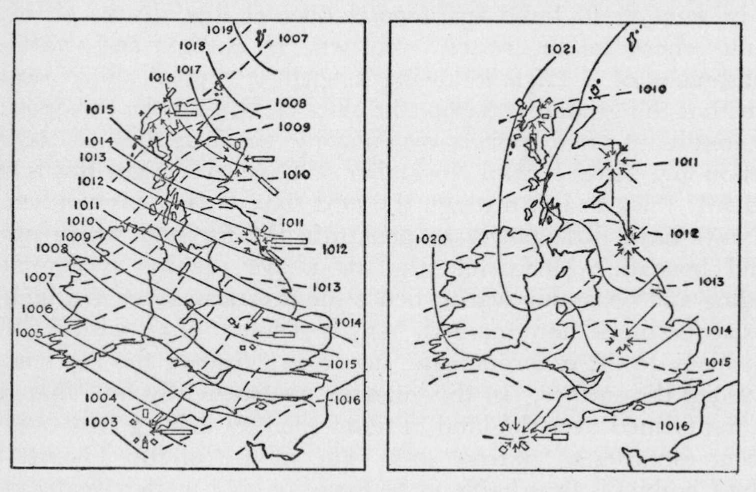

FIG. 25
Wind directions and mean pressure shown in millibars for February and August 1947. Normal shown by dashed lines

not yet fully explained their behaviour. Sometimes a large and healthy-looking anticyclone disappears unaccountably in a day or two as in mid-September, 1932. In occasional years a succession of anticyclones drifts over us throughout the summer apparently as detached portions of the Azores High. Moving slowly from west to east they fend off the Atlantic depressions for long periods, and Southern England experiences a very fine dry summer like that of 1921, or to a less extent 1933 and 1949. Out on the northern margin of the Highs, however, Scotland comes in for the stronger westerly winds associated with the passing depressions far to the north; in such years the Scottish summer is relatively cloudy and much cooler than that in the south.

Around the margins of our anticyclones there is a tendency for the surface air-streams to be slightly more convergent on the N.W. flank and on the S.E., with a corresponding tendency for slight divergence on the N.E. and S.W. flanks. Hence when a persistent anticyclone hovers to the north-westward, as in the dull cold May of 1923 and July-August 1931, these tendencies for cloud development particularly affect South-east England, especially as the air moving round the High has a long "fetch "down the North Sea. The comparison between Scotland and London is then reversed. Southern England especially is affected by the low stratus from the North Sea, and also by the cloud on the outer margins of depressions which find their way eastward in such seasons across N. France. Scotland, and notably Western Scotland in the lee of the mountains, has clear skies and sunshine. The report of the fine summer of 1931 in the Hebrides came in a year when the South Coast summer was particularly sunless and there was little money for holidays abroad, and undoubtedly the result was to attract many Londoners to Scotland in the next year. August 1945 and late June–early July 1948 gave for two weeks a somewhat similar experience. In August 1947, however, the High lay a little farther east and was more intense; clear skies predominated everywhere. In the summer of 1958 the situation was again like 1931.

In general, however, summer anticyclones more frequently spread across South Britain than elsewhere; they are accompanied by widespread small fine-weather cumulus developing inland in the afternoon, especially to the northward, in the light westerly current overlain by warmer subsiding air. It is rather in spring that we most often find anticyclones centred to the northward. One result is that in May the island of Tiree in the Southern Hebrides experiences on the average more sunshine than almost any other part of the British Isles.

The air moving gently round an anticyclone may fall by virtue of its origin in any of the classes already mentioned. But as it is moving slowly it has more time to become modified. Expressed otherwise, throughout the warmer months the sun has more opportunity to warm the land, and to modify whatever type of air is crossing the country in the direction of greater warmth and dryness. Hence with a normal summer anticyclone centred to the southward and a westerly wind, Eastern England has higher temperatures than the west. Rather rarely in winter when there is already a snow-cover, an outburst of severely cold continental-Arctic air is followed by calm, clear skies consequent on the rapid build-up of pressure over Britain. In such a case the surface air, already cold, falls very rapidly in temperature at night over the snow; and the occurrences of minima below zero in England, at least, can almost all be attributed to this sequence of events. In January 1940, –6° was recorded at Bodiam in Sussex, and –4° in Kent (at Canterbury); a century earlier, January 1838 gave minimum temperatures which by modern standards undoubtedly indicate more than 10° below zero in similar inland locations. Phenomenal dry warmth, such as the maximum of 77° recorded at Wakefield in March 1929, is again a concomitant of a gently moving stream of dry warm air from the south combined with clear skies; even more remarkable was the maximum of 75° in the London suburbs as early as the 9th in 1948, with 74° at Cromer and 70° as far north as Hull.

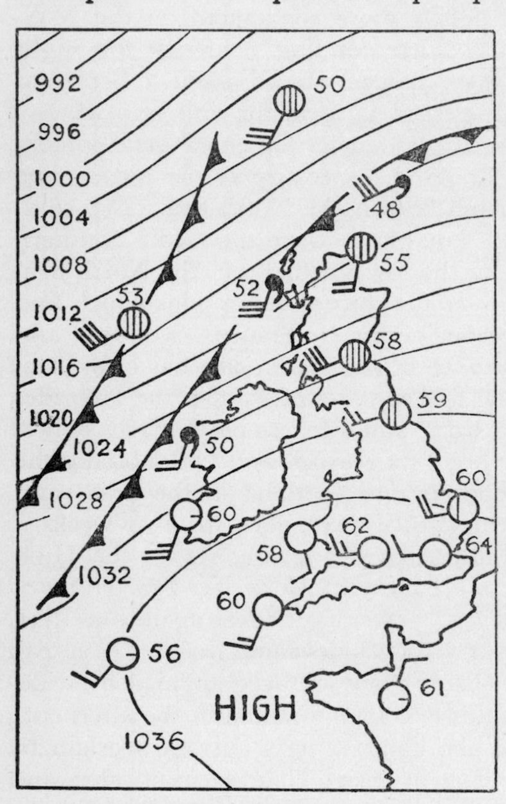

FIG. 26

Noon, 9 March 1948. Maximum in Cambridge 73° at 1500 hrs.; compare situation for 5 November 1938, p. 72, and 18 January 1947, p. 98 (Notation, p.5)

Incidentally the reasons for the lack of a sea-breeze at Cromer on this occasion in spite of the proximity of the North Sea at its coolest, are noted on p. 141. On 12th March 1957, 72° at Elgin is also noteworthy.

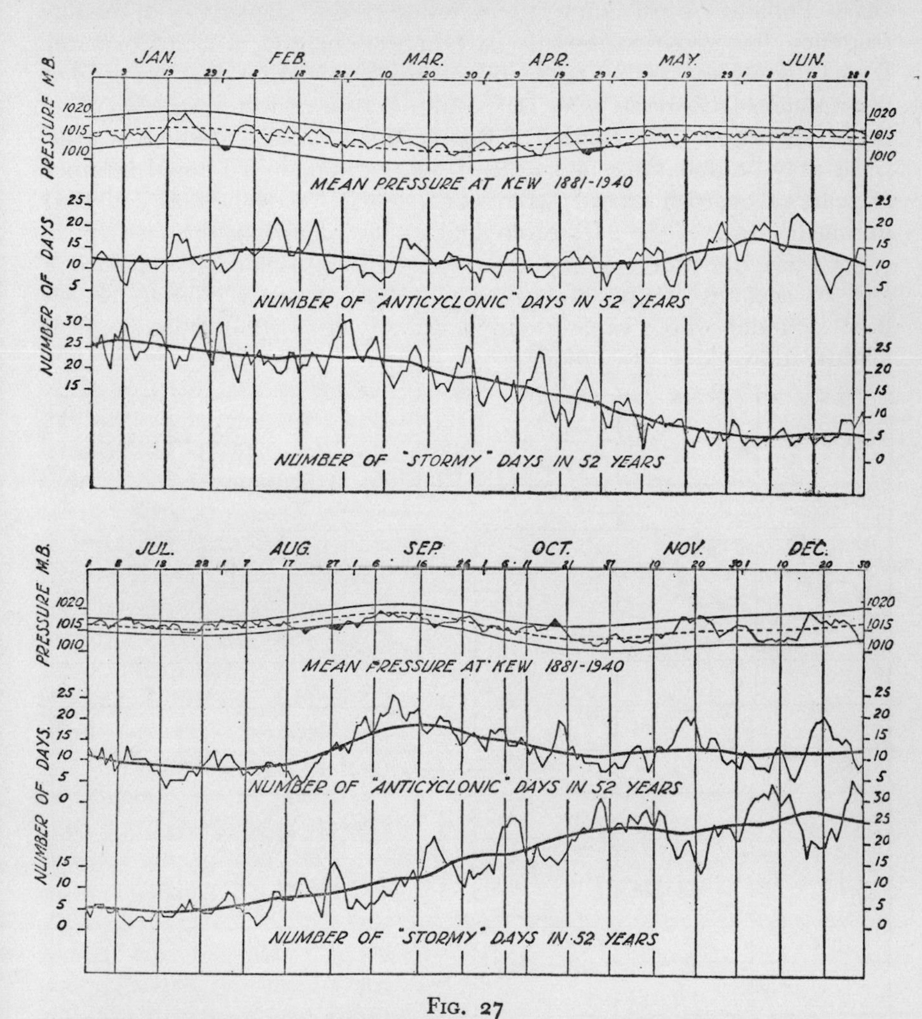

FIG. 27

Mean daily pressure at Kew, 1881–1940: frequency of 'anticyclonic' and 'stormy' days, British Isles, 1889–1940 (by courtesy of Dr. C. E. P. Brooks and "Weather")

In any discussion of the atmospheric circulation over the British Isles mention should be made of the extent to which spells of a particular type of weather tend repeatedly to occur at the same time of year. Folklore of all countries includes beliefs suggesting a regular tendency for abnormalities at certain seasons. In 1869 Alexander Buchan wrote his famous paper on *Interruptions in the regular rise and fall of temperature in the course of the year*, and much work has since been done in this direction. The reality of Buchan's six 'cold' and three 'warm' spells may be debatable as he based his findings on a limited number of years of Scottish records. But considerable evidence was collected during the war in the Meteorological Office regarding the recurrence of such 'singularities', as they have been called, in North-West Europe and an account of the results has now been given by Dr. C. E. P. Brooks, under whose superintendence many important climatological investigations have been made.

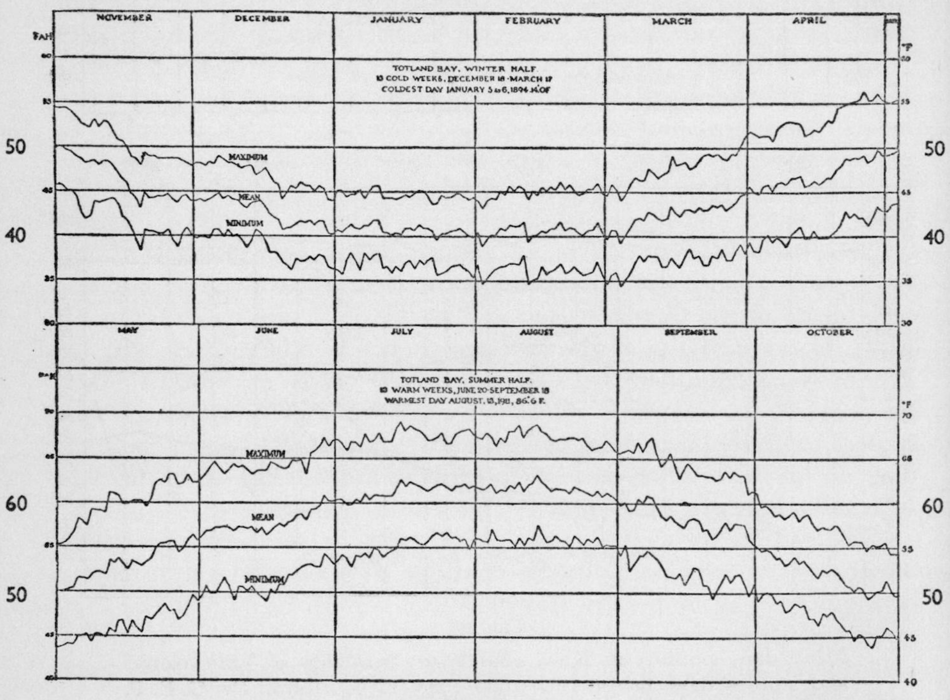

FIG. 28

The annual march of mean daily temperature at Totland Bay, 1886–1918
By permission, from *Q. J. Roy. Met. S.*, 1919. Cf. fig. 44, p. 147

Dr. Brooks' paper clearly shows (fig. 27, p. 89) that in South-East England the chance of disturbed weather reaches a peak in December from an early summer minimum; but there is also considerable variation in the chance of anticyclonic days in different months. Early June and the second week of September are well favoured in this respect. Some minor tendencies are also evident in many years, for example storminess in late November and early December, and again in the past 52 years disturbed weather has supervened sufficiently frequently during the week after Christmas to make another peak in the record. The tendency for cold anticyclonic weather towards the middle of February has been recognised by others (E. L. Hawke in *Buchan's Days*), and it appears in the annual course of the Totland Bay temperatures (p. 90). But it is interesting to note that Buchan's third cold spell associated on the Continent with the 'Ice Saints of May' (11–13 May) does not appear to occur with any consistency according to this later investigation. 'St. Lukes Summer', a brief anticyclonic spell in mid-October, occurs with fair frequency. It will be interesting to see how many of these singularities are accepted after we have acquired a further fifty years of observations; they are indeed interesting but several of them can scarcely be regarded as regular enough, as yet, to satisfy would-be holiday-makers. Further reference with regard to such spells can be made to a paper by H. H. Lamb.

A rather different approach to this problem, restricted to the evidence of abnormally high and low temperatures for the season, has been made in the *Quarterly Journal of the Royal Meteorological Society* by E. L. Hawke for the Greenwich temperatures (1841–1936) and by W. Dunbar for the Kilmarnock temperatures (1902–41); the tendency for occurrence of extreme values is by no means the same at the two stations and the incidence of extreme temperatures of Kilmarnock does not in this century appear to support the validity of the periods which Buchan originally derived for South Scotland.

The search for periodicities of one sort or another in meteorological events will undoubtedly continue to fascinate large numbers of inquirers. At the present time however, it must be emphasised that any claims that such tendencies exist must be treated with the utmost caution. At the very least we must have longer terms of records, kept with sufficient strictness to stand up to rigorous analysis.

SOME REFERENCES

Chapter 4

References to frontal phenomena and frontal behaviour abound in text-books and journals.

The *Daily Weather Report* of the Meteorological office may be watched with profit.

ADMIRALTY WEATHER MANUAL. London, H.M.S.O., latest ed.

BARTHOLOMEW'S ATLAS OF METEOROLOGY (1899). Edinburgh, old, but useful for purposes of general illustration as far as the British Isles are concerned.

BELASCO, J. E. (1945). Temperature Characteristics of Different Air Masses over the British Isles. Winter. *Q. J. Roy. Met. S. 71:* 351–76.
(1948). Incidence of Anticyclonic Days and Spells over the British Isles: *Weather, 3:* 233–42.

BROOKS, C. E. P. and MIRRLEES, S. T. A. (1930). Irregularities in the annual variation of the temperatures of London. *Q. J. Roy. Met. S. 56:* 375–88.

BROOKS, C. E. P. (1932). The origin of anticyclones. *Q. J. Roy. Met. S. 58:* 379–88.
(1946). Annual recurrences of weather: 'Singularities'. *Weather, 1:* 107–12, 130–34.

BRUNT, D. (1939). *Physical and Dynamical Meteorology.* Cambridge, University Press.

BUCHAN, Alexander (1869). Interruptions in the regular rise and fall of temperature in the course of the year. *J. Scott. Met. Soc.* n.s. *2:* 4.

CARRUTHERS, J. (1941). Some interrelationships of meteorology and oceanography. *Q. J. Roy. Met. S. 67:* 207–46; contains many valuable references with regard to the Gulf Stream, the effect of wind on sea disturbances, etc.

DIGHT, F. H. G. (1931). Thunderstorms of Nov. 2, 1930. *Q. J. Roy. Met. S. 57:* 101–03.

DUNBAR, W. (1942). Abnormally high and low daily mean temperatures, Kilmarnock, 1902–41. *Q. J. Roy. Met. S. 68:* 287–92.

GOLD, E. (1947). Weather Forecasts. *Q. J. Roy. Met. S. 73:* 151–85

HAWKE, E. L. (1937). *Buchan's Days.* London, Lovat Dickson.

LAMB, H. H. (1950). Types and spells of weather around the year in the British Isles. *Q. J. Roy. Met. S., 76:* 393–438.

SHAW, SIR NAPIER (1926–31). *Manual of Meteorology*, vols. 1–4. Cambridge University Press.

CHAPTER 5

SKY, TEMPERATURE AND SEASON: WINTER

. . Infant winter laughed upon the land
All cloudlessly and cold: when I, desiring
More in this world than any understand
Wept o'er the beauty . . .

SHELLEY

It is certainly a fact that so various are the effects of weather, light and atmosphere that the landscape in England, with its heights and distances, is never two days alike.

L. C. W. BONACINA

NOTHING is more refreshing or reassuring to the English mind than the contemplation on a clear day of the pageant of the sky from rising ground overlooking a wide stretch of country. The restful changeability of the scene thus viewed has the same fascination as that of a river. There is a quiet underlying sense of purpose, a deeply-felt flowing rhythm beneath the fleeting disorderliness and variety of aspect. Faintly-apprehended overtones and harmonies abound to a far greater degree than in the brighter and more simplified American scene with its harsher outlines, its cruder contrasts, its strongly emphasised seasonal tom-tom, mitigated only where higher latitude and wider ocean again begin to modify the behaviour of the air.

Rhythm can indeed be sought and felt in the roar of the winter storms, fitly expressed by Sibelius' incidental music to *The Tempest.* Under the howl of the winter wind the more northern dwellers in our island, in particular, recognise that such storms are a necessary and regular precursor of summer, and preparations still go forward. Storms in the north are not the capricious ill-tempered irregular devastations that beset the resigned peasantry of more southern lands; violent though they are, they must regularly be expected at some time in the colder months. Nevertheless, protection against them lies within the

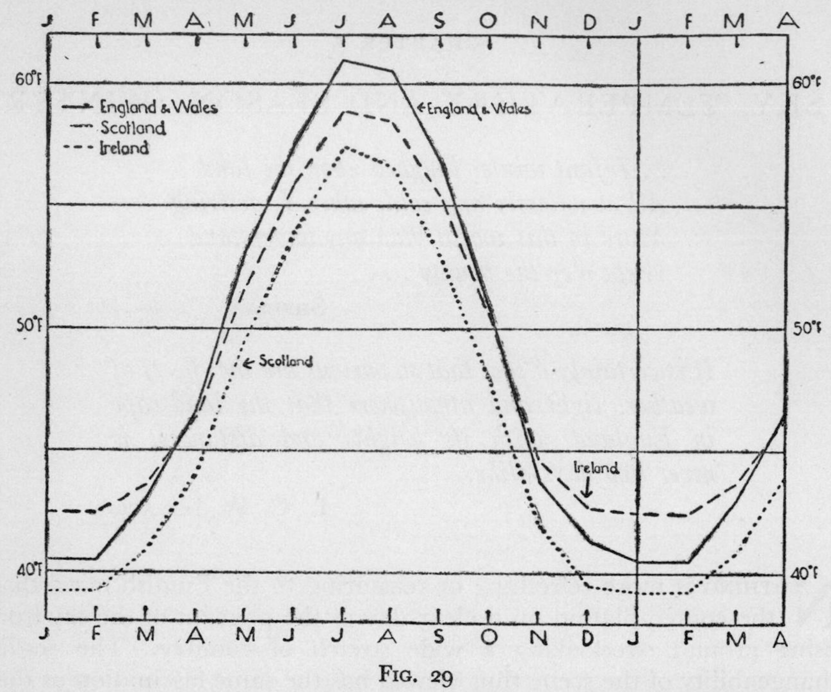

FIG. 29

Mean temperature 1901–1930, England, Wales, Scotland and Ireland (from J. Glasspoole, *Q. J. Roy. Met. S.*, 1941)

capacity of the intelligent individual of cautious outlook. Edinburgh's profusion of insurance companies can fairly be associated with the demands of the climate on the forethought of the Scottish farmers.

But discussion of such differences as there are between Scotland and England can be left to the reader. After the previous chapters we are in a position to explain why the Englishman who contemplates his weather out-of-doors sees and feels what he does. Let us consider

---

PLATE 11

DERBY: January morning. Going to work in a Midland city. Characteristic coppery smoke-haze in the morning inversion layer following a clear night and light snow-cover. Temperature 22°.

PLATE II

Cyril Newberry

PLATE 12

Cyril Newberry

the characteristic weather which he is likely to observe, month by month.

Winter Weather

The mid-winter month of January may be taken as typical; we can then expect a succession of deep Atlantic depressions to approach Britain. A new one appears off our shores almost every other day; but frequently they are occluded by the time they reach us. Polar air coming from the westward with a long fetch over the ocean gives much low cloud and afternoon temperatures of the order of 46° inland, falling to 38° at night; sea-surface temperatures in mid-winter range from about 50° off Cornwall to 45° in the Shetlands. (Cf. Fig. 14, p. 70). 'Moderate to fresh south-west wind inland, strong on the coasts', is the most usual outlook; if an occlusion passes, two or three hours steady rain is followed by a slight clearance but with little change of wind or temperature. With a shorter fetch on the Atlantic, however, west winds are colder and more unstable; bright intervals of from one to three or four hours' duration on the coasts are marked by a pale blue sky, and flying ragged cumulus over the sea which often builds up rather heavily over our hills and mountains. Temperature ranges between 35° at dawn to 45° by afternoon, tending to be appreciably lower in Scotland. Skies often clear a little towards evening with a rather pale sunset, broken near the horizon by the jagged tops of distant cumulus over the sea. In the evening, ragged low clouds still drift across the moon. If the wind dies down sufficiently there may be a touch of frost inland before dawn. This is a very prevalent winter type of weather; with slight variations depending largely on the strength of the wind and the degree of instability in the air moving over the sea. It is roughly true that a fast-moving current of maritime-polar air is more unstable than one which is crossing the sea slowly. Hence with a deep depression centred to the northward the stronger westerly winds commonly reach gale force on our exposed coasts; lashing showers of rain, not infrequently accompanied by hail, fall from the towering cumulus over the white-crested sea, greenish-grey

---

Plate 12
Allestree Lake, Derbyshire: fine anticyclonic winter day, March. Rough ice after earlier heavy snowfall. Temperature 34°.

in the changing light. Temperature rises by day little above 40°; little more than two thousand feet above, therefore, the air is close to freezing-point and our cloud-bedecked mountains receive fierce driving squalls of sleet and snow. The showery clouds march steadily

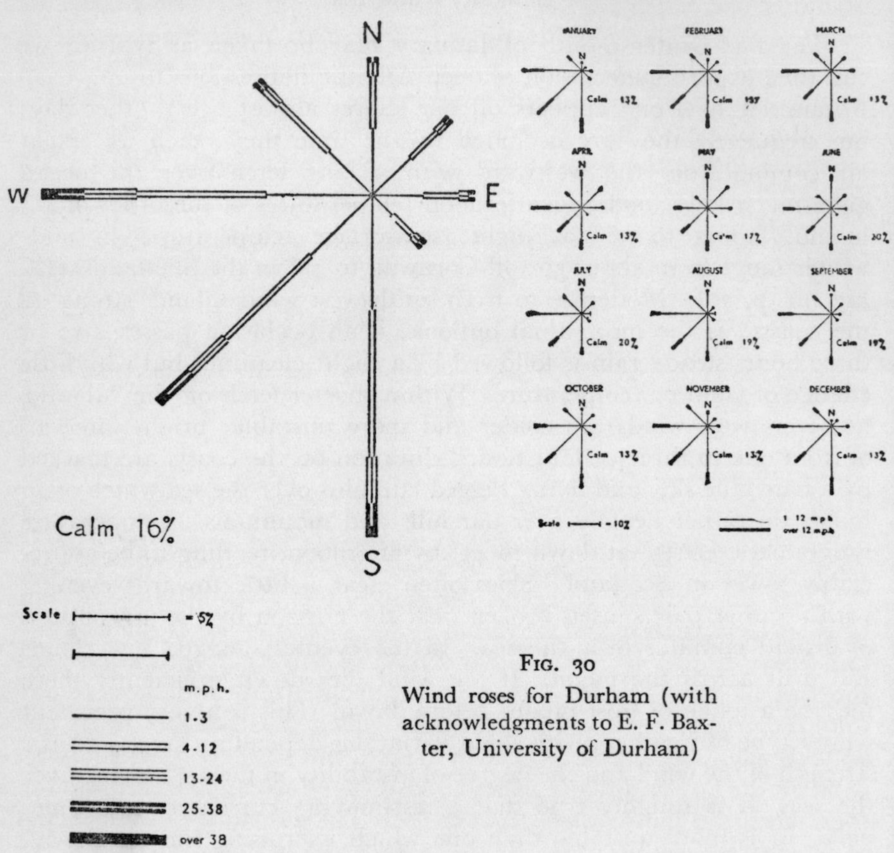

FIG. 30
Wind roses for Durham (with acknowledgments to E. F. Baxter, University of Durham)

across Britain, but as a general rule the showers are considerably less intense and frequent once the air has crossed the mountains.

If a vigorous depression approaches in which the warm sector is still present, the country may be flooded for a few hours with tropical air coming from the seas as far distant as Madeira. In this type of air the surface temperature exceeds 50° even in Scotland, and temperatures of 55° occur in S. England. Hurrying low cloud overhead, stratus or strato-cumulus is borne on a soft south-wester; almost every year, each winter month from November to February gives at least one particularly mild day of this type. Even at night the temperature often remains close to 50°. Tropical air of this warmth with a high humidity and extensive low cloud needs little further disturbance for rain to occur; and generally the arrival of such a warm current gives very heavy 'orographic rain' in our western mountain districts. Even on the mountain tops the air temperature rises well into the forties, and the incessant heavy rain quickly removes the covering snow, sometimes up to the level of the highest Scottish summits above 4,000 feet.

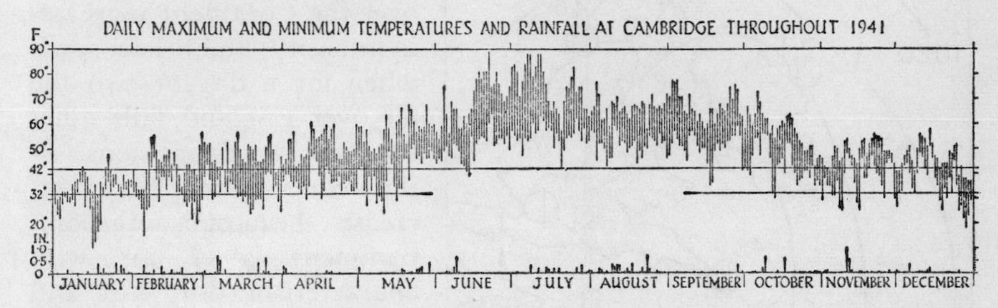

FIG. 31
Annual march of Cambridge temperature and rainfall

Over the country, however, the cloud-sheet gives little more than brief spells of drizzle and occasionally for short periods if the cloud is thin the sun may be visible. Mild days of this kind with occasional sun can be very pleasant given shelter from wind, and most of our southern resorts have their gardens in which it is possible to sit out on such days, although exposure to the soft, but strongly blowing damp wind on the headlands and capes may still produce some feeling of chill even when the thermometer is over 50°. The advantage of the resorts at which the sea lies towards the southward is that a good deal of reflected light comes off the sea during the mid-day hours. This adds to the impression of brightness near the coast even though the sun itself may be only intermittently visible through the thinner parts of the cloud. We have already noted that maritime-polar air which approaches this country from the S.W. after a very long fetch over the Atlantic differs little from tropical air as regards its surface temperature and it will readily be seen that this particularly mild version

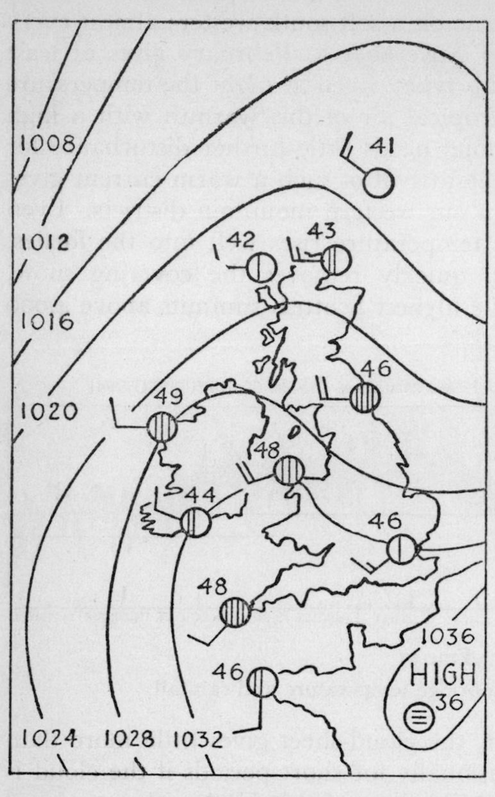

FIG. 32

Noon, 18 January 1947. Fine mild winter day; note fog at Paris. Very fine, clear and quiet interval in the Hebrides within the wedge of high pressure; see also Plate 9, opposite p. 78, taken on this day (see p. 5)

of *mP* is considerably more frequent on the south coasts of Ireland and of England, than, for example, in northern Scotland. (Cf. Fig. 15, p. 71).

In some part of January, however, we can generally expect that depressions will for some days be fended off when the high-pressure area over the Continent increases in intensity; and briefer spells when for a day or two the air over England falls quiet between two depressions. If the continental High increases, the most usual accompaniment is *cP* air with characteristic dull skies and a light to moderate south-east wind; we have already seen why, towards the margin of a winter high, dull overcast weather is wont to occur. On such days maximum temperatures fall below 40°; minima are close to freezing-point but do not generally fall below it unless the sky happens to clear; which it may do if the stream of air from the Continent becomes very slightly drier. More rarely, two or three sunny mild days occur, with light

---

PLATE 13

MARLEY COMMON, Surrey: birch trees after an ice-storm, March. Raindrops from warmer air at a high level fell through a stratum below freezing point, and froze on impact. The warmer air advancing from the south was unable to displace the colder air and while rain fell heavily nearer the Channel, north of the Thames there was very heavy snow.

PLATE 13

F. Goldring

PLATE 14

Cyril Newberry

frost and valley fog at night, if an anticyclone moves north-eastward from the Azores and is temporarily centred over N. France; the air coming round such a high is mild and rather moist and it is always touch and go whether low cloud does or does not develop in it. (Cf. Fig. 32 and Pl. 9). In some years several depressions follow tracks across the Midlands, up-Channel, or across France, usually when pressure is high over the Scandinavia-Baltic region. In this event cold continental air arrives north of the track of the centre, and precipitation may take the form of continuous snow for some hours. Under such circumstances it is by no means unusual to find that Southern England lies in a warm sector with temperatures approaching 50°, while Eastern Scotland is experiencing a heavy snowfall. Such a situation is illustrated by the adjacent chart.

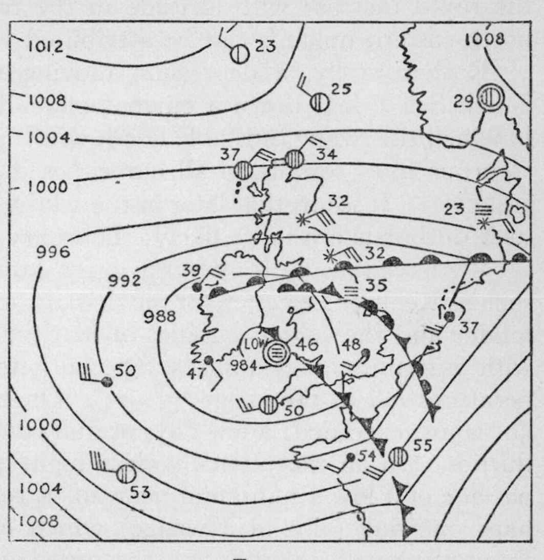

FIG. 33
Synoptic chart for 18h., 13 March 1947, showing that, with a low crossing the Midlands in March, southern England lies in a warm sector, whilst eastern Scotland is receiving cold continental air, giving snow towards the east coast (For notation, see p. 5)

Depressions crossing Great Britain in this manner sometimes move along the flank of the continental High towards the Baltic; but if the High is well-developed they slowly fill up in the North Sea. Squally cold raw north-easterly winds of Continental origin then prevail on our east coasts; these are raw on account of moisture and their rather low temperature, but after crossing the open North Sea normally give readings well above the freezing-point on the coast. Showers in

---

PLATE 14
WOODBRIDGE, Suffolk: nearly cloudless May forenoon. Anticyclonic weather with stable surface air near the North Sea coast.

such air may be of cold rain or sleet. From what has been said, however, it will be evident that air from the north-east on the margin of a low, with a surface temperature of say 37° at Tynemouth, will give much cloud and probably sleet showers; but at an altitude of 1,000 feet, temperature will be about 33° and heavy snow will fall. Much of the rapid increase with altitude in the frequency of snowfall on our north-eastern uplands can be attributed to this sequence of events.

Rather rarely, a depression moving from the Atlantic into the North Sea brings down a current of really cold maritime-Arctic air, in which the strong, unstable north winds give very disturbed skies and vigorous snow squalls on all our exposed uplands towards the north-east coast. It seems that later in the winter, about the end of February, such outbursts are more likely; February 1955 was notable.

We have thus built up our average January. At a Midland station such as Keele or Leicester, for much of the month air ranges between the milder and the colder varieties of maritime polar. It is rather windy with intermittent clearances especially towards nightfall, and temperatures fall in the range 35°–45°. On one or two mild misty days 50° is just exceeded; a few days of dull cold weather give values in the thirties. On an exceptionally clear night in the 'wedge' following the passage of a low a minimum near 20° is recorded inland; there is perhaps one day (also in a wedge) which can be called sunny almost throughout, with about six hours recorded duration. (Cf. Fig. 32, p. 98 for 18 January 1947, the wedge being over Scotland). Perhaps on ten mornings the minimum has fallen to 32° or below. Typically we should record by the end of the month that snow fell in small amounts on about four days, and on a few days during which the continental air prevailed, we should note it as lying on three mornings. Rain would have fallen on about sixteen days; mostly in rather small amounts but with two or three bigger falls of the order of half an inch when a rainy sector of a vigorous depression moved eastward across the country. Bright sun might have been recorded inland for forty hours of which three days might between them give fifteen; the rest would most frequently be recorded as one or two hours on a number of days, chiefly those with westerly wind blowing and clearances of some length between showers.

In the diagram below, based on data from Cambridge, we may observe the contrast between the trend of temperature in an exceptionally mild January (mean temperature 44·5° in 1916) and a very cold January 1940. In 1916 although the prevailing air supply was exceptionally mild the mean daily range differed little from that

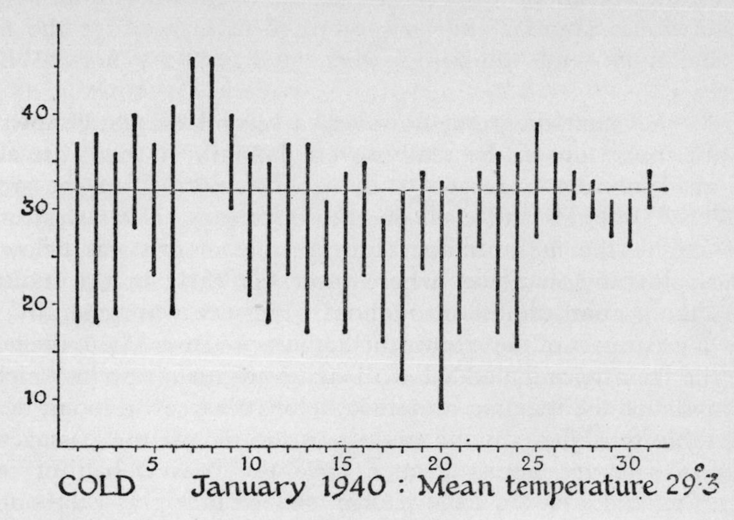

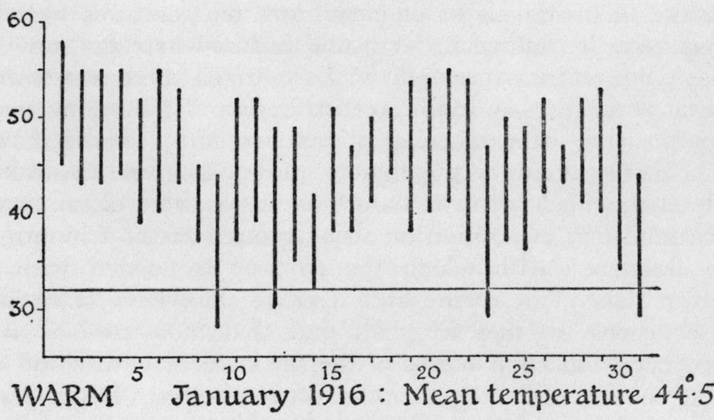

FIG. 34

Comparison of the daily maximum and minimum temperatures for January 1916 and January 1940 at Cambridge

which can be expected in a normal January when maritime-polar air predominates with a good deal of cloud. The chief characteristic of milder Januaries is that continental air scarcely reaches this country at all, and on several occasions tropical air with maxima over 50° may be experienced; hence in an exceptionally mild mid-winter month, the mean may reach 44° even in the eastern counties, with an average daily maximum approaching 50°. In 1916 at Cambridge the mean daily maximum and minimum were 50·8°, 38·2°; normal being 44·7°, 33·9°.

Below the main diagram, however, a second diagram shows the trend of temperature in the really severe January of 1940; the above means at Cambridge were 29·3°; and 35·8°, 22·8°. Over the greater part of the Midlands and even on the south coast as far west as the Isle of Wight, the mean temperature for the month was below the freezing-point and in places where snow fell early in the month it covered the ground almost throughout. Pressure was generally high and for a great part of the month the highest pressures lay to the northward; the country was flooded with dry continental air in which on some days even the maxima remained below the freezing-point despite considerable sunshine. In the middle of the month the passage of a depression eastward across France gave the coastal belt of South Eastern England a classical snowstorm (sixteen inches at Eastbourne); and on the following nights of clear skies temperatures well below zero were recorded inland in Kent and Sussex where the snow cover was deep. But on the same night in East Anglia there was scarcely a powdering of snow. The resultant distribution of minimum temperatures for the clear calm morning of January 20th, 1940 is shown on p. 169 (Fig. 54) as far as the figures allow. It gives an instructive example of the effects of urban and coastal location, of sandy soil in the Breckland and of position on rising ground in the Chilterns.

The frequency with which the country is flooded with moist surface air is the cause of the high average cloudiness; if we put all the observations together we shall find that most stations average between seven and eight tenths of the sky covered with cloud at the observing hours. For the months of November, December and February conditions are broadly similar. But as the average sea temperatures fall off Cornwall from nearly 55° in November to 48° in February, and correspondingly elsewhere (50°–44° at Wick, 50° to below 42° off Norfolk) the average temperatures prevailing during the onset of Atlantic air, whether tropical or polar by origin, steadily decrease. Hence in west-wind weather in February there is a greater chance of night frost occurring than in November. Tropical air with almost overcast skies in early November occasionally comes inland to give maxima over 60°; the exceptionally warm days and nights from November 20th–23rd, 1947 have already been mentioned (pp. 72, 73). Polar air coming by a shorter sea route in February often gives day-time temperatures below 40° and sleet may fall through it even on the coasts, especially in Scotland, with sharp snowfalls on high ground. In November, on account of the warmer sea surface, this is less likely to happen. Hence, at lower levels south of the Highlands the frequency of snow in February is about four times that in November.

To offset the effects of the cooler sea, however, by mid-February the sun is more powerful than at any time in November and in quiet clear weather the daily range begins to increase considerably. Moreover the air comprising the warm sectors of depressions which reaches our shores in November has a higher vapour content by reason of its higher temperature; November and December, therefore, can be expected to give heavier rains than January and February in a normal year. On the whole, December is the month in which the deep Atlantic lows most frequently affect our western coasts; in January and February when the Continental High is most likely to develop and spread, some of them may be fended off. Hence while great gales may occur in any winter month, the figures taken over many years show a slight maximum in December; and we can justify our view of dark December as the stormiest month. Such deep depressions with their floods of warm moist air surging from the wild Atlantic over our western mountains give immense falls of rain in certain well-known localities. There, convergence of the air streams in constricted mountain valleys enforces additional ascent of air to add to that resulting from the mountain barrier itself. (Fig. 41, p. 131).

Late February and onward into March is the season at which the sea is coolest, while in the Arctic February is often a cooler month than January. It is then that maritime-Arctic air sometimes descends on us with exceptional cold; this is the more marked in Scotland, as one might expect nearer to the source. Towards sunset the northern sky is often greenish by contrast with the remaining fragments of cloud; as the air falls quiet under the stars, with the aurora flickering above the northern horizon hard frost follows and the Highland valleys often experience their lowest minima of the year (−13° at Braemar, 1955).

Such outbreaks of cold air behind depressions following southerly tracks may also give a good deal of snow, not only in those eastern uplands where we normally expect it for orographic reasons, but also in unexpectedly mild locations towards the west.

Rather rarely, particularly heavy falls of snow are experienced for example in Western Ireland. These are attributable to exceptional developments in a stream of Arctic air. Normally as we have seen instability showers develop widely when such a cold stream debouches over a warmer sea. Sometimes the result of a cluster of such showers is that pressure is lowered sufficiently over an area of sea for the air to begin to follow the isobars round it; on the weather map we find a sinister little area of low pressure, perhaps in the region of the Faroe Islands. Like an eddy it moves southward in the general stream of air; within it, as in all low-pressure areas, there is considerable convergence and ascent of moist air, with consequent precipitation over a wide area. But in such a system there is nothing but cold air, and the result in general is an unexpectedly lively snowstorm. The blizzard of April 1st, 1917, in Western and Central Ireland; the heavy snowfalls of 16 May 1935 and 10 May 1943 in Lancashire; the fall of eighteen inches of snow in Suffolk on 27 April 1919 may be mentioned. Such depressions moving southward gradually draw in different air supplies and acquire more defined fronts; an example is shown for 10 May 1943 when the Northern English uplands received their heaviest snowfall of the 'winter'. (Fig. 36, p. 113.) Repeated instability showers with a strong north wind of Arctic origin gave very heavy drifting snowfalls in Northern Scotland in February. In sharp contrast a rare day of dry cloudless south wind gave 65° in London on Valentine's Day in 1961.

SOME REFERENCES

Chapter 5

BILHAM, E. G. (1938). *The Climate of the British Isles.* London, Macmillan.

BONACINA, L. C. W. (1937). *Q. J. Roy. Met. S.*, *63*, 483–90 Constable as a painter of weather.

(1939). ibid. *65*, 485–96. Landscape meteorology.

(1940). ibid. *66*, 379–88. Scenery, weather and climate.

(1941). ibid. *67*, 305–12. The scenic approach to meteorology.

HAWKE, E. L. (1937). *Buchan's Days.* London, Lovat Dickson.

CHAPTER 6

THE ENGLISH SPRING AND EARLY SUMMER

Bursts from a rending East in flaws
the young green leaflets' harrier
MEREDITH: *Hard Weather*

There was such deep contentment in the air
that every naked ash, and tardy tree
Yet leafless, showed as if the countenance
with which it looked on this beautiful day
were native to the summer.
WORDSWORTH

LONG AGO the remark was made in a March issue of *Punch* that "spring has set in with its accustomed severity". This expression indeed was used in a letter by Coleridge written at the beginning of May in 1826; but he in turn may have known that Madame de Sévigné used a similar phrase in 1689. Adopting the division of the year into four seasons of equal length, a division likely to appeal to most meteorologists in mid-temperate latitudes, it is undoubtedly best to consider March, April and May as comprising the spring. But surrounded as we are by the sea the rise of temperature with the lengthening day is slow and frequently suffers many setbacks; hence in Britain we have a very long season during which one after another of the familiar harbingers of spring appear. Many of our flowers derive from wild ancestors with an open deciduous woodland habitat, where they quickly responded to the increased light while the trees were still bare. Hence such flowers as crocus and daffodil, anemone and bluebell successively appear to remind us of the approaching warmer season. While in some years a venturesome crocus may appear a month early in mid-January the daffodil is equally capable of appearing a month late, in mid-April. March is indeed a particularly variable month, although not from the point of view of mean temperature; in that respect its possibilities are surpassed by December, January and February. The range of possibilities in March (and in Scotland, April) is however probably more effective for an easily-understood reason. Our prevailing vegetation characteristic of the temperate lands of Europe and parts, at least, of North America is such that growth begins and is maintained as soon as the mean temperature rises above a figure between 42° and 43°F. Climatological studies made in France gave this figure as 6°C. (42·8°F.). Arising out of such studies and that of others the Meteorological Office in Britain has for many years worked out and published values of "accumulated temperature" above 42° for each week, month and season. These figures represent the extent to which in any week or series of weeks the temperature has been in excess or deficit so far as plant growth is concerned, with a view to correlating the extent of the progress made by crops with the weather conditions.

It does not matter for the purpose of the present argument which figure we adopt around 42°; it is more important to realise that the mean sea-level temperature of March in England is 42·9° (period 1906–35) and of April in Scotland 44·2° for the same period. Hence it is immediately evident that small fluctuations of the mean temperature of these months will have a big effect as regards the progress made by vegetation. If March in England is severe, with a mean temperature of 38°, it is reasonable to assume that the average daily maximum will be about 44°, and the night minimum about 32°. Thus taken over the month as a whole the temperature will on the average only exceed 42° for about two hours out of the twenty-four. But in a mild March with a mean temperature of 46° and an average daily range from 52° to 40°, by far the greater part of every day and night will be above 42° when averaged over the month and the proportion of hours during which growth is likely to proceed without check may well be ten times that of the cold month.

Note too that we are here discussing the temperature of the air, taken by standard methods. Patches of ground beneath the shelter of a tree, or adjacent to a warm wall, do not radiate so freely at night,

---

PLATE 15
SHROPSHIRE PLAIN and Wrekin from Wenlock Edge: April. Cloudless warm anticyclonic April weather. Clear air free from smoke-pollution.

PLATE 15

Cyril Newberry

PLATE 16

Cyril Newberry

Cyril Newberry

and the minimum air temperature to which plants are subjected at night may be appreciably above that in a less protected situation. Thus we can find, especially in a cold month, marked variations even in the average garden in the response made by growing plants. If we extend our survey to cover the sheltered south-facing hillside nooks, the marshy frost-hollows with their coarse grass, the shady woodlands and the windswept ridges of this varied country of ours we shall find that the appearance of the first signs of spring and the progress of the season shown by vegetation depend to a remarkable extent on these micro-climatic differences, as we may call them. These are not as large in magnitude as in some countries elsewhere; but in the earlier spring months of March and April the differences in maximum and minimum temperatures from place to place arising from exposure, relief, soil and shelter are the more effective because they range on either side of several critical mean values (see also Chapters 8–10).

Spring sunshine is quite powerful and effective if the air is quiet. The sea, on the other hand, is at its coldest. To this we must add the fact that in a normal year there is still much ice in the Northern Baltic and that in the Greenland and Labrador seas the ice is generally more widespread and closely packed than at other seasons. When it is melting, a layer of cold and relatively fresh water from the ice lies on top of the salt water beneath by virtue of its lower density. We have already seen that in winter months the Siberian anticyclone dominates the surface circulation of the air over Asia and Eastern Europe. Towards spring it becomes less intense; but at the same time the extensive snow-covered Scandinavian Highlands and the icy Baltic remain as a relatively cold area by comparison with Europe to the southward. We find a well-marked tendency for anticyclones to develop and persist over the Baltic lands. We might justly attribute this in part to the fact that the movement of depressions depends on the existence of well-marked fronts. In the spring months air of Atlantic origin making its way into West and Central Europe, by this time generally quite free from snow, is often distinctly warmer than

---

PLATE 16

a. SHAP FELL, Westmorland: sunshine and shower, May. Cool westerly weather with showers developing at intervals over the Lake District in rather unstable maritime-polar air.

b. SHREWSBURY, Shropshire: warm anticyclonic April. Evening, by the Severn.

that which has been cooled by radiation over snow or ice farther north. Fronts therefore are often to be found somewhere across West Central Europe; and our islands often lie on their northward side.

Pressure is also (at sea level) generally a little higher over Greenland throughout the winter and spring, than it is over the open Norwegian sea or the North Atlantic. In spring this Greenland High frequently spreads across the sea to eastward. In such a case cold continental-polar, and even Arctic air reaches us persistently, in the form of the very common and familiar "east winds of spring"—it would be fairer to recall that their actual direction varies between N. and E. They share however the characteristic features of the continental air of winter already described in that they are unpleasantly cold and penetrating; as the proverb says "neither good for man nor beast".

Scandinavian and Arctic highs however are not persistent throughout and do not always become established at the same date. In many years the Atlantic depressions continue to develop on the polar front south of Iceland and to move with little or no interruption on the normal track towards the Norwegian coast. In this event cool westerly maritime polar winds again predominate with if anything slightly more stable conditions (owing to the cooler sea off our shores) than in mid-winter. A fresh blue sky with patches of small cumulus or cumulo-stratus, and a cool westerly breeze from the sea will be recalled by many as characteristic of March; the trees are still bare, but the sun is bright and distinctly warm out of the wind. The grass begins to grow, larks sing, the cattle are turned out for a period each day; daffodils flower here and there, crocus is abundant, the willows are turning green. Daytime maxima reach 50° or just above. At night the thermometer may fall to below 35°, but in many places, especially where the air is still moving, there will certainly be no sign of ice. The mean temperature of course is still close to that of the sea surface from which the air is coming; it lies in the neighbourhood of 43°–46°.

In such circumstances there is in March a decidedly significant difference between Cornwall, where the sea temperature by now is down to 48°, and the Outer Hebrides where it is 44°. The shorter distance travelled by the maritime-polar air to the Hebrides and the cooler sea combine to render the normal west-wind day as described above about five degrees cooler in the north. This means that at 2,000 feet on the Brecon Beacons where the probable afternoon temperature is about 39°, sleet will not be observed in a passing shower. But at 2,000 feet on the N.W. Highlands the day-time average will be about 33°; and frequently the ground at this level will be white with snow whenever the normal cool westerly weather prevails. This is reflected in the figures; the mean frequency of occurrence of snow-cover in March at 2,000 feet is about five times as great in Sutherlandshire as it is in South Wales.

Tropical air if it reaches us comes from the sea at its coolest, and hence tends to carry a good deal less water-vapour than it will for example in October. As tropical air masses, either directly or after occlusion, often give us our longer-duration rains we tend to get less rain in March, and still less rain in April than we expect in January or February; a further cause of this decreased rainfall lies in the slightly greater stability of the 'showery' west wind above-mentioned, and in the frequency with which anticyclones develop to the northward, as we saw. All these factors together make the average April one of the driest months of the year; although the frequency of measurable rain throughout the spring months does not decrease at all so markedly as the quantity, by comparison with autumn and winter.

March is then a variable month in which as a rule we can generally expect several days of dry, cold north-easter, with some strato-cumulus cloud as before near the east coast, especially towards the southern North Sea where a north-easter has a longer 'fetch'. There will be a number of days of cool west-wind weather; if a cloudy warm sector crosses us, the air temperature (52°–53° by afternoon) is no higher than in December or January. But if the sky happens to clear for a time, even with the wind blowing freshly from the S.W. the sun may give us an afternoon maximum inland of 56°–57° (Midlands) on one of those mild March days, and at night temperatures will scarcely fall below 45°. The dry cold days give (with continental air) maxima of the order of 45° inland, and minima if the sky clears of about 30°. The west-wind days we have already mentioned.

Minor fronts in the prevailing maritime air-stream give much of the rain; but in this respect March varies greatly. In a year with a vigorous Scandinavian high and disturbances crossing the country well to the southward along a well-developed front between the cold continental air and that of the Atlantic, there may be a good deal of snow, especially in the north-east. Moreover the existence of high pressure in the north means that in almost every March, and quite often in April, there is one real outburst of maritime-Arctic air, giving the snowy 'instability showers' we have already described on all the exposed coasts and uplands, sometimes as far south as the E. Kent hills. By reason of the less distance the air has had to travel over the warmer sea, such showers of snow are very normal in E. Scotland where they have acquired the name of 'the lambing storms'. The frequency of snow-falling and of snow-lying varies more noticeably from south to north in March than in any other month.

Generally in the clear night skies and calm air following such a bout of north wind severe frost is recorded at least once in the month.

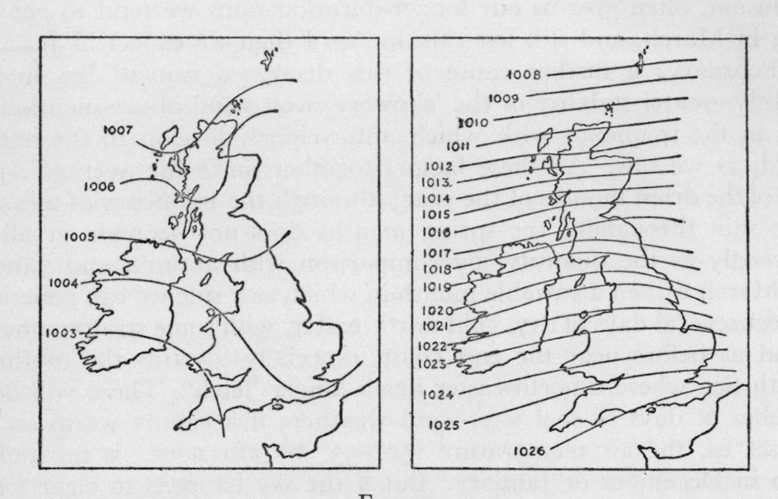

FIG. 35

Comparison of the mean pressure at 7 a.m. for March 1937 and March 1938 (in millibars)

In the Midlands the average extreme minimum for March is about 23°, only a degree or two warmer than that for January or February.

The most notable departures occur in March when anticyclonic weather is really persistent. We may illustrate very well by reference to the very cold March of 1937 and the very warm one of 1938 in the above charts. Given a warmish air supply off the margin of a High over France and dry ground extraordinarily high day-time maxima

---

PLATE 17

PRINCES STREET, Edinburgh: early May. Springtime: almost calm, slight haze, very light air from S.E. Cautious retention of coats by older Scotsmen.

Cyril Newberry

PLATE 18

Cyril Newberry

Eric Hosking

above 70° sometimes result, although the ground beneath the surface is still relatively cold and little warmth is conducted to the surface on clear radiation nights, so that night frost still occurs. 74° in Northumberland in 1957, 73° at London in 1961, are notable.

With regard to April air masses practically the same remarks may be made as for March; but by this time, whatever happens at night, the sun is strong enough by day to send the maximum temperatures well up the scale into the fifties in the South of England. Towards the coast the effect of the cool coastal sea breeze on afternoon temperatures begins to be noticeable on quiet days, especially adjacent to the North Sea. Moreover we find that at this time of year air crossing the cool sea either from the warmer land of the continent to the S. and S.E., or from the warmer ocean away to the S.W., tends to become decidedly stable in its lower layers; it may give extensive cloud, but there is little chance of precipitation, in contrast to the conditions of November or December. As we have seen the Scandinavian High quite frequently develops, and an early April north-easter across the cool North Sea is very little warmer than that of March; but as the month wears on the chance of such chilly surface air derived from Scandinavia steadily decreases, at least in a normal year.

Frost still occurs at night in April given clear skies and a cool air supply; and if the air is of maritime-Arctic origin, coming down behind a depression in the North Sea, the resultant night frost may still be quite severe. At habitable levels the chance of a day-time maximum below 32° is now very remote even in the North. Snow can still be expected to fall, at inland stations in Southern England, on one day in a normal April usually in the form of a shower; it is distinctly unusual in the south for snow to cover the ground. In Scotland, an April day with snow-cover must regularly be expected over the greater part of the low country; and at high levels the frequency is still considerable. Notice again how the south-north gradient

---

PLATE 18

a. THE CLYDE ESTUARY from above Greenock, Renfrewshire: late afternoon, May. Warm dry south-easterly wind, slight build-up of cumulus over distant hills: depression to westward of Ireland.

b. GORPLE RESERVOIR, Yorkshire West Riding: early summer on the gritstone Pennines near Burnley, June. Fine weather cumulus with light N.W. wind (hence excellent visibility); cotton grass seeding; characteristic June anticyclone well to northward in the Atlantic.

of mean temperature is greatest in March and April. The most rapid change in the feel of the season can be got at that time, a fact which the Edinburgh spring visitors to the sheltered South Devon coast regularly perceive and enjoy; the ultimate cause lies in the appreciable difference between the sea temperatures and the fact that they lie close to the critical value for plant growth.

April then like March is a mixture of dry cool weather with a good deal of sunshine, somewhat showery cool westerly weather with long fair intervals, and now and then the more extensive cloud and rain of a transient depression. The soft moist growing weather beloved of the farmer is fairly frequent, associated as a rule with a slow-moving current of humid air round the margin of a High over France or Germany. Such air is frequently 'returning maritime-polar'; it may be moderately stable. A typical soft April day of this type gives afternoon temperatures of 58°, and probably clearer skies and more sun than the broadly analogous type of day in March.

Dry clear warmth in an April High moving over the country may give us a very large day-night range of temperature; in Southern England, about every other April gives one day with a maximum just over 70°. In Scotland the narrower extent of land *i.e.* closer proximity to the sea rarely permits such high temperatures even in quiet sunny anticyclonic weather. To get really warm days in Scotland in April it is usually necessary to postulate a High over the region of Germany with a drift of dry and fairly warm air of Continental origin all the way up to Scotland over land; even then, April maxima over 70° occur rather rarely. By contrast the fearful effects on temperatures of the deep snow-cover which covered the north early in April 1917 may be mentioned. The night after the snowfall was clear and calm; radiation was rapid, and at Penrith the screen minimum fell to 5°. Many places in Scotland and North England recorded below 10° and there were some unofficial reports of minima below zero. As the majority of Aprils since 1930 have been warm we tend to forget the possibilities.

In May it is still possible for night temperatures to fall dangerously low, given clear nights. This is one of the most fundamental features of our climate, of the greatest importance to fruit-growers, and also to all who are interested in the production of early vegetables.

The general conditions are much the same as in April; in some years depressions continue to follow the normal Atlantic routes, in others they are temporarily checked by the development of a large High, quite frequently over the Icelandic-Norwegian seas. Sometimes however the Azores High begins to spread northward, and fine clear dry west-wind weather is established for a period in S. England, with stable air and little chance of rain. The High may gradually spread right over Britain.

If an anticyclone in which the air as a whole has been derived from warmer regions is centred over these islands with clear skies, the May sun gives high maxima; 80° is occasionally reached in S. England before the 10th of the month. In the north and in Scotland however 80° is very rarely touched before the end of May. Further the ground is often dry after a fine April, and hence radiation on clear nights is effective. But anticyclones may equally well spread over us from farther north. In the event of a cool dry northerly or north-easterly current spreading over us associated with an advancing high behind a depression which has moved eastward, day-time maxima are governed by the fact that the mean surface temperature for the whole day can differ little from that of the North Sea over which the air has come. Hence while the centre of a High still lies to the northward, the cool north-east wind in S.E. England is likely to give a maxima little above 55°, and minima fall well below 40° at night over most of the country. With minima at representative country stations of the order of 35°–38° many localities in valleys and hollows, especially with

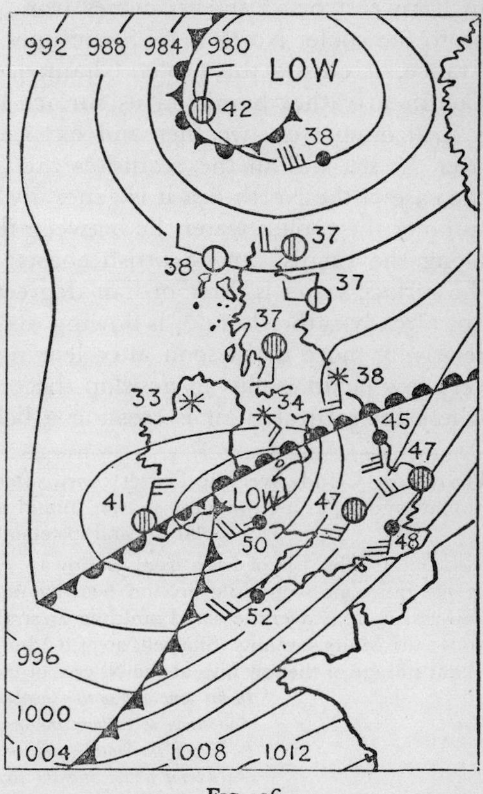

Fig. 36
Snowfall in May; 0700 hrs., 10 May 1943

lighter soils, may experience frost. In a later chapter we shall see how important this sporadic incidence of May frosts can be in a normal year.

Arctic air is still capable of reaching us and giving snow showers down to sea level in Scotland and N. England; as we saw in the previous chapter. But with the more powerful sun and the consequent warming of the land, such a cold air stream generally warms up sufficiently in the surface layers to make May snowfalls distinctly infrequent in the South, and snow-cover even for a brief period is very rare.

On the other hand we can recognise the advent of a new and very characteristic type of weather on and near the coasts, especially that of the North Sea. If pressure is fairly high over the region of Germany and low towards Iceland or the Bay of Biscay the air stream reaching us from east or south-east comes from the relatively warm continent on to the cooler North Sea. Sometimes, coming from the south across France, it crosses the cooler Channel. If such a current leaves the Continent rather humid in its surface layers, as it may well do after a spell of showery weather and extensive damp ground, it is cooled over the sea and all the requisites are present for 'advection fog'. In the case of the North Sea it is generally found that in spring and early summer the coolest waters lie between the Humber and Aberdeenshire along the English and Scottish coasts; on the Dutch-German coasts the surface water is three or four degrees warmer. Hence we find that air, already rather humid, is flowing over a cooler sea and so becoming relatively more moist soon after leaving the Dutch coast; and fog or very low cloud begins to develop, becoming quite thick on our coasts where the saturated air is crossing a belt of still cooler surface water.

PLATE V*a*: Clouds over the Dorset coast, afternoon, mid-March, 1925. Cumulonimbus and a passing thundery shower inland to eastward; unsettled with unstable polar air and excellent visibility.

b: The Isle of Man from Galloway, September 1937. Excellent visibility in the quiet air with little ground heating underlying extensive high cloud. Air flowing smoothly over the island building up strato-cumulus with base about 1000 feet above the Manx summits (Snaefell, 2034 ft.) but unable to ascend to greater heights. Slight mirage of the low hills at the N. end of the islands over the relatively cool sea.

"*In his lone course the shepherd oft will pause*
And strive to fathom the mysterious laws
By which the clouds, arrayed in light or gloom
On Mona settle, and the shapes assume
Of all her peaks and ridges."

WORDSWORTH

PLATE V*a*. *C. J. P. Cave*

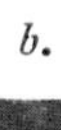

b. *Robert M. Adam*

Robert Adam

PLATE VI: Loch Maree: maritime-arctic air in early April leaving overnight snowfalls; approaching shower-cloud visible to the right.

The air-mass leaving the Continent is thus cooled to a depth which is often of the order of 1,000–1,500 feet. By the time the air reaches our coasts it is virtually saturated, with consequent fog formation throughout this depth. Here then are our classic advection fogs; but we must note that with a slight increase in wind speed and consequently increased surface turbulence, the base of the fog is found slightly above the surface, perhaps from 200 to 500 feet. The process has already been described in Chapter 4.

The result is the formless thin grey-morning fog, or very low cloud, sufficiently conspicuous in Edinburgh and E. Scotland in the spring and early summer to acquire its local name of 'haar'. 'Sea-fret' is commonly heard farther south; in Lincolnshire and Yorkshire 'sea-roke', 'roke' or 'north roke' are heard. (Icelandic *reykur*, Danish *røk*= reek, smoke or fog). In some years with a persistent High there may be several days when it lasts throughout, e.g. May 1935. In general however it shows some sign of lifting and disappearing during the daytime, especially if there is a clear sky above. For, as the sun rises during the forenoon, radiation penetrates the low cloud and reaches the ground from which the surface air over the land is warmed; moisture is absorbed, and after a time the low cloud disappears. Two other factors aid the process. As the land warms, the sea breeze increases in speed; this may give more turbulence over the land surface, mixing the saturated air with that which is unsaturated above. Off the coast it also leads to some slight descent of the dry air above, which may be sufficient to absorb some of the moisture in the surface layer. Moreover, as soon as direct sunshine begins to penetrate to the ground and can thus warm the ground well, the fret generally disappears quickly. If on the other hand there is a slowly spreading cloud-sheet at a higher level, obscuring the sun, the surface sea-fret remains very obstinately and the gloomy damp greyness is indeed depressing. In general such a spreading cloud-sheet, at a higher level, is the precursor of a front; eventually the sea-fog beneath it will slowly be removed after the onset of rain, followed by the arrival of a different type of air. Interesting experiments can be carried out by any amateur with two or three thermometers during the dissipation of fog with a clear sky above (cf. Fig. 3, p. 27).

The rather sudden appearance and disappearance of this extremely characteristic and dismally chilly veil of low cloud has for long been observed and in many parts of the coast is proverbially associated with the tide. In 1750, Dr. Thomas Short of Sheffield commented on the 'tide-weather' of Lincolnshire. "This sort of weather will change with the tide" is commonly heard; in general however such a statement is not justified, as anyone who notes the times of high and low water and correlates them with the variations of the sea-fret will soon observe. Locally however, for example, in the Essex estuaries where the ebb reveals a wide stretch of sand, the drying of the sand and the ease with which its surface warms by radiation from above probably does play some part with regard to the moisture of the adjacent surface air, and hence the local variations in the thickness of the cloud. The forecasting of the spread of night radiation fogs in certain sandy estuaries is affected by the extent to which the tide covers the sand; and it is not unreasonable to suppose a similar effect with regard to the advection fog so characteristic of the late spring and early summer. Patches of cold water may also play a part, as is well known by airmen near the Solent. Such associations with the tide however can at best be very local; and it would be wrong to force such conclusions on the rocky North Yorkshire coast, for example, in a region where the tide recedes but a short way from the cliffs.

Coastal fogs of a similar type are by no means unusual in the region of the mouth of the Channel, notably in June and sometimes in July or even later. A light warm southerly wind coming round the flank of a large continental high from across W. France is frequently accompanied by widespread fog or very low cloud along the South Devon and South Cornwall coasts. It may spread up the west coast to such peninsular regions as that of the Lleyn in N. Wales. More commonly however as we go northward the wind increases and the result is the extensive low stratus at 1,000–1,500 feet accompanying a mild but curiously relaxing south to south-west wind in the Lake District, for example, or in S.W. Scotland. Anyone who has tried to walk on such a day in Eskdale in early June will appreciate the way in which a singular lassitude discourages one from climbing into the formless stratus above, although from this type of cloud in the stable surface air nothing more than a slight drizzle occurs even in the mountains. No doubt if our mountains were high enough it would be possible to climb above the cloud into clear dry air.

The sense of lassitude is probably associated with the fact that inland in the valleys the surface air movement is slight; on the coast wind speed may be of the order of 8–12 m.p.h. Secondly, the surface air is humid and evaporation from the body is slow, especially with little air movement. By June the sea surface temperature on our west coasts is well into the upper fifties, the air temperature in the day probably lies somewhat above 60° and there is a curiously oppressive sense of warmth from overhead due to radiation through the cloud, not merely from the invisible sun but from the overlying warmer mass of air. In a later chapter it will be seen that such temperatures with high humidity and little air movement are much more discouraging to exertion than if the air were saturated and 10° cooler. The frequency with which such conditions occur in the summer months towards our west coasts goes far to explain the relaxing quality of the climate, especially in those inlets and deep valleys among mountains in which the movement of air is restricted.

All these early summer effects as we have seen arise primarily from the flow of warm air over cooler seas. More, however, should be said of June. By this time the Arctic ice off Greenland has largely dispersed; the sun is at its most powerful and throughout the month is overhead at noon in latitude 20° or more north of the equator. Associated with this northward displacement of the region in which incoming radiation exceeds that which is out-going we find that the great mid-Atlantic or Azores High tends to take up its most northerly station, while at the same time the warming of the land produces, on the average, lower pressures to the eastward. Britain in June has a very fair chance of lying within the influence of an Azores high, or within that of the extensive masses of quiet air called anticyclones which appear to detach themselves from it and to move slowly north-eastward over the north-west coasts of Europe. In such massive anticyclones the air above is warm, dry and clear, partly on account of the subsidence which is necessarily associated with such highs. Surface air generally moves from westerly points, from a sea which is still rather cooler than the land. Hence the air as it flows over the land is warmed, and the chance of extensive cloud formation within it is decreased. Inland the strong radiation encourages convection, and there is still in general enough moisture rising from the surface to provide small cumulus clouds in the day-time. But, on account of the subsidence of warmer and dryer air at higher levels the vertical growth of cumulus is generally checked at no great height. We thus get the very characteristic fine-weather cumulus associated as regards place with inland locations, increasing a little towards afternoon and dying down at night, perhaps disappearing entirely, perhaps remaining as fragmentary stratus. (Cf. the photograph on Pl. 18b, p. 111). The sky above is a clear blue; many hours of bright sunshine are recorded, and indeed an anticyclonic spell in June can scarcely be matched for sheer enjoyment in any other climate in the world.

On the long run of years the first few days of June appear to be one of the most favourite 'spells', especially in north-western districts; the great regard in the north for the Whitsuntide holiday owes much to this tendency.

Visibility is generally good wherever the sea is not far away. The clear unpolluted smoothly-flowing oceanic air from west or north-west gives sharp outlines. In Scotland, where the oceanic air is particularly pure, the brilliance of the colouring in fine June weather with a High to the westward is unforgettable. Farther inland however, where the air has had a longer land travel, the rising currents from the surface produce considerable haze, part of which can be attributed to dust from the surface and part to the smoke of our industrial areas. For example, in a summer anticyclonic spell a light westerly wind at Cambridge is often associated with brilliant skies above, but visibility at the surface of less than three or four miles in the afternoon. This can no doubt be partly attributed to the gradual shift of the morning's smoke from the Midland cities, associated with the surface turbulence and rising currents to a height of between 3 to 4,000 feet as a general rule.

The frequency with which such anticyclones develop, the decreased activity and speed of movement of Icelandic and other depressions, and the prevailing stability near the coast of surface currents moving from a cool sea on to a warmer land, all combine with the length of day to make June the most sunny month of the year. Indeed even when we allow for the length of day and take out the proportions of bright sunshine for each month, June almost everywhere shows the highest average percentage of possible bright sunshine; though in some parts of the South it is slightly surpassed by May.

REFERENCES

Chapter 6

MANLEY, G. (1935). Some notes on the climate of N.E. England. *Q. J. Roy. Met. S. 61:* 405–10.

SHORT, Thomas (1750). *New Observations on the Bills of Mortality*. London.

CHAPTER 7

HIGH SUMMER AND AUTUMN

Less fair is summer riding high
in fierce solstitial power
Less fair than when a lenient sky
brings on her parting hour
WORDSWORTH

FROM ALL that has been said it will be clear that no sharp line can be drawn between the seasons. On rather rare occasions a persistent anticyclone to the north of us may prevail into June; cool north-easterly winds from Scandinavia cross the chilly North Sea and continue to give extensive day-time strato-cumulus cloud towards the east of the country. At night the decreased turbulence and clear skies, following a day with a maximum temperature in the fifties, allow of rapid cooling; and towards the north, inland valley-bottoms can generally expect a night or two with minima close to the freezing-point in the earlier part of the month. From mid-June to late August however, it is pretty unlikely that even the worst-favoured inland location will experience a screen minimum as low as 32°; though they have occurred as the table on p. 257 shows, e.g. at West Linton. West Linton lies in a broad upland basin 10 m. S.W. of Edinburgh, surrounded by bare grassy uplands. With the break-up of much of the Arctic ice and the warming of the Arctic land-masses, it is scarcely possible from June to early October to make any distinction of maritime-arctic from maritime-polar air.

Throughout the summer months it is still true that we are more often than not under the sway of air masses which have been for some time over the Atlantic. Westerly winds prevail; depressions continue to move along the polar front, that is from Iceland to N. Norway. But the whole circulation tends to be less lively. Depressions are not so deep; winds are much less strong as a rule, shown by the fact that

June and July have fewer gales than any other months. As a corollary, depressions move less rapidly; not infrequently in a bad summer they persist for several days in unfavourable locations. One of the worst possible situations with regard to S.E. England arises when slow moving or stationary depressions repeatedly find their way into the southern North Sea (August 1915, August 1946, June 1948).

But still the streams of tropical and polar air flow over us; slower-moving and less different in temperature than in winter, but still capable of producing frontal cloud and rain. Summer is then a season at which in principle at least the weather is similar to that of much of winter. However, the air is warmer and where saturated can hold much more moisture. Hence if rain falls in summer it generally falls more heavily although not as a rule for such long periods, as in winter. This is reflected by the statistics of rainfall; the amounts in July and August are larger although the number of days with measurable rain is generally less than in January-February. Slow-moving humid air masses are also very liable to produce the characteristic summer thunderstorm, of which two main types can be recognised. The first of these is the series of scattered thunderstorms due to heating of a somewhat unstable air-mass over the land. The second is rather of frontal origin, and is very characteristic of the break-up of a summer heat-wave. So much indeed is this noticeable that Englishmen will still lend their amused support to Charles II's opinion:—"The English summer consists of three fine days and a thunderstorm."

Let us consider a familiar sequence. On a July day a depression is filling up to the East of Iceland, while pressure is high towards the Bay of Biscay and South West of Ireland. Cool westerly winds give the familiar strato-cumulus with occasional bigger cumulus and showers among the mountains of Scotland. In England towards the Midlands the weather is fair, and the warming of the surface layers of air in East England lowers the relative humidity; fine weather cumulus and a pleasant westerly breeze prevail; the cumulus tending to have a higher base and to decrease in amount towards Eastern England. Along the Channel coast the wind sets slightly off the sea during the day, giving stable air on the coast and long hours of sunshine. Afternoon temperature reaches 70° on the South coast, 74° in London, 67° at Blackpool, 62° in Skye. A clear night follows, with minimum temperatures down to 50° inland.

Next day the anticyclone spreads north-eastward; winds are still W. to N.W. but decrease in strength. Clear sunny weather prevails over most of the country; the day is hot inland, and coastal sea-breezes are well-developed. Probably 80° is exceeded in London and the Thames Valley, and 76°–78° in our other Midland cities; a little small-sized cumulus is found here and there towards afternoon particularly in belts a few miles inland from the coasts.

On the third day the anticyclone has moved so that its centre lies over the North Sea. In W. Scotland a gentle humid current from the S.W. fulfils the conditions described in the last chapter and gives relaxing damp warmth. But in most of England the warm Continental air supply with a light S.E. wind gives clear skies and a rapid rise of temperature to something like 85° to 87° inland in the South, and probably 80° in the North. A truly hot day, with in the evening continued warmth among the buildings of the cities; and on the succeeding night temperature scarcely falls below 60° even in the country. Next morning the sultry air warms up rapidly; but it is probably more humid. For as the anticyclone retreats eastward and the south wind freshens the moist air begins to advance again over the country from the direction of France; sometimes it may even come from the Mediterranean. Converging towards this warm stream of air however there is probably a cooler stream from the Atlantic; the two meet along a north-south boundary or front, which moves slowly eastward.

At such a slow-moving front it is by no means uncommon to find great instability. Some of the advancing cool air is held up by friction at the surface, while above it moves more freely; hence over a broad belt of country we find that while the surface air is still hot, the air above is much cooler. We have already seen that under such conditions rising masses of humid air are extremely unstable; expressed otherwise, cumulus once it begins to form at any point grows very rapidly upward and may easily reach the heights at which it acquires the title of cumulo-nimbus, that is a height at which ice crystals form, shown by the fibrous appearance at the top. We have already seen that once this stage is reached violent thunderstorms are liable to break out; they develop very widely from south to north across country and after they have passed, the heat is dispelled by the advent of the pleasantly fresh and cool Atlantic air, as a rule many degrees cooler than that before the storm. Normal S.W. to W. conditions with maritime-polar air then resume their sway.

Frontal thunderstorms of this type occasionally last for some hours in S.E. England, following exceptionally hot weather. Not uncommonly they are associated with the front formed when cooler air lying over the North Sea is being overridden by hot, humid air from France. It has been observed that afternoon maxima over 90° in France are very usual precursors of such storms. The warm humid air takes some hours to spread northward before reaching a sufficient height for the necessary instability to develop; hence in association with slow-moving fronts they frequently develop in the evening hours as well as the afternoon, and may continue after midnight, as Londoners who recall the great July storms of 1923 and 1945 will recall. (Cf. a paper by C. K. M. Douglas and J. Harding in *Q. J. Roy. Met. S.*, 1946). London complained of a similar storm in July, 1565.

In the main however the predominant theme is that of the west wind with the maritime-polar air; sometimes with occlusions and an hour or two of slow-moving warm rain with little wind; sometimes with widespread layer cloud when the south-wester of 'tropical' origin blows. Occasionally very unpleasantly oppressive warm moist air of this type gives intermittent sunshine and temperatures over 80° with a wet-bulb of 70° or so in eastern England. At the same time in Cornwall the very low stratus is scarcely broken at all and maxima of the order of 66°–68° prevail. The steady forward movement of tropical air under these conditions gives very heavy orographic rain on our western mountains. A notable example occurred on 29 July 1938 when over seven inches fell at Buttermere. (Cf. Fig. 42, p. 137).

Yet in a dry season quite vigorous fronts may pass with practically no rain at all in the Eastern counties. An outstanding instance befell on 16 September 1947. By 13h. temperature at Cambridge was 84° with a strong wind, a falling barometer and increasing cloud; an hour later the temperature was 70° and the sky was again clear. A cold front of some note had passed yet very little rain occurred. It is probable that this event was partly attributable to the intense drought; there was practically no ground moisture and even in the warm southerly air-stream but little cloud developed (Fig. 37, below).

Occasional years give persistent anticyclonic drought as in 1921; if the anticyclones move rather to the north, a hot dry east-wind summer results, as in 1911 or 1887. A shorter spell, lasting for a week, of such weather marked the opening of the Olympic Games at the end

The Times

PLATE VII: Heavy seas on the Sussex coast in a December gale. Bright interval following the passage of a minor front in a deep winter depression.

PLATE VIII*a*. *James Paton*

b. *Fox Photos, Ltd.*

of July 1948; on four successive days maxima surpassed 90° in the South, and with the north-westward spread of the warm air there was a rare occurrence of a maximum of 90° in Scotland, at Prestwick on 29 July. By contrast, other years give us bad summers when for several weeks slow-moving depressions from the Atlantic make their way towards the Channel and on to the Continent. The rainy sectors of such lows traverse S. England, each one giving cool easterly to north-easterly winds and rain for many hours; while as the depression moves slowly onward, minor fronts give additional rain and cloud in the rear. August 1912, 1924, 1931 and 1946 may be mentioned as examples. In August 1945 for two weeks persistent North Sea cloud spread over S.E. England in association with an equally persistent anticyclone in the region of the Hebrides. As we might expect, this August was one of the dullest on record near London, while at Stornoway it was one of the finest and most sunny for a number of years.

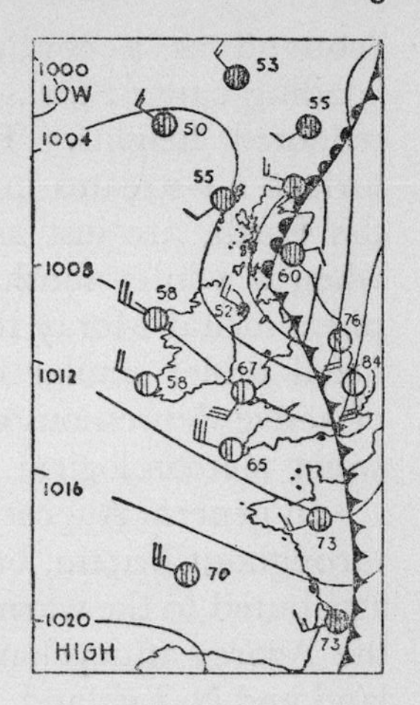

Fig. 37
1200 hrs., 16 September 1947

More recently, August 1948 gave persistent heavy rains in Southern England, associated with a slowly-moving low off Southern Ireland and a very humid cloudy air stream with minor fronts within it. With quite a minor front between humid air from the direction of France and a cooler air-stream from the North Sea, very heavy rain fell for several hours on the night of 2–3 August; in all nearly three inches fell in the Cambridge district. Later in the month, a depression whose centre crossed England gave an appalling deluge of rain north of the centre. With the rainy sector lying to the north-ward we should expect a continuous fall for several hours in any case. But on this occasion the easterly surface wind off the North Sea gave exceptional falls on all the eastward-facing hills. Nearly three inches fell in the Harrogate district, and upwards of four inches in South-eastern

Plate VIII*a*: Rays of Aurora borealis: from Abernethy, March 17, 1949.

b: December 1946. Stormy seas breaking on the Scilly Isles. The resemblance to the shape of cumulus will be noted, and the relatively bright light compared with inland at the same season.

Scotland in a continuous downpour lasting many hours. Severe flooding carried away railway and road bridges, rivers rose to unprecedented heights. The results accompanying convergence of the surface air-streams into constricted valleys, and into estuaries such as the Forth, are just as evident when the wind is easterly as they are when a winter southerly gale besets Snowdonia (cf. Fig. 43). The phenomenal Moray floods of August 1829 on the Findhorn and Spey, befell under similar circumstances. There is no doubt that a series of active depressions crossing Southern Britain in August is one of the worst meteorological events that can occur in our climate.

In general August is appreciably wetter and more cloudy than July throughout Britain, but especially in W. Scotland. This can in part be attributed to the warmer sea, and the diminished frequency with which the Azores anticyclone spreads far enough northward to affect Scotland and N. England. At the same time, in association with the warmer sea and also, we may presume, the lack of Arctic ice, it is rather rare to find a well-established High to the northward. There is little to impede the eastward movement of Atlantic lows, therefore, in a normal year, although these do not move quickly and are not in general of great depth as yet. We have generally to wait till October for the first conspicuous results of the sharp increase of the temperature gradient between the rapidly-cooling Arctic and the still-warm Atlantic, in the shape of more active depressions and the stronger winds of autumn on our exposed coasts.

In association with the warmer sea and the slightly less powerful sun, August is on the average more humid inland; the surface wind from the sea has a higher water-vapour content. At the same time with a less powerful sun the land is not quite as warm; hence we find August tends to be more cloudy than July. Harvest goes forward under varying skies, and in the north the greater proximity to the normal route of Icelandic lows means that there is almost always a good deal of intermittent rain, with consequent difficulty for the farmer. It is unusual, in the South at least, to experience low minimum temperatures at night, partly on account of the more humid air, and the dampness of the ground after rain which is rather frequent in a normal August. In Scotland however, the cool north wind behind a passing depression may give a night of clear sky and a sharp foretaste of autumn before the end of the month; occasionally temperatures below 32° are recorded in the latter half of August over a wide area. This occurred for example towards the end of August 1946 when a calm night of clear sky and relatively cool air befell in a feeble ridge of high pressure, behind a deep depression which had moved along the English Channel.

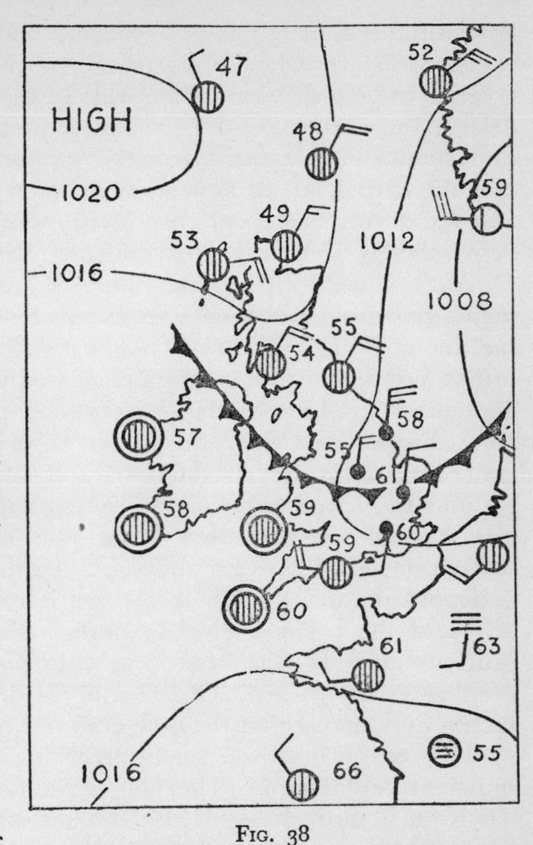

Fig. 38
0600 hrs., 3 August 1948. Slow-moving cold front with very heavy rain in the Midlands overnight (nearly 3 inches at Cambridge)

September does not in general continue the steady process of decline which appears to set in with the rather cloudy west wind of July, through the increased cloud and humidity of August. We have already seen that from time to time the advent of anticyclones of exceptional persistence leads to months of a very different type from the normal, the most remarkable example in recent years being August 1947. In September the march of the Atlantic lows is frequently arrested for some time, and one or more spells of fine, quiet dry weather are so regularly experienced in September that in general it is a much drier month than either August or October all over Great Britain.

Exactly why this should occur in not yet clear. Moreover, the character of the month of September has tended to change; since the late Victorian era it has tended on the whole to be rather drier than previously. As it is, some Septembers are characterised by much windy south-west to westerly weather, and the story of the maritime-polar air is repeatea for yet another month. With little ice in the Arctic however the polar air is not especially cold by comparison with the warmer seas over which it must travel to reach us. As in other months, it tends to give showers in the west of the country, and to be a little drier to the eastward with longer periods of sunshine. This is extremely important for the Scottish harvest; on the higher ground towards the east the spring, as we have seen, is later. Harvest is therefore later; and the drying of the oat crop often owes more to evaporation in the September wind than to the sunshine. (See Pl. XXIII*b*. p. 286)

Anticyclonic days in September are very characteristic. The longer nights give more time for the temperature to fall as a result of out-going radiation. Following a rainy August the ground is not uncommonly rather damp, moisture having penetrated to some depth. In the warm September sunshine by day evaporation into the air is still considerable. Evaporation from the surface however results in more moisture rising from below if it is available; and thus we can see why with quiet air and the lengthening nights the conditions are favourable for the formation of dew and mist. On a quiet evening as we have seen the surface air soon falls to saturation point adjacent to the ground and dew is deposited; and the whole process is emphasised when the surface layers of the soil are already damp. Hence too the Englishman's cautious attitude with regard to sitting out on a warm September evening in his garden. Occasionally a dry warm September follows such a dry August that the dew does not readily form.

It is to be observed that longer nights in themselves are not a sufficient explanation. The fall of temperature between sunset and dawn on a short June night is on the average greater than that in September, or indeed at any time of year unless there happens to be an exceptional snow-cover. This is largely because the fall of temperature is mainly during the three hours or so after sunset, and also the ground in June is in general drier. Dry ground loses heat from its surface more rapidly, as we have seen in the instance of sandy soils (p. 171). The air too is drier, allowing more radiation to escape from the ground.

In the quiet air of September valley-inversions very readily form at night. In no month are the average differences of temperature more

---

PLATE 19

ABERYSTWYTH, Cardiganshire: quiet July afternoon on the Welsh coast. Strato-cumulus and cumulus in a light humid south-west air stream; cloud slightly broken in the lee of Pembrokeshire after passage over the land.

PLATE 19

Cyril Newberry

Julian Huxley

Julian Huxley

acute between favoured hill sites and valley-bottoms inland. (Cf. Chap. 9, pp. 166–8). For while there is rapid cooling of the ground on a clear evening by radiation, this leads as we have already seen to continuous "ponding" of the cooled air, especially in undulating country. No cyclist homeward bound on a fine September evening will fail to recognise this.

But the air at higher levels remains warm; and the inversion boundary between the cool surface air and that at higher levels tends to subside at night, partly as the result of the slight outward movement of the land-breeze all round our coasts and partly owing to the increased density of the cooled surface layers.

Later in the night therefore the air on a hill finds itself losing heat by radiation, both from itself and from the adjacent ground. But at the same time the subsidence of the warmer air aloft means that inward radiation from the warmer air-mass above to the ground is to some extent balancing the outward loss. Hence we find that at higher level stations the temperature does not fall, later in the evening, so rapidly as in the valleys below. In a month such as September the upper air is frequently quite warm and dry, fulfilling the needful conditions. The minimum temperatures attained at hill stations, especially where the ground retains some warmth from the sun, may be many degrees above those in the nearby valleys.

Accordingly, it is in September that very marked variations in the incidence of early morning frost commonly occur. These particularly interest the amateur gardener; but they are perhaps not so economically significant to the fruit-grower as those of spring. The table on p. 168 shows that over 15 years the greatest differences between the minimum temperatures on the flanks of the Malvern Hills and in the Severn valley tend to occur in September.

With warm subsiding air aloft the stirring-up of the surface air is hindered; hence quiet days in September are often hazy, especially inland in the Midlands and South. Sometimes the light dry easterly drift off the continent round a large anticyclone is associated with a good deal of haze; rising currents carrying dust or smoke are checked

---

Plate 20

a. Vale of Ock, Somerset: July afternoon. Summer shower from heavy approaching cumulus and cumulo-nimbus.

b. The same shower passing away.

by the existence of the subsidence inversion-layer at a height of the order of 3,000 feet. (Cf. Pl. XII*a*, p. 159). We have already seen that the subsidence-inversion is commonly developed in large anticyclones detached from the region of the Azores.

It is rare at this season to find the cold air from the Arctic bursting out sufficiently to build up a high-pressure region of the colder type such as we experience in the spring. Nevertheless, such things have happened; in recent experience the most memorable example befell on September 19–20, 1919. On the morning of the 20th snow lay down to 800 feet or thereabouts at many places in Scotland and N. England, and even at Princetown (1,360 ft.) on Dartmoor.

In any year however, a wet windy September may occur, with active fronts frequently crossing the country and continuing the rain and wind of such an August as that of 1950. We are a very long way as yet from full knowledge of the factors which determine why in some summers, the Azores High rarely spreads towards us, and instead, vigorous depressions follow tracks much to the southward of normal expectation. The South Devon rains in autumn 1960 were exceptional.

October is a definite autumn month which in some years gives us quiet anticyclonic spells like September; but more generally a series of active Atlantic depressions with well-marked fronts begins to predominate. There follows the normal sequence of weather, in which the several air-masses affect the country in greater or less degree from south to north, of warm dull days with maritime-tropical air, much low cloud and heavy orographic rain (Figs. 39, 40, below); of brighter west-wind days with maritime-polar air in its several possible varieties. Generally from the rapidly cooling Arctic a foretaste of winter arrives; occasionally we can again recognise the unstable maritime-Arctic air, giving the first snow, hail and sleet squalls of winter and the first upland snow-cover. The air flooding over the country ahead of these vigorous October depressions is still from a warm sea, and gives heavy and prolonged rain ahead of warm fronts (Figs. 39, 40); while polar air is now more unstable and prone to give showers. The effects of exposure to wind and rain become apparent in the woods. In the north of England and Scotland, leaves fall fast with the wind. It is only in sheltered inland valleys after an anticyclonic spell that the glory of the autumn woods can really be seen, for example on Tweedside. But the South of England lies farther from the track of many lows, and hence the leaves generally give an attractive display of colour especially following the first frosts of a dry autumn. Over the greater part of the Midlands the first morning with a minimum below 32° in the screen can be expected early in October.

November is also fairly to be described as autumn, rather than winter, in a normal year. Radiation fogs tend to be both more frequent and, with the feebler sunshine, more persistent. Both in November and December many depressions tend to follow tracks north-eastward along our western coasts, so that cloudy days and strong winds or gales are frequent, with very heavy orographic rains in all our western districts. Inland to the eastward however, the wind is often much lighter, especially when pressure tends to be higher towards the Continent. We have already seen that the lightly-blowing humid southerly air-stream under these circumstances is very liable to give mist and fog, especially in the Midlands, whether through radiation at night or advection over land already cooled. (Cf. Fig. 4, p. 31 for 22 October 1937 and Fig. 67, p. 256 for 27 November 1948).

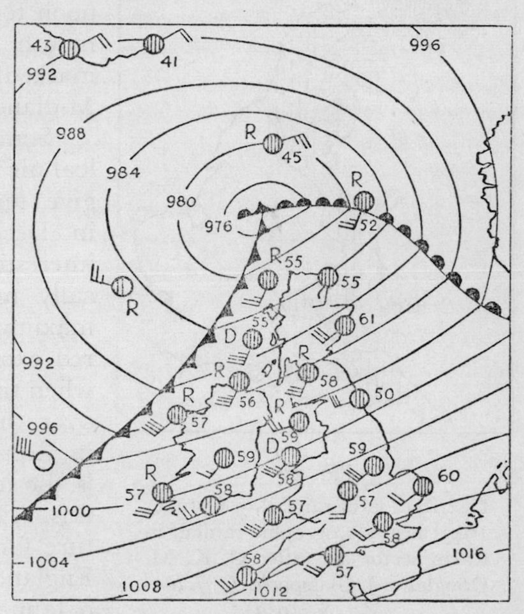

Fig. 39

1800 hrs., 5 October 1943. R, rain; D, drizzle (after C. K. M. Douglas & J. Glasspoole, *Q. J. Roy. Met. S.*, 1943)

The map for 27 November 1948 is interesting, showing the situation during one of the most persistent London fogs for many years. Surface temperatures were little above freezing-point in East Anglia, but were nearly 20° warmer a thousand feet above the ground. The surface air with its fog drifted northward beneath this inversion as far as Tyneside and East Scotland, lifting there into low stratus cloud.

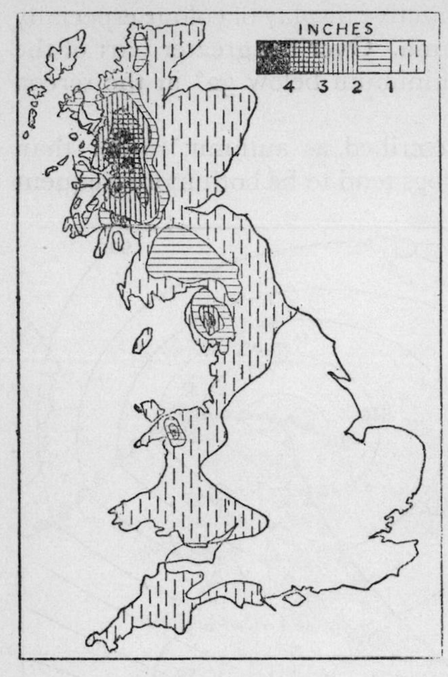

FIG. 40
Distribution of rainfall, 5 October 1943; heavy orographic rainfall in warm sector air (after C. K. M. Douglas & J. Glasspoole, *Q. J. Roy. Met. S.*, 1943)

To the westward however a stronger southerly breeze prevailed ahead of a minor cold front off Scotland. This air, descending over the mountains, became warmer still along the Moray Firth; and the noon temperature of 59° at Lossiemouth accordingly contrasts very markedly with the dismally cold Midland fog.

Somewhat rarely maritime tropical air reaches us in November to give temperatures over 60°. But in December and January it is interesting to observe that practically no example of a day-time maximum above 60° has ever occurred except in the lee of mountains when the air-stream was descending over them; 65° at Achnashellach in Ross-shire on 2 December 1948 is the most remarkable example of this. The highest January temperature for the past hundred years in England was 62·5° at Durham on 9 January 1888. Sixty degrees has been recorded at Aberdeen on Christmas Eve (1931); 63° at Aber in 1916, in N. Wales in the lee of Snowdon in January 1929, and at Rhyl in 1916; 61° at Wrexham in January 1944; 62° in Dublin, also in 1888. Each of these instances affords a reminder that with a moist air current under stable conditions, generally round the margin of an anticyclone with warm subsiding air at some higher level, the air will descend on the lee side of a mountain range in much the same manner as the Swiss föhn; and our highest November maximum on record (71° at Prestatyn in 1946)

PLATE 21
STUDLAND BAY, Dorset: fine and nearly cloudless August evening on the south coast, with slight cirro-stratus hinting at a distant depression.

PLATE 21

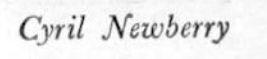

PLATE 22

Cyril Newberry

occurred for similar reasons. Although it can be shown that with our small mountain ranges the effect must be slight, it is just recognisable in the distribution of our highest mid-winter maxima, and perhaps more frequently on the North Wales coast than elsewhere. (Cf. also Chapter 13).

Severe cold in November is somewhat unusual; the range of variability of the monthly means (about six degrees on either side) is considerably less than in December. Indeed December resembles January and February in the contrast it can afford between a stormy 'Atlantic' month such as that of 1934 (mean temperature about 7° above normal, Fig. 41) and that of 1890 or 1878, when the Continental anticyclone built up exceptionally early and the mean temperature was 10–12° below normal over most of the country. It is in December that the sunshine duration is lowest, especially inland, so that the relative brightness of our coasts is then most noticeable. Plate VII, page 122 will bring to mind that even stormy December can give its own unsurpassed magnificence where the cliffs stand forth against the gale.

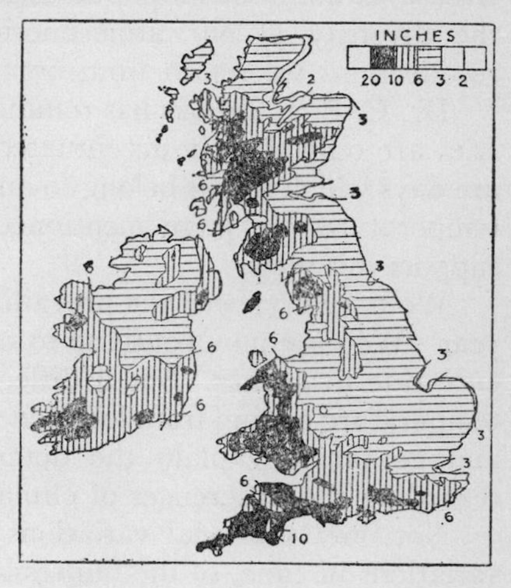

FIG. 41

Rainfall, December 1934; a very mild south-westerly month (from *British Rainfall* by permission of Her Majesty's Stationery Office)

So concludes the year; from many points of view it can be summarised as the story of the possibilities of maritime-polar air, now moving quickly, now stagnating; now coming over a cool sea, and five months later over a sea twelve or fourteen degrees warmer. In the far North it tends to be more unstable at all times; in the extreme south-west it may almost assume the stable characteristics of the humid

---

PLATE 22

HUNTINGDONSHIRE: July afternoon, before harvest. Cumulus clouds showing varying degrees of growth, but as yet little threat of rain. Moderate westerly wind, afternoon maximum 74°.

tropical-air which it closely resembles after its very long sea travel. With relatively small overall range of mean temperature there is not only a wide range of possibilities. There is also a particularly noticeable variation in the intensity of the light from winter to summer by contrast with many other populous lands. All these factors contribute to the diversity of our atmospheric effects and the impressions and recollections we retain from our travels.

Dr. C. E. P. Brooks has reminded us that our seasons in the British Isles are only slightly accentuated, so that all through the year there are days which might belong to any month. And certainly the extreme temperatures we have mentioned for the various months go far to support this view.

We have reviewed the prevailing vicissitudes in time, in an average year We may now continue to examine the variations in our experience due to *location*. Small differences in wind-speed, humidity, and temperature arising from location affect our perceptions very markedly and go far to explain the determination with which many argue regarding the differences of climate from place to place.

Not only the local variations in place, but also the year-to-year variations in time, of the impression given by our climate, owe a very great deal to the extent to which the air remains in vigorous motion. In some months, or even for the greater part of a year as in 1949, it is as if the whole circulation of the atmosphere over N.W. Europe were slowed down, by contrast with other spells of great liveliness. Further, the less rapidly the air-mass moves, the greater the chance that modification in the direction of greater cold or greater warmth will take place even in our relatively small islands. Looking farther back in time, we are beginning to suspect that sometimes there have been not merely single years, but whole centuries during which the general vigour of the circulation and the frequency of 'change-of-air-mass' over our islands tended to be less than at present. It will be sufficient for the moment to point out some contrasts between the mean temperature of quiet summer and winter months, inland and by the sea, and the means for windy and cloudy unsettled months of the same name. Such contrasts, if maintained, would go far to explain the slight but significant changes in our past climate demanded by archaeologists, palæo-botanists and even some historians; and the figures below will suffice to remind the reader that the range of variation between quiet and breezy summer months is more effective inland. Some aspects of this variability have been discussed elsewhere by the present writer.

July 1921—quiet, anticyclonic and sunny:
Inland—Nottingham mean 66·2°, 5·2° above normal (77·9°, 54·5°).
Coast—Skegness ,, 62·8°, 3·8° above normal (70·5°, 55·0°).
Rain 70% below normal, Sun 40% above, for East Midlands generally.

July 1922—cool, unsettled and cloudy:
Inland—Nottingham mean 56·9°, 4·1° below normal (63·9°, 49·9°)
Coast—Skegness ,, 57·1°, 1·9° below normal (63·5°, 50·6°)
Rain 60% above normal, Sun 25% below, for East Midlands generally.

Figures in brackets are mean daily maxima and minima.

REFERENCES

Chapter 7

BROOKS, C. E. P. (1929). *Climate*, p. 21. London, Benn.

DOUGLAS, C. K. M. and HARDING, J. (1946). The thunderstorm of July 14–15, 1945. *Q. J. Roy. Met. S. 72:* 323–31.

MANLEY, G. (1951). The range of variation of the British climate. *Geogr. J. 117:* p. 43–68.

CHAPTER 8

LANDSCAPE FEATURES AND THEIR EFFECT ON WEATHER—PART 1

. . . Humid evening, gliding o'er the sky
in her chill progress, to the ground condens'd
the vapours throws. Where creeping waters ooze;
Where marshes stagnate and where rivers wind,
cluster the rolling fogs, and swim along
the dusky-mantled lawn.

THOMSON: *The Seasons*

THE impression of the scenery of Britain gained by the passing traveller undoubtedly owes a great deal to the changing light; and his emotional response to any view may be totally different if it is seen under differing conditions of weather, even by the same individual on the same day. Moreover, the individual traveller may chose that aspect of the scenery which gives him the most satisfaction; and his degree of contentment not only depends on his personal feelings and make up. It also depends on what he wishes to see in accordance with the spirit of the time in which he lives. Hence the one time popularity of *Monarch of the Glen* studies of the Highlands, arising from early Victorian, and indeed Germanic, romanticism. Contrast the urbane and reasonable calm of the 18th century landscape under a hazy blue sky; and the restfulness of the fashionable 20th century Hebridean water-colour responding to the mood of the Eriskay love lilt in which so many of the modern generation wish to see those withdrawn islands.

So far we have laid the emphasis on the several types of air we may expect to sweep over our islands during the year. We have seen how these different air-masses originate; and we have noticed the effects arising from the tracks they follow en route for our shores. Slight differences are also to be noticed from season to season in the characteristic weather resulting from the movement of any one air-mass across us. In particular, that most frequent type we call maritime-polar tends to give showers more frequently throughout the colder months, although with much the same amount of cloud-cover as is present during the warmer half of the year.

When the air reaches our shores the type, character and thickness of cloud developed owes much to the features of the landscape over which the air continues its course. In this chapter we shall consider these effects, together with those associated with varying clarity of the air.

The important thing to remember about so much of the weather we experience is that it depends on very slight variations in the humidity of the surface layers of air, up to five thousand feet or thereabouts. If we analyse the variations of relative humidity with height we shall find many occasions when, at one place, the R.H. is 95% between say 3,500 and 4,000 feet and no cloud exists. A few miles away the air at the same level is just saturated at the same level and a continuous cloud sheet lasts for many hours. Over the year as a whole the average temperature at 4,000 feet over the Midlands is about 35°. Air at this temperature with 97% R.H. has only to be cooled 0·8°F. to bring it to saturation. This simple illustration will serve to remind the reader that one of the most characteristic features of our climate is the ease with which low cloud develops and spreads over the sky, especially in hilly districts. Many of the photographs show this characteristic low cloud, sometimes detached (Plates 2, 5, pp. 3, 50) or more continuous (Plates 19, XIIb, pp. 126, 159) whose formation has been discussed in an earlier chapter.

But here we touch on a vital factor in our changeable British skies; changeable not merely in time, but also over quite short distances in space, as many motorists know. Our mountains rarely attain 4,000 feet; south of the border few reach 3,000 feet; yet they play an important role on many days and not only with regard to the formation of cloud in the ascending currents on the windward side.

It is to be recalled that an inversion is said to exist when the temperature in a layer of air a few hundred feet thick remains steady or rises with height instead of falling. (Cf. Chap. 3, pp. 27, 49). If the cloud sheet is extensively developed to windward and the base is fairly low, and at the same time a well-marked inversion is present at no great height above the mountain summits, rising air on the one side is accompanied by descent on the other. This will readily be perceived if we consider what happens to a small mass of air following the streamlines over a mountain range. With an inversion layer above, it will have risen into a relatively warmer environment than itself, and so will tend to sink (cf. Fig. 9*a*, p. 51).

Sinking air is however warmed as it descends as a result of compression by the air above; and air which is being warmed becomes drier, as we have already seen. Hence it is very common to find that when an extensive cloud-sheet is present on the windward side of one of our hill ranges, the cloud is much more broken up on the lee side, and may even vanish entirely. This explains why there are many days when a motorist crossing the Pennines or the Welsh mountains finds that the amount of cloud on one side is very different from that on the other. In a springtime anticyclone with an easterly wind Yorkshire may have no sunshine at all while Southport has twelve hours a day; while conversely, Shropshire is often sunny when Aberystwyth has grey skies. For such a break-up of a cloud sheet to occur it appears that in general the height of the top of the inversion should be not more than about three times that of the obstacle beneath.

Nevertheless there are many occasions when no inversion is found above our hill ranges until a much greater height is reached; in such cases the hills play little part in the extent and thickness of the cloud. They may however still play an important part as regards the amount of rain, most of which is derived from the elevation of surface air currents. The diagram below illustrates the amount of rain which fell in an intensely humid and warm current of tropical air in July 1938; all over the country it was cloudy, but the rain varied very greatly in amount to windward and leeward of our hills.

Rainfall varies enormously in amount over the British Isles as everyone knows. Moreover even among our mountain ranges local variations in rainfall are very considerable. The frequency and rate of ascent of moist air depends not only on the presence of the hills but on the degree to which air currents converge into certain valleys leading to additional forced ascent in constricted channels. Excellent examples are found in the Lake District and North Wales; the rainfall map for Snowdonia illustrates the effect of convergence up the Glaslyn valley, and that for the Lake District illustrates the effect of the convergence of air-streams entering several radiating valleys on the knot of mountains at the head of Borrowdale and Langdale. The 'wet patch' round Princetown on Dartmoor lies at the head of the valleys running S.W. towards Plymouth.

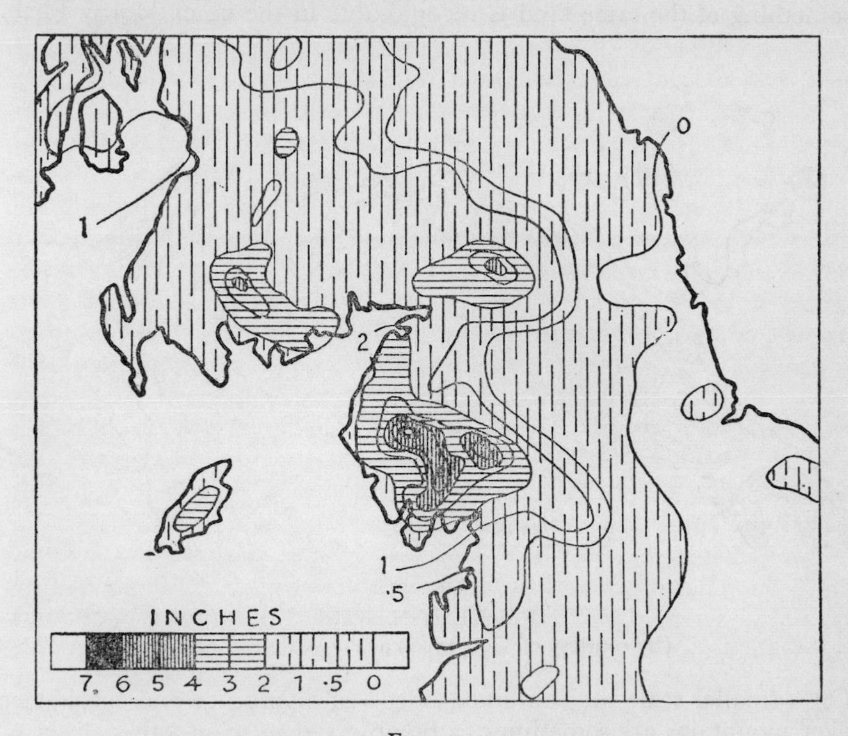

FIG. 42

Rainfall, 29 July 1938; heavy orographic rainfall in warm sector air (from *British Rainfall* by permission of Her Majesty's Stationery Office)

Similar effects lead to a very rapid increase of rainfall up Loch Linnhe, and to the exceptional falls in the group of hills lying just to the westward in S.W. Inverness-shire. It is considered that in a small area at the head of Glen Garry, the average annual rainfall exceeds 200 inches.

The converse effect is also found. In many places horizontal spreading out or divergence of air currents leads to subsidence and hence, as we have seen, a tendency for clearer skies. The best example is provided by the relatively high sunshine durations in Fife and

Angus. The divergent trend of the Grampians and the hills to southward is often very effective when the wind blows from westerly points; something of the same kind is recognisable in the inner Moray Firth.

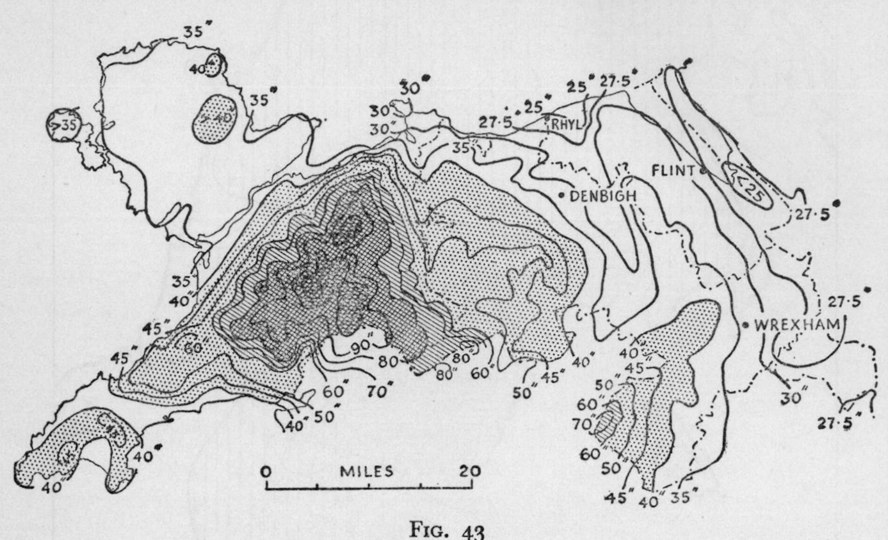

FIG. 43
Rainfall of N. Wales
(by courtesy of Geographical Publications Ltd.)

On a smaller scale the combined effects of divergence and subsidence over mountains are sometimes to be observed in many other districts, round the head of Morecambe Bay for example in north-easterly weather, and to some extent at Plymouth. On the windward side, extensive strato-cumulus or even stratus at 3,000–5,000 feet prevails; to leeward of the hills only occasional fragments of cloud are to be seen. Any motorists driving about the north of England, Wales or Scotland soon becomes familiar with these effects especially in summer. Descent accompanied by divergence of the surface air-streams not only leads to decreased low cloud but also to decreased rainfall. Perhaps the best example of this is at the small island of

---

PLATE 23
BEELEY, Derbyshire: September. Fine-weather cumulus, in a westerly air-stream. Temperature 67°.

PLATE 23

Cyril Newberry

PLATE 24

B. A. Crouch

Fidra in the Firth of Forth, where the lowest rainfall of any Scottish station is found (21·8″); similar effects obtain in the inner Moray Firth, at Nairn for example.

In more unsettled showery weather the changes in the light between flat coasts and mountain districts are especially noteworthy and become more marked when the sun is low. This arises partly from the reflection of light from the nearby sea, and also from the much greater thickness of the cloud building up over the mountains.

Further, the characteristic tones of the grasses in wetter districts are darker. The heavy green of *Molinia*, which predominates in the hay crop in wetter districts, contrasts markedly with the lighter *Agrostis*, and above the cultivated fields dark moorland predominates. Limestone districts owe much of their relative brightness to the lighter colour of the prevailing vegetation. Among our higher hills the variations in the light-intensity between the sunny intervals and those during which the heavy shower-clouds predominate are very much more noticeable than in the plains. Cloud shadows are not only more frequent but more intense; this is well borne out by the contrast between Plates 1 and 26 (on pages 2, and 175).

The intensity with which the gradations and contrasts of colour in the landscape appear to our eyes depends a great deal on the quality of the light reaching us. Hence contrasts are sharper when the air is relatively transparent, *i.e.* visibility is good. The impediments to visibility in Britain are of two kinds; the liquid particles in the atmosphere in the form of mist, fog, drizzle or rain (to which we may add falling snow for convenience), and the solid particles or dust. Solid material in the form of excessively minute particles (of the order of 1/1000 of a mm. in diameter) is largely swept up from the earth's surface into the air whenever the air is turbulent; it is obviously more marked when the ground is dry. In dry clear summer weather it is quite usual to find that the visibility decreases towards afternoon by reason of the diffused solid particles in the lowest 3 or 4 thousand feet—giving the characteristic golden haze effect of a hot summer afternoon. Some indeed is often carried over from the Continent, and pilots of

---

PLATE 24

BUTTERMERE, Cumberland: rain and hail shower, October. Characteristic showery weather in mountains with cool S.W.–W. wind in autumn, rather unstable maritime-polar-air. Cf. Plate 16a.

aircraft are familiar with the yellowish-brown ground haze above which in warm weather they commonly climb at 4,000 feet. This development of day-time haze owes much to the development of instability in the lower layers of the atmosphere during the day. By early afternoon in the warmer months, as we have seen, the surface air can generally be expected to rise in temperature as it moves inland from the sea. The mean daily maximum in July at Woolacombe in N.W. Devonshire is 65·7°; a short distance inland at Barnstaple 67·5°, near Exeter 71·3°, Reading 71·4°; it may be added that Camden Square (London) gives 73·5°.

Thus many occasions will befall when the surface air is stable on the coast, but some miles inland it is warmed up sufficiently to be unstable and to give rapidly rising thermals or convectional currents. These rising bubbles of air, carried along by the wind, give rise to the well-known day-time haze of warm summer weather. The longer the land track and the more powerful the sunshine, the greater the haze. For this reason fine weather with a light westerly to northerly wind is commonly marked by excellent visibility over the sea and near the west and north coasts; but while from Skye one may see the blue Outer Islands all day long, in Cambridge, on the same day and in the same air-mass it will probably be difficult to see any object more than three or four miles distant.

If however, with a light south-west to westerly wind in the warm months the sky is overcast the visibility even far inland often remains very good; turbulence in the surface layers is checked. Driving up the Watling Street beyond Atherstone the hills of Charnwood Forest stand out clearly; from Dunstable Downs the crests of the Cotswolds are seen. Further north, where low cloud sheets are more often found one is frequently reminded of Belloc's lines:

"*The men who live in the North Country*
I saw them for a day
Their hearts are set on the waste fells
Their skies are fast and grey.
From their castle walls a man may see
The mountains, far away."

Stand on the battlements of York and see under the pale grey stratocumulus of a west-wind September noon the long low line of the Pennine Hills to the north-westward; or recall the days when from Shrewsbury the Welsh border was watched for signs of movement under a similar sky. Not that East Anglia is free from such grey skies in early autumn, as Wordsworth says of the way to Cambridge in 1789:

> "*It was a dreary morning when the wheels*
> *Rolled over a wide plain o'er hung with clouds,*
> *And nothing cheered our way till first we saw*
> *The long roofed chapel of King's College. . . .*"

Generally speaking visibility is better when after rainfall the air is subsiding and convection is limited; hence as a rule it is excellent in a narrow wedge of high pressure between two passing depressions. Further, when the weather is warm and sunny inland sea-breeze phenomena become well marked on our coasts. The existence of a sea breeze implies that some slight descent of air is proceeding just off shore. Not only is the air relatively free from impurities but the sky in a strip along the coasts tends to be rather freer from cloud; hence in summer it is commonly found that the visibility is better and the sunshine duration slightly higher on the coast, than inland. Over the year as a whole, for example, Lowestoft averages 1,720 hours of bright sunshine, Norwich 1,600. Any visitor familiar with the weather at a coast resort will not fail to have noticed, too, how hazy the air is on occasions when in summer the wind continues to blow from the land. Nowhere is this more noticeable than in Lancashire. The occasional day of stifling, brassy heat at Blackpool accompanied by a south-east wind from the smoky inland towns and an afternoon temperature in the eighties provides for many a disappointing contrast with the more normal fresh westerly breeze, bright sky and an afternoon temperature of 66°–68° in August.

It may be asked why on such a warm day the surface sea breeze does not set in and undercut the hot wind off the land, inasmuch as the sea is still cool. But if the air at greater heights is still very warm, as it sometimes is when a deep stream of warm air spreads from the continent, normal sea breeze circulation across the coast cannot establish itself. Any cool air spreading inland from the sea cannot rise from the surface as the air above is too warm by comparison; hence no circulation is set up and the heat is equally great on the coast or inland. The temperature of 93° at Bournemouth on 16 August 1947 offers a good example; also 74° at Cromer in early March beside the cold North Sea (p. 88).

Indeed the brazen sky now and then in summer accompanying the hot south-easter at Blackpool, or for that matter the north-east wind of Central European origin at Brighton is a reminder that to the natural dust in Britain's atmosphere, derived by normal processes from the earth's surface, we must add the man-made smoke. Dust haze and smoke are far more effective than most town-dwellers realise, in cutting down the quality of the light and detracting from the colour of the landscape. A well-known artist at Lamorna Cove once described to the writer how he had one summer been invited to paint in Derbyshire; "but", he said, "when I got there the atmosphere was yellow. There's no light for painting at all." This was said on a Cornish April day when he had stopped work for similar reasons—because of the north-east wind—the rare wind which, turbulent due to ground heating on a spring day and travelling over a long stretch of land to West Cornwall, brings the haze which takes almost all colour out of the landscape even in the far west. In western Ireland the wind from the east is known as 'the black wind'; though it is probable that this refers to other evils beside diminished visibility.

But all over England and Wales, over part of Scotland and even in Ireland, the effects of the man-made smoke are at times apparent. On many days when the Yorkshireman of a former day looked out from his castle walls in the clear air under a silver-grey sheet of slightly rippled stratus, he now finds an undoubted brownish cast; and this is equally true of Lincoln. In the Lake District, even the Glasgow smoke reaches Skiddaw on occasion; and in particular the south wind ahead of a slow-moving warm front often brings an unpleasing additional gloom from Lancashire.[1] In Cambridge the smoke of London is at times perceptible under similar circumstances and it has been known to affect visibility at Norwich. For the existence of extensive layer cloud

[1] Compare here Wordsworth's poem "written on the mountain Black Combe" the gloom which beset a "geographic labourer", Col. Mudge of the Ordnance Survey, in 1813.

PLATE IX*a*: Daffodils in early May in the high valleys of the Scottish border. Characteristic cumulo-stratus associated with anticyclonic weather and a light N.E. wind towards the North Sea coasts.

b: Near Baildon in the West Riding of Yorkshire. The fine weather cumulus of a fine day in early summer; the turbulent westerly breeze of the Pennines shown by smoke; rather dark-coloured quarries and grey stone by contrast with Plate XIX*a*, p. 234.

PLATE IX*a*. *The Times*

b. *C. H. Wood*

PLATE X*a*. *Thomas H. Elwell, New York*

b. *The Times*

frequently implies that just above there is an inversion of temperature through which the rising air from the cities cannot ascend; hence the smoke may be borne at a higher level many miles from its source. Some years ago a well-known meteorologist plotted the visibility at various places during a motor run across Central Scotland; he showed how the Glasgow smoke haze spread eastward along the foot of the Ochils and seriously affected visibility on the opposite coast sixty miles away. With light winds the smoke from industrial areas sometimes follows quite a narrow track. The writer has stood on a nine hundred-foot summit in Durham in a light north wind, with the Cheviot clearly visible to the north fifty miles distant, yet, due east, the visibility was less than two miles—the result of the smoke haze drifting off Tyneside and the Durham coalfield. Pilots will remember the difficulty with a similar north wind and splendid visibility elsewhere, of landing on West Lancashire air-fields in winter when smoke haze from Preston was obscuring the ground below.

Smoke is indeed a gloomy subject; for while there are those who will rhapsodise over the hazy City sunsets across the Thames, there are many more who deplore the vanished glories of a sunny little Manchester from which the green and gold Pennine slopes were visible on many April mornings two centuries ago. It is possible that the whole country is affected more than we think. Early last century the Ordnance Survey sighted the Welsh mountains from Bardon Hill in Leicestershire. Ralph Thoresby espied the shipping in the Thames as he rode over Harrow Hill in May 1702. Celia Fiennes saw the Isle of Man from near Chester; Defoe (1726) was informed that from the Cheviot the view extended to the Tyne, and George Smith saw the cliffs beside the North Sea from Crossfell in 1747. But Wordsworth noted the London smoke haze from Hampstead Heath.

In many parts of Britain smoke from the towns spreads in certain well marked directions. Studies of the frequency of surface wind in

---

PLATE X*a*: Unsettled summer afternoon over the Gareloch looking towards Dunbartonshire, June 1945. Humid westerly wind; extensive low cloud breaking a little in the lee of the hills. Cumulus beginning to tower above the low cloud, threatening showers with the possibility of thunder.

b: Harvest in County Durham, looking south-eastward from Plawsworth to the hills of East Durham. August afternoon with characteristic tendency to unsettled weather supported by the distant cumulus with a hint of cumulo-nimbus over N.E. Yorkshire. Clouds show signs of flattening out over the sea.

quiet weather have been made which point to a tendency for the wind on many winter nights to set slightly from a southerly point near the flanks of the Cheshire Pennines. Perhaps this helps to explain why in more recent years the development of the suburbs of Manchester has taken place to the southward where the air is cleaner and purer; in an earlier generation it was more fashionable to live on the rising ground to the north and north-west. Generally speaking the surface relief influences to an appreciable extent the movement of air whenever the wind resulting from the pressure gradient, *i.e.* the spacing of the isobars, is less than force 3 or about 10 miles an hour. Canalising of the flow of air along valleys and the like then begins to become notable as regards the spread of smoke haze, especially at night and in the early morning when there is commonly an inversion a few hundred feet above the surface so that the smoke cannot rise.

It is important to recognise that over uniform country smoke, especially in the day-time, is distributed upwards as well as sideways by turbulent eddy motion in the wind. Hence in favourable circumstances wind can quickly remove nine-tenths of the smoke from surface air. Recent studies of atmospheric pollution in Leicester show that in practically every type of weather the highest concentration of smoke at street level was in the centre of the city, but that most daylight is cut off between a half and one mile down wind from the centre. There is also a horizontal spread with distance, but in more hilly country this may, as we have seen, be restricted by the topography. Details are given in a recent publication of the Department of Scientific and Industrial Research.

It is evident that if unpleasant smoke and fumes are to be liberated this should be done in the middle of the day when in general turbulence is most active. At night turbulence in the surface layers is often checked by the development of inversions; in such an event the smoke cannot be carried upwards. Accordingly, a light drift of air is sufficient to affect suburbs to the leeward. Dwellers in north-west London were particularly affected by this during the prolonged spell of fog at the end of November 1948, accompanied as it was by a light south-east wind (Fig. 67, p. 256). Farther up the country the conditions were often worse; and the unpleasant acrid murk in Wharfedale on that occasion will long be remembered as the penalty of living to leeward of the great Yorkshire coalfield and the belt of large towns on its north-western edge.

Smell plays its part in our impressions and one might fairly remind the airmen of the West Riding Squadron during the war of the proverbial saying that in a light west wind the pilots could smell their way home all the way from the Yorkshire coast. In Hampshire Gilbert White himself recognised the haze "with somewhat the smell of coal smoke" at Selborne, and that such a "blue mist" was always accompanied by a north-east wind. For the meteorological conditions affecting the olfactory powers of such animals as deer, reference may be made to Dr. Fraser Darling's work in this series on the Natural History of the Highlands.

A good scenting November morning in Leicestershire owes much to the combination of damp clay, humid air and little turbulence associated with the soft grey cloud sheet of a typical quiet autumn day on the flank of the Continental anticyclone, when a distant Icelandic low brings the gentle south-wester across England. The melancholy smell of cabbages on dull afternoons in early December on the Gault clay outside Cambridge will also be recalled; similar phenomena are no doubt equally familiar to Oxford men, as they are throughout the clayey Midlands and can almost be described as an integral part of our own impressions, if not of the scenery.

Generally speaking, polar air gives the best visibility though much depends on the length of its land travel. Towards the north-west Highlands and islands the frequency of excellent visibility in the pure unpolluted air with a long sea travel is greater than in any other part of Britain. In 1878 a Scottish amateur meteorologist recorded that over twenty-one years Lochnagar (45 miles distant) had been visible on an average of one day out of four from Aberdeen. The deep blue distances of a September afternoon in Argyllshire provide the traveller with yet another enlivening contrast when he recalls the hazy golden light over the stubble of Norfolk in the same air-mass. The characteristic blueness of the more distant Scottish hills, especially in the Highlands, has been remarked by all travellers. Indeed the very name *Cairngorm* means 'blue hill'. Seen from the great Inverness-Perth road on the long descent to Carrbridge, from the hills of Affric or from the outskirts of Aberdeen the appropriateness of the name is very evident throughout the warmer months.

The visible solar spectrum is composed of light of different wavelengths from red through green and blue to violet. As it traverses our atmosphere the shorter wave lengths are scattered in all directions by the molecules of which it is composed, and by the excessively minute particles of dust which float in it. In a pure atmosphere free from dust the scattering is almost all of the blue and violet, so that for example we perceive the colour of a clear sky to be blue; and at high altitudes the sky becomes noticeably darker as less light is scattered. It may be added that if the larger particles we call dust are present—even they are so small that there may be many thousands in a cubic inch of air—the scattering applies to all the light traversing. An atmosphere with a good deal of dust in it appears much more white. Hence the sky becomes whiter near the horizon, partly because one is looking through a greater thickness of air, partly because in the layers close to the earth's surface there is more dust.

Now if the air is clear and free from dust the scattered light, giving the same impression of blueness, is as it were superimposed by the atmosphere between the observer and the hills. Indeed we find that we even tend to estimate distance by the blueness of the distant hills. Hence the characteristic blue distances of the Highlands in clear weather are principally to be attributed to the purity of the air; in general this has had a very long travel over the Atlantic, and then over a country in which little dust is raised, partly as a result of the relatively frequent rains. Prolonged dry weather, especially if the wind is south-easterly with a long fetch over the land produces a characteristic haze in which objects even at a mere five miles distance appear almost colourless. In Skye we are then reminded of the similar effects seen on sunny days in Cornwall with a north-easter, mentioned on page 142.

Optical effects in regard to landscape have been the subject of a well known work by Minnaert. In this country Mr. James Paton, of the University of Edinburgh, is a meteorologist and physicist who has recently given a most attractive account of some of the less known effects; the above paragraphs are largely based on his article.

Topographical Effects on Wind, and the Results

We have already indicated some of the more obvious results of the canalisation of the movement of surface air-streams by hills and other features. This becomes especially noteworthy where a broad water surface, as in the Firth of Forth, gradually increases in width between the hills. Friction is less over the water, and whenever the isobars allow for the development of a wind from between south-west and north-west the set of the wind in the Forth tends to be from westerly points and the strength is a little greater than it would be if there were no hills. Buildings and other obstacles play their local part; Scotswomen may be reminded of the frequency with which the wind hurries the shoppers in one direction or the other along Edinburgh's Princes Street on breezy winter afternoons. Similar effects are found in a great many estuaries and channels. The pleasures of yachting off Cowes are enhanced by the canalisation of the sea-breeze up the Solent on a warm August afternoon.

FIG. 44

The Solent and Spithead. Totland Bay (p. 90) is due south of Hurst Castle

It has already been shown that the wind tends to descend hill slopes more forcibly if an inversion-layer is present above the summit at some height which should not be too great; one may say, roughly three times the height of the obstacle surmounted. The best example of this process in Britain is provided by the 'helm wind' of Crossfell and the neighbouring Pennine slopes in Cumberland and Westmorland. On days when the north-east wind prevails it is often found to blow with exceptional strength down the treeless slopes of the Crossfell escarpment facing the Eden valley, and over the fields and villages lying at the foot. A little way farther to the south-west, however, the air is nearly calm. Accompanying the strong wind on such occasions, a characteristic wall of cloud is seen lying along, or just above, the summits of the Pennines. Its top is generally smooth in outline. Parallel to this cloud, known as the 'helm'— a word probably (though not quite certainly) expressive of the helmet-like nature of the cloud— there lies, at the same level but four or five miles distant to the south-west, another line of clouds sometimes continuous, sometimes broken up into fragments. This is called the 'bar'; it is seen as a cloud stationary with respect to the ground but in vigorous motion within itself. The phenomena are illustrated in Fig. 45 below.

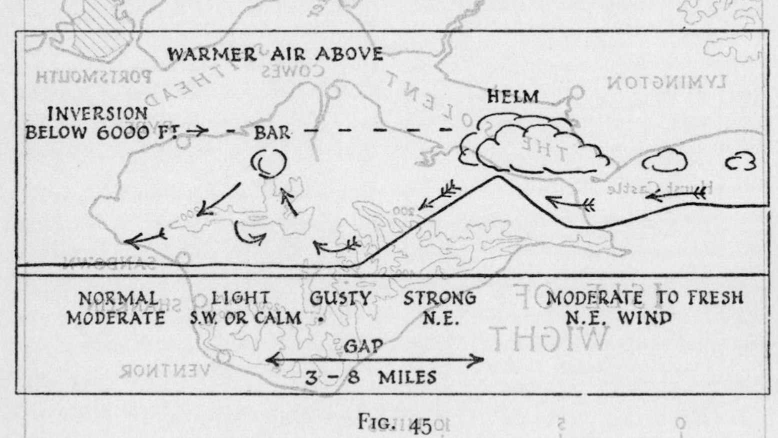

FIG. 45

Normal helm wind; bar formed at crest of 'standing wave' to leeward of the escarpment

It is evident that the escarpment lying transverse to the surface wind acts like a submerged weir in a stream of water; a standing wave is set up just below the weir. High above this surface wave, stationary lenticular clouds sometimes appear (Pl. XXI*b*, p. 270). The clear sky between helm and bar is associated with the descending air; where the air again reaches the same level as the helm, cloud is again present as the bar. This bar is continually forming in the rising current on its eastern side, and at the same time dissipating as the air descends on the west. It often shows rotary motion, a horizontal eddy being formed. Various modifications and partial developments occur which go to show that the upper edge of the helm, marked by an inversion, should not be more than 6,000 feet above the sea, the average height of the range being 2,500 feet, falling 2,000 feet to the valley. Moreover the phenomenon does not develop unless the prevailing strength of the north-easter at say, Tynemouth exceeds about 15 m.p.h. It is also

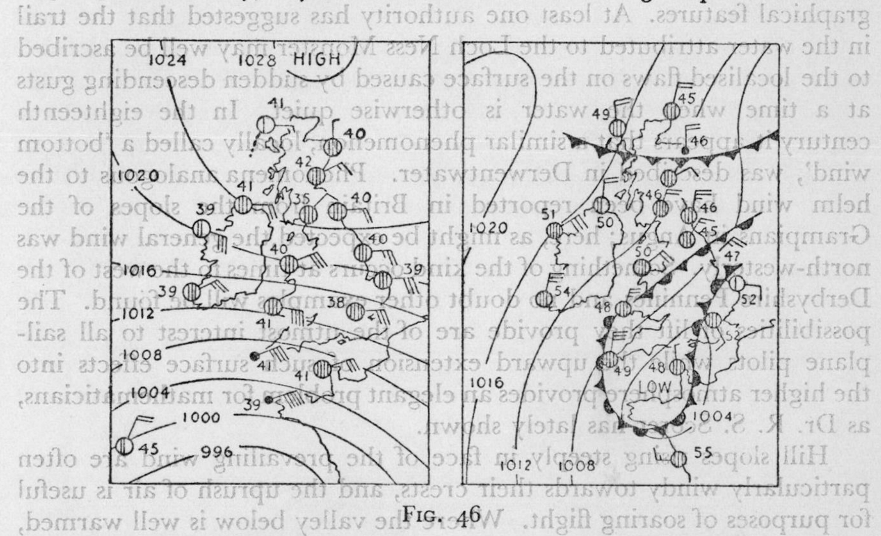

FIG. 46

13 hrs., 29 January 1939 and 18 hrs., 17 May 1939. Moderate to fresh E. to N.E. wind across N. England; situations giving rise to helm wind on the Crossfell escarpment (For notation, see p. 5)

evident that if the westward slopes of the escarpment were much steeper the flow would not be smooth, and places near the foot of such a steep slope would intermittently experience severe gusts. Something of this kind is known to occur in the outer harbour when an east wind (the 'Levanter') blows at Gibraltar, where the steep-sided Rock presents a north-to-south barrier 1,400 feet high and directly in the path of the wind.

Now if we extend these ideas more widely it will be evident that many places among the mountainous Scottish coasts will be subject to variable squalls descending over the crests and giving disturbed patches of water, on occasions when the wind is strong and especially when the summits are capped by cloud. Much depends on the length and steepness of the ridge over which the wind blows, as well as the prevailing strength of the wind. If the wind is only light or moderate, up to ten or twelve miles an hour, it seems that as a general rule the flow of the air would tend to become canalised along the foot of the ranges rather than across their crests. Many squally and disturbed patches of water, many windy and gusty corners among our valleys and hills, can be found which owe much to the surrounding topographical features. At least one authority has suggested that the trail in the water attributed to the Loch Ness Monster may well be ascribed to the localised flaws on the surface caused by sudden descending gusts at a time when the water is otherwise quiet. In the eighteenth century it appears that a similar phenomenon, locally called a 'bottom wind', was described in Derwentwater. Phenomena analogous to the helm wind have been reported in Britain from the slopes of the Grampians in Angus; here, as might be expected, the general wind was north-westerly. Something of the kind occurs at times to the west of the Derbyshire Pennines and no doubt other examples will be found. The possibilities of lift they provide are of the utmost interest to all sailplane pilots while the upward extension of such surface effects into the higher atmosphere provides an elegant problem for mathematicians, as Dr. R. S. Scorer has lately shown.

Hill slopes rising steeply in face of the prevailing wind are often particularly windy towards their crests, and the uprush of air is useful for purposes of soaring flight. Where the valley below is well warmed, a thermal effect may be added to that of the prevailing wind; at gliding sites such as the Dunstable Downs and Bradwell Edge in the Peak District of Derbyshire much use is made of this in summer. But all these local variations in wind speed play an additional part in moulding our scenery. Quite apart from questions of soil, it becomes very difficult to establish trees in windy sites; animals and men too often avoid the wind, or save their energies by adopting the prone position. Hence the establishment of many of our villages in just those locations where in conjunction with other needs, greater shelter was afforded; Telscombe, near the crest of the South Downs between Brighton and Newhaven, offers a good example, in contrast to what many deem the breezy exposure of Peacehaven nearby.

Hence, too, still another factor leading to diversity of scenery. Nowhere in England is the effect of wind better seen than on the north-east coast. The bare cliff-tops of Durham and Cleveland, so well known to those who visit Whitby Abbey or the bleak ugliness of Blackhall Colliery, contrast emphatically with the thickly-wooded denes, with Brignall Banks in Teesdale celebrated by Scott, or with the woodlands of Eskdale, where the Esk has cut so deeply into the North York Moors. In Banffshire the north-east corner is nearly treeless; but inland the splendid woods of Glenlivet at 500–800 feet give emphatic evidence of the value of shelter. For centuries our travellers have been even more struck by the contrast between the treeless windy uplands of Cornwall and the lush vegetation wherever there is protection, especially along the inlets on the south coast such as the Helford river; and in South Devon, Dartmouth and Salcombe are similarly favoured.

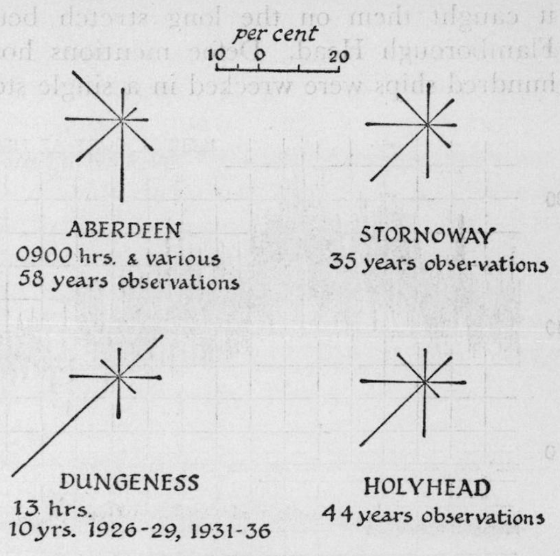

FIG. 47
Observations of wind direction; means of the monthly percentages of observation from each direction made at various coastal points

Towards the British coasts we may often recognise in the distortion of the trees the effect of the dominant wind from the sea, as distinct from the more prevalent south-wester. This becomes very evident on the Northumberland coast, and even in the suburbs of Sunderland the ash trees are woefully stunted through the combined effect of wind and cool air. W. V. Lewis has distinguished the importance of the dominant north-east wind with a long fetch over the sea in giving rise to powerful waves whose onset has done much to mould the features of the North Norfolk coast and elsewhere, although it blows far less frequently than the south-wester. It seems probable that our impressions of coastal landscapes are also affected by our subconscious recognition of the dominance of the gales from the sea. The sight of the bent trees along the Lincolnshire marshland and the landmark of Winterton Church tower north of Yarmouth brings to mind the stories of the Cromer lifeboat, and for how long the north-easter was justly dreaded by the Newcastle colliers on the way to and from London if it caught them on the long stretch between Winterton Ness and Flamborough Head. Defoe mentions how in one year (1692) two hundred ships were wrecked in a single storm.

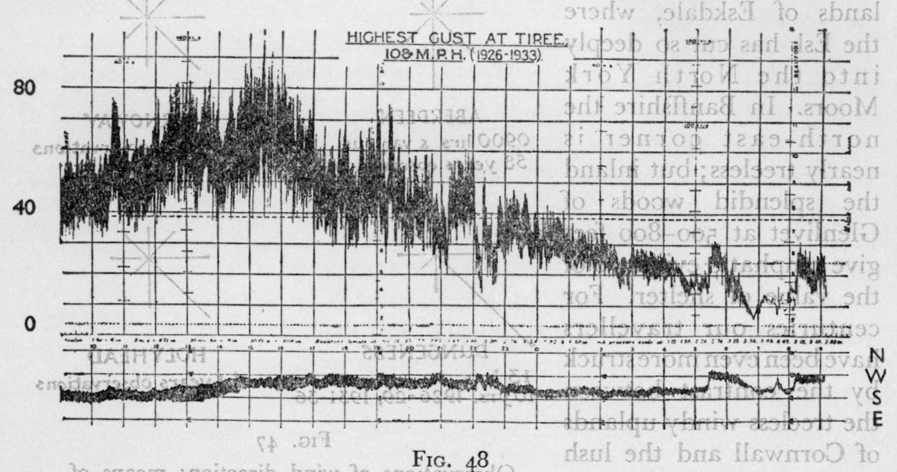

Fig. 48
Highest gust at Tiree, 28–29 January 1927. 108 m.p.h.

Lee slopes as we have seen can receive steady winds, strong gusts at intervals, or no wind at all, depending on the overall force of the wind and on the degree of steepness of the slope. Sometimes back eddies develop (Fig. 9) and we can well understand now the development of the rapidly varying swirling masses of cloud among the irregular passes, peaks and corries of our heavily glaciated mountains. These effects have become familiar in the work of many artists. English and Scottish painters in particular have been perennially fascinated with the expression of the lively and complex rhythm of the movement of the air amid the mountains as well as that of the sea along our rock-bound coasts. Even our instrumental records indicate at times a rhythmic fluctuation in wind speed, illustrated for Bell Rock in the Firth of Tay by Mr. Ernest Gold in a Presidential Address ("Wind in Britain") to the Royal Meteorological Society in 1936.

It would be appropriate to add a note on the occasional extremes of wind experienced inland. Great gales on the coast blow with a higher average speed, as the anemograph trace associated with such a gale at Tiree shows. On this occasion the highest gust reached 108 m.p.h., while for one hour the average speed of the wind was 66 m.p.h. (force 11).

Inland in a great city a really windy day typically gives a result such as that at Kensington. Although the highest gust reached 70 m.p.h. the average wind speed for a whole hour was not greater than force 6 (about 28 m.p.h.).

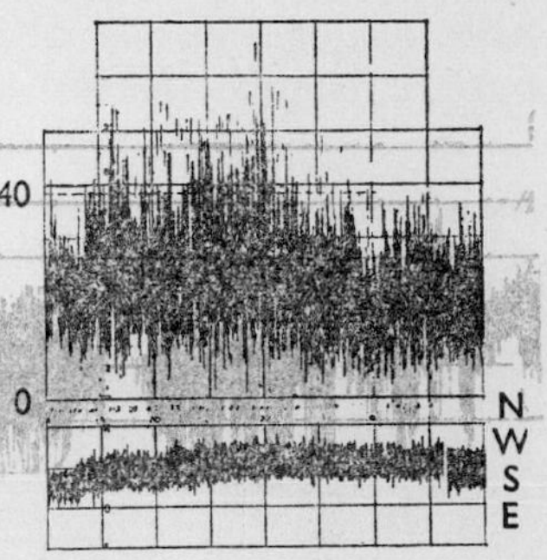

FIG. 49

South Kensington: a strong, very gusty westerly wind, 18*h* 12 January to 2*h* 13 January, 1930. Highest gust at 70 m.p.h.

By reason of surface friction the mean speed is much lower, but occasional gusts still reach, or nearly reach, the violence and strength of those at sea. It is these gusts which do the structural and other damage. For the growth of our largest trees some degree of shelter is generally necessary, such as that provided by an extensive surrounding stand of well grown forest or sometimes by neighbouring hills.

Severe gales are of more frequent occurrence during the months when most of the trees are bare, as Table X in the Appendix will show.

Gustiness for the reason mentioned above is much more marked inland than at sea; the winds over open plains and uplands are also less gusty than in wooded and cultivated lowlands, but are still not so steady as at sea. For a second factor enters into the development of gusty winds. Under conditions of cool air flowing over ground which is being rapidly warmed, for example on a sunny spring afternoon, the air near the ground is very unstable; warm bubbles rise on the least provocation and are replaced by chilly gusts, which feel the colder as the air under such conditions is generally rather dry. Hence a most characteristic type of day especially in East Anglia; the bright warm sun, especially hot in sheltered corners, but the gusty chilly north-easter finding its way round the houses and copses at intervals, and giving rise to appreciable soil erosion in the Fens after dry weather.

Here the black peaty soil is absorbent of radiation while at the same time it is very fine; hence great turbulence develops with the strong heating of the ground, sweeping up the fine material. It will be noticed that the worst damage occurred in late April, 1943—the same season of year when, for similar reasons, the worst dust storms of the Middle West have occurred.

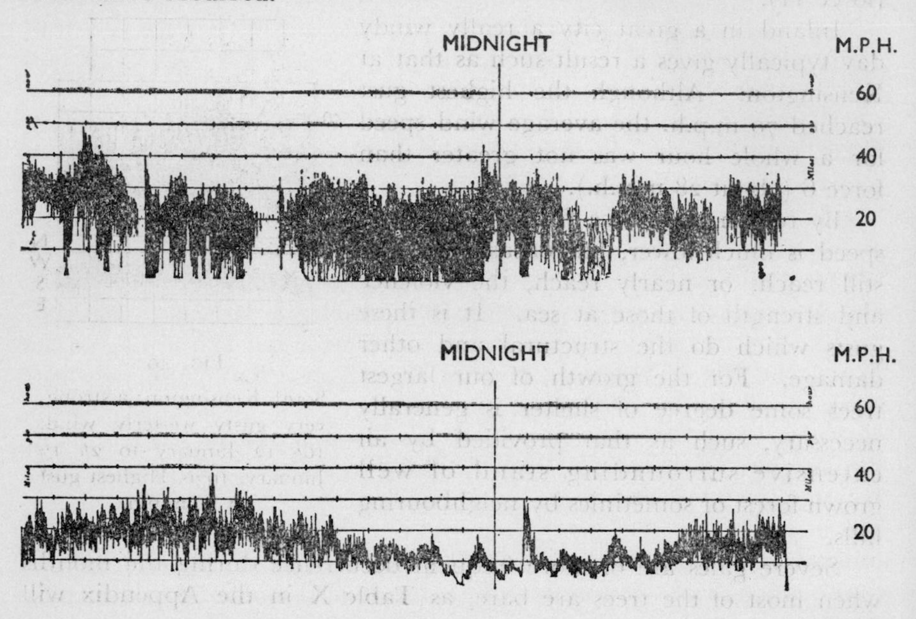

FIG. 50

Anemograms from St. Ann's Head, Pembroke, and Cardington, Bedfordshire, 5–6 June 1944, from 9h. to 9h. "D-day"; strong westerly winds, decreasing in the night (by courtesy of the Director of the Meteorological Office)

One may comment on the statement often made by Southern visitors, how few large trees are found north of a line from Mersey to Humber, except in particularly sheltered places. As the climatic differences in respect of temperature and sunshine are small it seems probable that this is partly a matter of wind, partly of less fertile soils, for it is indeed true that in the aggregate the North is more breezy and more cloudy, especially in summer. The impression is probably enhanced by the fact that so many of the most frequently visited parts of the North, whether for business or pleasure, lie in the uplands. Moreover, in counties such as Durham the clearance of woodlands in the past was particularly active, notably in the eighteenth century. The effect of soil however must not be forgotten. Quite sizeable beeches are found on the limestone near Penrith a thousand feet above sea-level; nearby, other soils are nearly barren of timber. Of the various trees so characteristic of the English scene the oak is perhaps the most celebrated, but farther north on colder soils and on higher ground it is the ash which attracts the attention.

"The oak and the ash, and the bonny ivy-tree" sang the homesick north-country maid of Elizabeth's day; and the ancestral stock of many an accomplished English family is derived from a strictly-built hill farm in a wide and windy country of stone walls, whose straight and sombre lines in autumn were only interrupted by the struggling ash trees and the hurrying cloud above. Of the Welsh border country and even of parts of high Leicestershire much the same may be said.

Far down to the westward in kindlier Devon and Pembroke, in the Lleyn, the Isle of Man and even in West Cumberland, the high turf banks dividing the small irregular fields are most noticeable features of the scenery. Climate again has played its part; in the damper west, the keeping of animals has always been more prominent; pasturage and hay are more abundant. Both grass and stock benefit from shelter, and in a country where hedges and trees are less easily established the turf-grown banks with their wealth of wild flowers have survived from time immemorial. The extended enclosure of the open Northern hills by means of stone walls was a later matter; much of it began in the seventeenth century, and continued thenceforth until after 1800. Readers may make whatever they wish of the fact that they were largely contemporary with the evolution of Cartesian geometry, the Fahrenheit thermometer and the well-tempered clavier. The erection of stone walls rather than turf banks has indeed an extremely conspicuous effect on the landscape, but interpretations in terms of climate would beg too many questions to be essayed here. That the greater windiness and wetness of the north and west, the uplands and exposed seacoasts has influenced in numerous ways the characteristics of the country that we now see is evident to us all. The immediate remark of a Bradford sixth-form girl returning from a first term at a University in the South of England was "how much more colour there is"; but grit-stone and smoke, laburnum and brick play their part as well as the climatic factors and the prevailing northerly aspect of her birthplace. (Plates 14, IXb, pp. 99, 142).

Thermal Effects: Sea and Mountain Breezes

Whenever there is an appreciable difference in the surface temperature across a sharp boundary local air movements are set up. English houses become far more draughty in winter, when the temperature difference on either side of ill-fitting windows is large. By far the most conspicuous of these thermal effects are the sea-breezes resulting from the sharp difference of temperature set up across a coastline in warm sunny weather. The land being so much warmer than the sea a shallow local flow of air sets in almost at right angles to the coast if the weather is calm; typically on a hot summer day it begins to be noticeable as a light breeze about 10 a.m. and may continue till 6 p.m. or so. It is generally strongest, and penetrates farthest inland, in mid-afternoon. The opposite effect—the surface flow of cool land air towards the warmer sea—occurs in the later hours of the night towards dawn. A not uncommon feature of the British scene, therefore, was the setting forth of the local fishermen in the earliest dawn of a summer morning, returning with the fish soon after 10 for quick sale to the diligent seaside housewife; something of this activity is indeed still found.

Sea breezes of this kind are quite shallow, about 500 feet being common. This is easily to be observed on some of our industrialised stretches of coast, as the diagram below will show. On that summer day the afternoon maximum at Durham was 73° with a west wind, force 3; in an open car the sea breeze began to be felt, undercutting the west wind, about two miles inland from Seaton Carew just south of Hartlepool with its smoking chimneys; and on the beach the temperature was 58°. It will be evident from the diagram that the inflowing air from the cool North Sea attained a temperature corresponding with that of the prevailing westerly wind about two miles inland. About 6 p.m. the smoke of the chimneys was no longer being carried inland, but rose and spread for a time rather erratically; by 7 p.m. the steady drift of smoke seaward was noticeable. (See Fig. 51). It is evident that the sea breeze often sets in towards the coast from some miles to seaward. Watchers of the shipping from the cliffs above Scarborough and Whitby will often observe how the smoke of the steamers making their way between Forth or Tyne ports and London spreads out horizontally some hundreds of feet above the ship, hanging in persistent wreaths at that level which indicates the upper limit of the cool sea air.

Sea breezes however, sometimes combine to a greater or less degree with whatever prevailing wind exists as a consequence of the overall trend of the isobars on a given day.

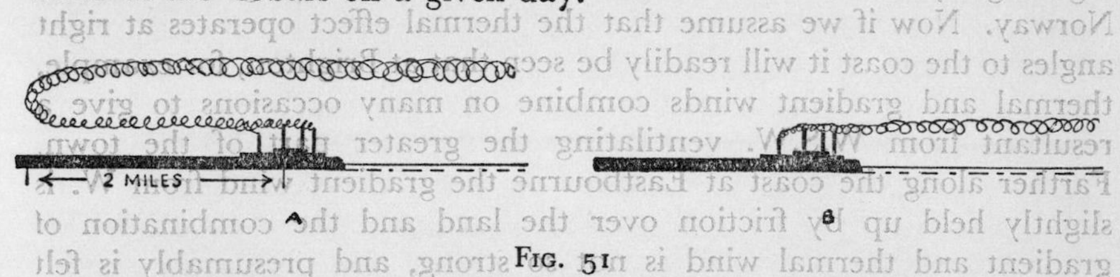

FIG. 51

A July sea breeze at Hartlepool on the Durham coast

A. Mid-afternoon; depth of sea breeze on coast about 500 ft. Westerly wind inland at Durham and overlying sea breeze, shown by movement of smoke. Sketched in 1935. B. Early evening; sea breeze ceases, westerly gradient wind prevails

Moreover, sea breezes do not invariably develop or spread far inland in the hottest weather. In order that the sea breeze may spread inland it is necessary for the warmed air to be able to rise, and this will take place more freely if the lapse-rate is large; that is, if the air falls off in temperature fairly rapidly with height. When the weather is very hot and clear the air even at 2,000 feet may still be at a temperature of 80°. In order that any surface air flowing from the sea can rise freely it will have to attain a temperature of 91° or more at sea level; the rate of fall of temperature with height will then just exceed the "dry adiabatic", 5·4°F. per 1,000 feet. Hence we find that on occasions when a deep warm current of air is spreading from the Continent, the temperature in the streets of Brighton or Eastbourne can be just as high as in London. The baking crowds beloved by the newspaper photographers roast themselves in the glare as close to the cool water as possible, to catch whatever feeble puffs of air they can. Since observations began, the highest values for upper air temperatures at South Farnborough in such air-masses lie in the neighbourhood of 70° at 5,000 feet; such values imply that in theory the afternoon temperatures at ground level might attain 97° even at the seaside, a figure well in keeping with experience (93° at Bournemouth, August 1947; 92° at Margate, August 1932; 91° at Southport, July 1948).

The varying impression of bracing or relaxing qualities associated with our seaside resorts probably owes much to the extent and character of the daily exchange of air between sea and land in quieter weather, especially in summer. Throughout many summer days the prevailing pressure gradient favours westerly winds with an Azores high slightly to the south west, an Icelandic low moving east towards Norway. Now if we assume that the thermal effect operates at right angles to the coast it will readily be seen that at Brighton, for example, thermal and gradient winds combine on many occasions to give a resultant from W.S.W. ventilating the greater part of the town. Farther along the coast at Eastbourne the gradient wind from W. is slightly held up by friction over the land and the combination of gradient and thermal wind is not so strong, and presumably is felt over a smaller area during fewer hours of the day. At Bournemouth an effect similar to that at Eastbourne appears to arise from the shelter provided by the Isle of Purbeck and is probably more marked; and the net result of the relatively diminished vigour of the movement of air from the sea may be that which leads many to consider Bournemouth as 'relaxing and sleepy', while others appreciate its 'pleasantly sheltered' qualities. There is no doubt that the local complexities of air movement and exchange along our sea coasts deserve further study; no precise assessment of the qualities which distinguish the air of our seaside is yet possible. The distinction between Hoylake and West Kirby, Margate and Ramsgate, St. Ives and Penzance can however be reasonably attributed to the varying frequency and vigour with which the sea air undercuts the air off the land at each place in quiet weather. There are many naval meteorological officers accustomed to dealing with the problems of particular harbours and coastal airfields who will readily provide useful explanations of numerous happenings at our holiday resorts.

The general regard in which the bracing qualities of our east coast are held owes much to the sea breeze from the cool North Sea. Accordingly, the contrast in afternoon temperature between Yarmouth and Norwich, Scarborough and Hull, Whitley Bay and Hexham in

---

PLATE XI*a*: Cumulus cloud over the Channel, "Armada weather" July forenoon; fresh S.W. breeze.

b: Village cricket at Wombourne, Staffordshire. Early afternoon in fine anticyclonic weather, July.

PLATE XI*a*. *R. M. Poulter*

b. *The Times*

PLATE XII*a*. *B. R. Goodfellow*

b. *Robert Atkinson*

sunny warm summer weather is particularly noticeable, the sea-breeze undercutting from the opposite direction the prevailing warm land wind in the manner illustrated above at Seaton Carew south of Hartlepool. On the west coast, gradient wind and sea breeze combine their efforts more frequently; the change in the afternoon maxima between Bolton, Wigan and Southport, or between Preston and Blackpool, is not in general so marked as between Norwich and Yarmouth. Precise comparisons are not easily made owing to minor local site factors but the following figures will serve to illustrate.

APPROXIMATE MEAN DAILY MAXIMUM FOR JULY:
WEST AND EAST COAST COUNTIES.

Blackpool	65·7°F.		Yarmouth	66·8°F.	
Southport	66·5		Lowestoft	66·9	
Hutton (Nr. Preston)	67·0	(10)	Norwich*	70·2	(18)
Bolton*	67·9	(25)	Geldeston-Bungay	70·2	(10)
(*adding 1° for correction to sea-level)			Cambridge	71·4	(50)
			(*adding 0·3° for correction to sea level)		

Distances from sea in miles in brackets.

It follows however, that if the gradient wind is from the E. or N.E. and thus reinforces the sea-breeze from the North Sea the effects may be felt much farther inland. The map below illustrates the distribution of mean daily maxima at places close to sea level, for July 1934, a very warm fine sunny month with a considerable number of days during which light easterly winds prevailed. In the warm dry August of 1947, similar effects along the coast of Northumberland were even more marked. Farther north the average afternoon temperature at Nairn, Inverness and Fortrose was appreciably lower than at Banff or Lossiemouth, evidently because of the tendency in this exceptionally warm month in the Highlands for the sea breeze to be drawn in to the

PLATE XIIa: Lliwedd from Crib y Ddysgl, Snowdon. "Autumn anticyclone." Afternoon, October 14, 1945, looking south. Light E. wind; smoke haze reaching 2500 ft., Cader Idris in distance.

b: The low stratus of an autumn day in the hilly Outer Hebrides: Harris, October, 1947. Remnants of the jetty of abandoned whaling station.

inner Moray Firth. Normally Fortrose and Nairn at the observing stations tend to be slightly warmer in the afternoon than Banff.

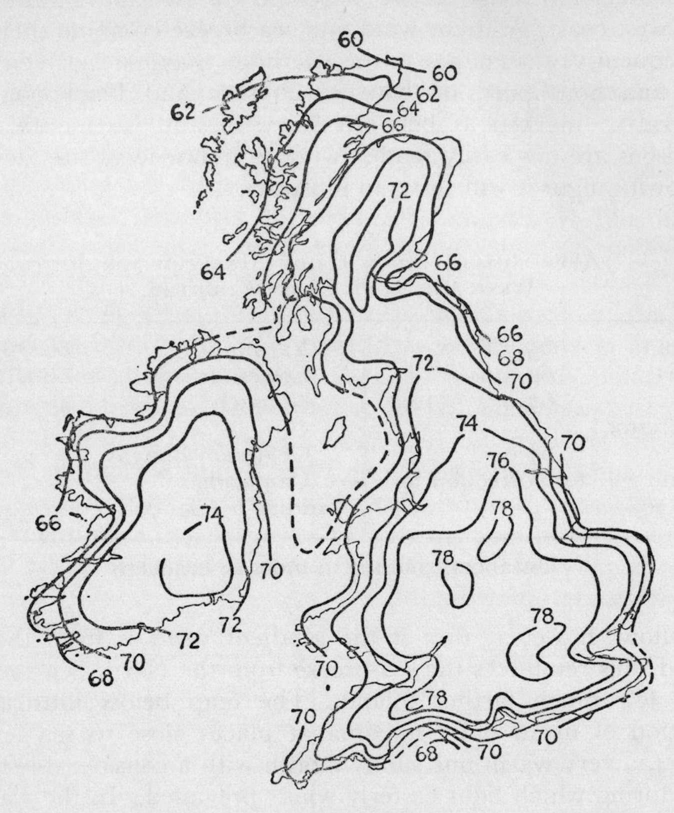

Fig. 52

Mean daily maxima over the British Isles, July 1934. Isopleths at intervals of 2°F.; a fine warm month with marked sea breeze effects, especially on east coast

If the sea breeze and gradient wind combine, it is occasionally possible on warm days for a slight but perceptible effect to reach Cambridge, 50 miles inland from the Wash. The freshening wind is accompanied by a slight fall of temperature and increase in humidity lasting an hour or so. On hot summer afternoons with an easterly breeze a certain freshness about 5 p.m. in London may possibly be attributable to the same cause. Similarly the afternoon breeze from the Severn estuary may penetrate some distance towards Gloucester; but effects such as these are infrequent.

It has already been shown how, in summer, the decreased convectional activity and the slight descent of air near our coast on summer days result in slightly greater freedom from cloud and a tendency for sunshine duration to be, on the whole, about 10% greater on the coast than inland.

Lakes in the British Isles are scarcely large enough to have noteworthy effects on wind, although those which lie longitudinally in Scotland and N. England resemble the estuaries, forming channels along which the wind moves more freely. The winter north-easter along Ullswater, and the sharp hail squalls sweeping down Loch Shin in Sutherland from the north-west are extremely perceptible if one happens to be there at the right time. The slight local air movements arising round the shores of lakes and large rivers on calm nights however are significant; in the same manner as on the seacoast, they provide just sufficient air movement close to the shore to mitigate the incidence and severity of frost on clear nights, so long as the waters remain unfrozen. But the effect of the freezing over of a lake such as Derwentwater can be suspected by comparing the figures of minima on exceptionally cold mornings at Keswick, adjacent to the lake, and at Penrith some miles distant.

	Keswick	*Penrith*	*Keswick: Derwentwater*
8 May 1938	29	23	Lake open
21 Jan. 1940	0	0	Lake frozen
26 Jan. 1945	5	11	Lake frozen
29 Oct. 1946	30	27	Lake open
30 Oct. 1949	30	27	Lake open

Some examples of minima on cold mornings at Keswick and Penrith, Cumberland.

Local breezes of another kind arise wherever there are hills; especially if the hills are bare of woody vegetation. On a clear evening the loss of heat by radiation from the surface of the earth is particularly marked around sunset. Radiation from the tops of exposed hills and plateaux leads to the cooling of the adjacent layer of the air below that at the same level away from the hill-top. Hence the cooler and denser air tends to slip downward at once, and as radiation continues and more air is cooled, a quite perceptible shallow current of air gravitates down the hillsides, gradually undercutting the warmer air below. Such winds are known as 'katabatics' by the meteorologist; the Westmorland farmer at the foot of the Crossfell escarpment knows them as 'the fell wind' which often sets in as a gentle breeze on a clear quiet evening towards sunset. In country of gentle relief such movements are slight; shallow currents near the ground can be detected by means of cigarette smoke in places such as the chalk downs. Sometimes, for example in the Chilterns, the smoke from autumn bonfires can be seen creeping down the hillsides until sometime later in the evening the valleys are filled with a cooled mass of air from which radiation continues; not uncommonly the result as many will at once recall, is a valley fog early next morning.

Katabatic flows in our country of gentle relief rarely attain any notable intensity, but they serve to reinforce the land breeze on occasion along the Scottish sea lochs, and they may play some part on the Northumberland and other coasts. They appear too to play some part on quiet nights in regard to the spread of smoke haze from cities such as Manchester, Sheffield and Leeds, as we have seen.

All thermal effects of this kind are reinforced when the ground is snow-covered. Freshly fallen snow includes a great deal of air and hence, like dry sand, it is a bad conductor of heat from the subsoil. Moreover, snow not only forms a good reflector of any radiation received during the day; a snow surface is also a very good radiator at night. Hence we find that the land breeze is more marked on a clear night when the land is snow-covered, and the Tyneside smoke is carried far out to sea as the Norwegian skippers know. Descending air-streams from the mountains are more vigorous and the consequent ponding of cold air is more marked. On a small scale indeed we reproduce in Britain all the effects which become really noteworthy along the Greenland and other polar coasts. Here however the katabatic becomes a roaring gale down the fjords, although its origin is attested by the fact that it is often not more than a thousand feet deep and fresh-fallen snow remains undisturbed on the higher ledges of the coastal mountains.

CHAPTER 9

LANDSCAPE FEATURES AND THEIR EFFECT ON WEATHER—PART II

The torpor of the year when feeble dreams
Visit the hidden buds, or dreamless sleep
Holds every future leaf and flower . .

SHELLEY

THE TABLES and maps above (pp. 159–60) have already drawn attention to the nature and amount of the temperature difference to be expected on warm summer afternoons between seaside and inland stations. We appear to be particularly perceptive of small variations of temperature and humidity around 60°–70°, for reasons to be discussed in Chapter 15 (p. 291). Small variations around the lower forties are important with regard to the progress of vegetation in spring. Small variations in the lower thirties are very significant as they often determine whether delicate plants, or crops at critical stages of their growth, are to be damaged by frost. At still lower temperatures small differences in location may determine whether peach trees for example are killed to the roots or not; the lethal temperature for a number of familiar cultivated plants lies between 25°F. and zero (olive 20°F., peach about 0°F., macrocarpa about 23°F., for example). Even winter wheat may suffer considerably if it is insufficiently protected by snow, as in Eastern France early in the winter of 1946–47.

We may review the character and amount of local variations in temperature. These small but significant variations in air temperature, from day to day and from place to place, can be largely ascribed to those factors which lead to decrease or increase in the freedom of movement of the surface air at times when inward or outward radiation is active. Quiet clear sunny weather with little air movement is often associated with the passage of a large anticyclone, and the clear skies by day and night result in a wide daily range of temperature.

Inland position with less air movement than by the sea is also conducive to a greater daily range. Valley locations, again leading to restriction of air movement give greater daily ranges than neighbouring hill tops.

Conversely, heavily clouded skies, vigorous air movement in more stormy types of weather, exposed hill-top location, are all conducive to decreased daily range. If we add to these any other features allowing for free air movement, such as lakes, we shall find that adjacent to their shores the daily range of temperature is slightly less than in similar locations lacking such opportunities.

Within extensive woodland, although the movement of air is impeded, the result of the impediment to inward and outward radiation in clear weather is a decreased day-to-night range of air temperature compared with the country outside. Small copses and belts of trees are however more common in Britain; together with hedges, they impede air movement but do little to shade the ground as a whole. Hence the irregularly wooded character of the verdant countryside we see from any hill-top leads in many instances to increased range in some fields while elsewhere some protection is afforded from extremes, depending on the degree of shade and other factors, notably minor irregularities of the local relief.

Topographic barriers of every kind from mountains to stone walls hinder the free movement of surface air. Further, in regions of irregular relief extensive ponding of cold air in all hollows becomes very noticeable. On a clear evening the mechanism is evident; from the neighbouring higher ground there is a shallow gravitational flow of air which is often at a maximum a little after sunset, as we have already seen. Gradually the warm valley air is undercut; it rises and spreads. After the cooling of the uplands has continued for some hours sufficiently to maintain the katabatic flow, downward radiation from the warmer air which has been spreading at higher levels begins to balance the loss of heat from the surface and hence we do not as a rule find that the katabatic flow, resulting from the cooling by radiation of the uplands, is maintained throughout the night; it often stops before midnight, unless the summits are thickly covered by snow in which case outward radiation predominates for a longer period.

Readers must not assume that these katabatic flows represent vigorous breezes. They are often very shallow, only a few feet at most, and very gentle; but the aggregate effect is considerable over a large area. To quote from a paper by E. L. Hawke with regard to the frost-hollow at Rickmansworth "on quiet clear autumn evenings when garden bonfires of damp leaves are burning, it is common to see during the hour after sunset, rivers of white smoke slowly winding their way down into the northern strip of the valley from points in all four quadrants of the compass . . . the estimated speed of the air flow seldom exceeds about 2 m.p.h. It does not take long for a lake of cold air 30–40 feet deep to accumulate."

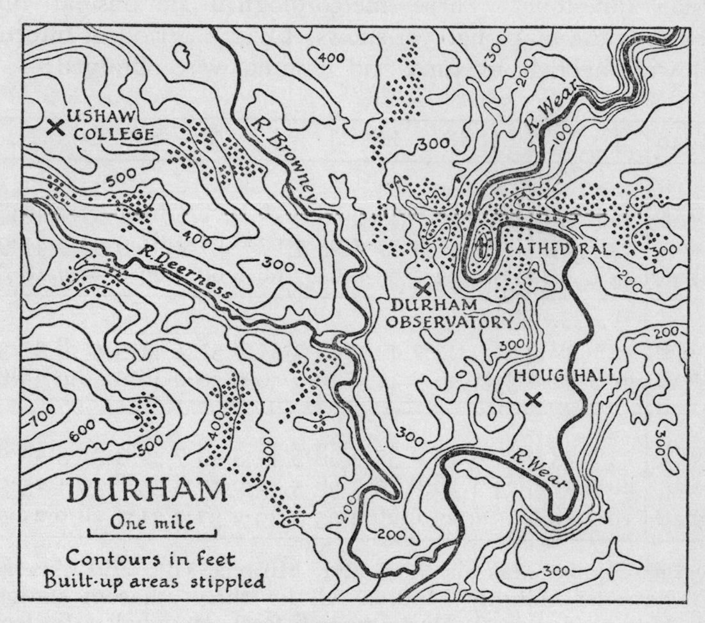

FIG. 53
Location of Ushaw, Durham and Houghall meteorological stations

As the "ponds" of cooled air accumulate, they in turn continue to lose heat to the clear sky above; within them, again, the coldest air sinks to the bottom. By dawn the result is that the valley-floor minima are sometimes many degrees below those on the neighbouring hills. These differences, as one might expect are enhanced when not only the summits but also the valleys are deeply covered by snow. The present writer estimates that the difference of temperature between hill and valley over 150–200 feet, following a calm clear night with a deep fresh snow-cover, is likely to be nearly three times as great as it would otherwise have been. Examples are given below.

The extent and character of these effects on minimum temperature arising from differences of relief within a broad valley can be very well shown in the neighbourhood of Durham. Here the river Wear has incised its course sufficiently to make a valley often half a mile or more in breadth within the larger valley, six or eight miles broad, lying between the flanks of the Pennines and the East Durham upland of magnesian limestone. Three meteorological stations at different levels exist as the map (fig. 53) shows. Over a period of fifteen years the following average maxima and minima were obtained:—

1925-1940	J	F	M	A	M	J	J	A	S	O	N	D
A												
Ushaw °F.	42·4	42·6	47·3	50·8	56·5	63·0	66·6	66·5	61·1	53·2	46·8	42·4
Durham	43·1	43·7	48·3	51·6	57·1	63·7	67·8	67·2	61·9	54·7	47·9	43·1
Houghall	44·0	44·7	49·3	52·5	57·9	64·7	68·6	68·1	63·0	55·7	48·8	43·9
B												
Ushaw °F.	32·6	32·9	35·3	37·7	42·3	47·2	51·8	51·1	47·5	41·8	37·3	33·9
Houghall	31·1	32·0	33·7	36·8	41·0	46·4	50·9	50·0	46·0	40·0	35·6	32·1
C												
Ushaw °F.	23·2	23·8	26·3	29·2	32·4	38·5	44·5	43·7	38·3	31·3	28·8	25·2
Durham	20·9	21·5	24·1	26·6	29·2	36·5	41·6	40·3	35·6	29·1	25·9	23·5
Houghall	16·9	18·6	20·3	23·8	26·4	33·3	39·2	37·3	31·7	26·6	22·7	19·7

Average Daily Maxima (A) and Minima (B), and Average Extreme Monthly Minima (C) for three adjacent stations, Ushaw (594 feet), Durham (336 feet), Houghall (160 feet). Period 1925–1940. Mean Minima at Durham are omitted as the observations refer to different terminal hours (21h. not 9h.) and are therefore not strictly comparable.

Looking over these figures it is evident that throughout the year the average daily range of temperature is lower on the hill-top than in the valley, and the effect of relief is more marked with regard to the minima. These are lower in every month than on the hill. But not all nights are clear and calm. Many are windy and cloudy, and if on a cloudy night with, say, a strong west wind blowing, the minimum temperatures are compared it will be found that they conform closely to that we should expect from the known expansion-rate of air rising and falling in its turbulent course over the surface. The adiabatic lapse rate, considered in Chapter 3, is about 5·4° per 1,000 feet; on a windy and cloudy night therefore Ushaw, 434 feet above Houghall can be expected to be 2·3° cooler, a value borne out by the observations.

But inasmuch as the valley station is generally 1·6° cooler than the hill-top, and as clear skies predominate on about one night in three, it follows that to make up for the windy and cloudy nights which are warmer in the valley, the minima on clear nights must average about 4° cooler in the valley than on the hill. This is closely confirmed by the observations, as the average extreme minima show.

The exceptional effect of outward radiation on clear calm nights with a deep snow-cover impeding the conduction of heat from the soil below is shown by the values of the screen minimum temperature on occasional winter nights, some of which are quoted below:—

	1928 Mar. 4	1929 Feb. 16	1940 Jan. 21	1941 Jan. 5	1941 Feb. 26	1942 Feb. 8	1945 Jan. 24	1947 Feb. 23	1947 Mar. 4
Ushaw, 594 ft. °F.	22	12	12	19	19	18	17	18	14
Durham, 336 ft.	18	11	3	8	14	14	9	14	5
Houghall, 160 ft.	8	−1	−4	−6	3	7	2	2	−6

Exactly similar effects with much the same magnitude are found in many other districts. At Malvern, overlooking the Severn valley from the flanks of the narrow north-south ridge of the Malvern hills, the incidence of frost is much less acute than at Perdiswell near Worcester or at Droitwich, or for that matter even Cheltenham on the slightly higher ground towards the foot of the Cotswolds.

Perdiswell, incidentally is the Worcestershire County Agricultural station; it lies two miles from Worcester, about a mile from the river Severn and forty feet above it. It is however, important to recognise in all these discussions that much depends on the size of the open hill-top from which radiation leading to downward flow of cool air proceeds. Moreton-in-the-Marsh (450 ft.) lies higher than the station at Malvern, but is in a valley between wide stretches of open though cultivated Cotswold plateau; it is accordingly far more liable to frost by virtue of its position. Bromyard (393 ft.) is also a valley site liable to frost, a fact which appears to have been well known in

1926–1940	J	F	M	A	M	J	J	A	S	O	N	D	*Mean Difference.*
A °F.													
Perdiswell	33·4	33·7	34·9	38·5	43·1	48·3	52·5	51·6	47·6	41·6	37·6	34·2	+2·6
Malvern	35·3	35·5	37·4	40·7	45·4	51·1	55·4	54·8	51·2	44·8	40·2	36·1	
B °F.													
Perdiswell	20·8	22·4	22·3	26·1	30·9	37·1	41·7	40·4	33·0	26·6	23·9	21·1	+6·1
Malvern	23·9	26·6	26·8	32·4	36·6	44·3	48·9	47·8	42·0	34·4	30·9	24·9	

Average Daily Minima (A) at Perdiswell, near Worcester (94 feet) and Malvern (377 feet), and Average Extreme Monthly Minima (B), 1926–1940.

the seventeenth century as it is mentioned in a treatise on Herefordshire orchards written in 1657.

Now there is a long period of the year, notably in autumn and spring during which in Great Britain the arrival of one of the cooler air-masses may well result in a clear night with widespread minima under 40°. Under these circumstances the local geographical factors play an extremely important part. For example if at Ushaw and Malvern the minimum falls to 38° there is a considerable chance that at Perdiswell or Houghall the air temperature will fall to 32° with consequent damage to many plants at a critical stage. Early in May the Worcestershire fruit trees are in blossom; later in the month those at Durham, together with the newly-sown potatoes, are vulnerable. Over a period of years the average length of the season free from frost (minimum air temperature below 32°) is about eight weeks longer at Malvern than at Perdiswell; and there is evidence that the particular location chosen for the Droitwich climatological station is slightly more liable to frost than Perdiswell, five miles to the south-westward. These general features of the Severn lowland are supported by other stations such as Prestwood, Stonehouse and Defford with the Cotswold upland airfield at Little Rissington for comparison.

All over the country these effects have been recognised by the intelligent farmer long ago. Kentish orchards thrive better on the slopes; around Cambridge a local company makes extensive use of the slightly rising ground near Haslingfield. In Worcestershire the belt of orchards girdling Bredon Hill may be mentioned. In Scotland the slopes of the middle Clyde valley are more favoured than the frosty upland basins of Peebles-shire. In many other districts however the favourable sites for cultivation of fruit can be recognised at once by anyone with a little experience of the manner of incidence of frost; coupled with a recognition that it is equally necessary to avoid the higher exposed and windy summits.

It has already been pointed out that among the factors discouraging the ponding of air is the proximity of lakes, estuaries or the sea. Sloping ground in such positions is often particularly free from frost unless there is considerable opportunity for drainage of air from uplands in the interior. Pendennis Head, small in area and surrounded by the sea on three sides, has fewer frosts than Falmouth Observatory a mile away although the latter lies 167 feet above the sea and on a favourable slope for air drainage. Around the Observatory Falmouth in turn is slightly less liable to frost than Fowey, and both are much freer from frost than Bude, where as any map will show the air can drain downward from quite a wide area. On the Devonshire coast, the Sidmouth station, at the mouth of a narrow wooded valley, shows slightly less tendency for extremes of frost than that at Seaton, lying at the mouth of a broader, longer and more open valley with a nearly flat floor.

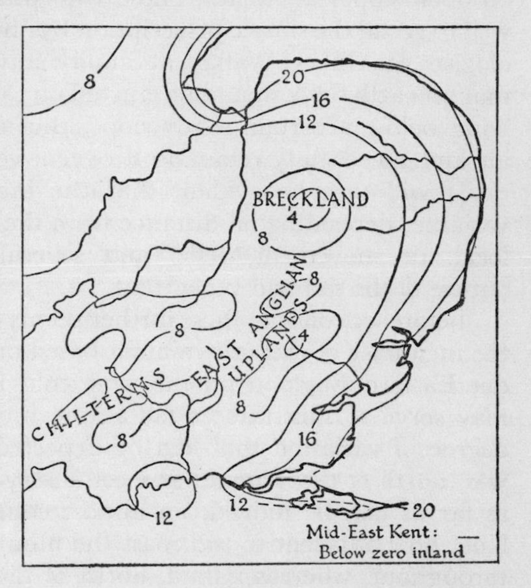

FIG. 54

Approximate distribution of minimum temperatures in degrees F. for East Anglia on 20 January 1940, with isopleths sketched at 4° intervals. Inland minima in the Weald of Kent fell considerably lower owing to the presence of a deep snow-cover, although at Biggin Hill (567 ft.) the minimum was 12°, of the same order as that above 500 ft. on the Chilterns. In the Rickmansworth frost-hollow zero was recorded

Almost all over Britain the local relief is appreciable. Officers of the Forestry Commission have noted some remarkable frost-hollows among the undulations of the central Weald of Sussex. Elsewhere severe frost occurs in the boggy hollows among the Eden valley drumlins of Westmorland. Moreover, quite small barriers—even walls—may serve either to divert or to impound the gentle downward streaming of the cool air on open slopes at night. Those who wish for early skating would do well to recall the simple prescription by one of our most eminent meteorologists, Sir Gilbert Walker, at Simla many years ago. He recommended that an earth bank should be erected on the border of some tennis courts lying below a certain shady slope, thus damming up the flow of cool air and successfully retaining it over a very shallow flooded surface.

It will thus be evident that the incidence of frost is extremely variable, depending on distance from the sea, degree of cloudiness and local air movement, relief and several other factors, notably the nature of the soil and vegetation.

Before we consider these further points the Figure (54) above showing the incidence of extreme minima based on the official records throughout Eastern England on the very cold morning of 20 January 1940 may serve to illustrate as far as the limited observations permit the degree of variation that can be expected on an occasion when there was, north of the Thames at most a very thin powdering of snow and, as far as can be judged, an almost completely calm cloudless night. Note how adjacent to the coast the minima were very much the same throughout, whereas inland, north of the Thames, they ranged down to 20° colder in exceptional hollows, notably that at Rickmansworth.

On the same night however East Kent and S.E. Sussex lay under a deep snow-cover. Although the coastal minima at Dover, Margate and Hastings were very similar to those in Norfolk, inland minima were very much lower. In hollows inland, they ranged down to –6°, for example at Bodiam in Sussex, with –4° at Canterbury. It is evident that when the Downs are snow-covered very low minima indeed can occur in such locations; in January 1947, –6° was again recorded, at Elmstone in E. Kent. In East Yorkshire, very low minima are recorded from time to time at the foot of the Wolds. This helps to explain why the minima reported from Bridlington in winter are much lower than those at Scarborough.

For the morning of 20 January 1940 the isopleths may be conjectured to have run somewhat as shown. It is however rare to find a night on which the prevailing conditions over an extensive area are sufficiently uniform for relief and other factors to play their full part, but with the aid of the data in the Monthly Weather Report many will find it interesting to attempt the construction of similar maps for such extreme nights as 3 March 1947.

This map will serve to draw attention to other factors governing the incidence of extremely low minima. The first of these is the character of the soil. The surface of the earth loses heat by radiation at night, but this is to some extent offset by the conduction of heat to the surface from below. The surface soil layer, *i.e.* the top two or three feet, varies considerably in conductivity. If the soil is coarse-grained and includes a good deal of air, it is a poor conductor. Hence sands and sandy soils by virtue of their low conductivity and rapid drainage give surface temperatures on clear nights well below those obtained on more normal clayey loams which not only contain less air but are better conductors by reason of their longer retention of water. Incidentally a deep fresh snow-cover contains a great deal of air and is a very poor conductor indeed of warmth from the ground below. Hence as we have seen exceptionally low minima can be expected when the sky clears following a heavy snowfall. (Table p. 167).

In the map above it will be seen that the lowest minimum in East Anglia occurred on the Breckland near Thetford. This is an area of particularly light, well drained sandy soil and on clear nights minima upon it are generally about six degrees lower than on gravelly loam at Cambridge, unless the sand has just been wetted by heavy rain or else there is a continuous snow-cover. The same factors operate elsewhere; at South Farnborough minima tend to be rather lower than at other inland Hampshire stations because of the situation in a broad valley with a rather light sandy soil.

1941 May	4th	8th	9th	10th	11th	16th	18th
Lynford	15	18	19	17	15	24	27
Cambridge (Univ. Farm)	28	30	29	29	27	31	38

The effect of sandy soil at Lynford on the Norfolk Breckland.
Night Minima at Lynford and Cambridge with a Scandinavian anticyclone following dry weather, May, 1941.

1933–1941	J	F	M	A	M	J	J	A	S	O	N	D	*Year*
A	2·2	2·5	2·6	2·7	3·1	3·1	3·1	3·7	2·8	2·4	1·9	1·9	2·7
B	6·0	6·1	5·8	6·8	6·1	7·5	6·6	7·6	6·1	5·9	4·7	5·4	6·2

Departures (°F.) of the Average Daily Minima (A) and Average Extreme Monthly Minima (B) at Lynford (99 feet) *below* those at Cambridge University Farm (78 feet), 1933–1941.

Relief and dry, well drained soil combine to provide in a small valley in the Chilterns near Rickmansworth one of the most exceptionally frosty locations yet known in the British Isles. The extreme minima apply, it should be noted, within a shallow layer of air on the floor of a dry valley in the chalk, surrounded in the main by bare grassy slopes rising 100 to 150 feet above. The area concerned is very small, about half a mile long and perhaps a hundred yards wide; but there is no doubt that similar exceptionally cold pools occur here and there in neighbouring valleys. Phenomenally low temperatures at all times of year have been recorded at Rickmansworth; quite commonly on clear nights minima are 10°–12° below those of nearby London suburbs. Our knowledge of the possibilities of British frost hollows owes a great deal to the enterprise of a long-standing Honorary Secretary of the Royal Meteorological Society, Mr. E. L. Hawke, in establishing and maintaining a record on the floor of this remarkable valley, in which dahlias may be lost in mid-August whereas a quarter of a mile away they survive until November.

In such a constricted valley temperatures by day tend to rise higher than elsewhere; and on a calm clear and dry day, 29 August 1936, the exceptional daily range of 50·9° was observed, with a maximum 85°, minimum 34° within ten hours. A daily range of this magnitude is almost worthy of the Sahara.

Still another factor in the incidence of frost, considered as air temperature below 32°, is the nature of the vegetation cover. Studies at a number of stations have shown that the lowest minimum temperatures tend to occur over long coarse grass. Bare ploughed earth, many will be surprised to hear, does not in general give such low minima. By and large it would appear that centuries of drainage and cultivation have on the whole tended to improve the climate. The occurrence of low minima over long coarse grass is probably because the grass-blades themselves radiate heat, and their cool surfaces chill the air between the grass blades while impeding its movement; also the grass stems are not good conductors of warmth from the roots below. In many districts the ill-drained boggy hollows are often the first to experience the effects of frost in autumn, and are very slow to warm up in spring.

From what has been said it will be evident that local variation in minimum temperature on severely cold mornings can be very large; and that really low temperatures often appertain merely to small patches of ground. Hence much discrimination must be used in estimating the liability to frost at any particular place in comparison with nearby meteorological stations.

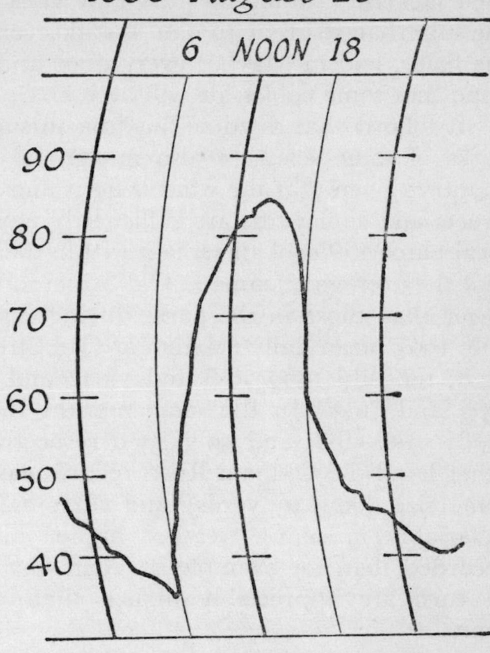

FIG. 55
Thermograph trace showing one of the greatest daily ranges on record in England; minimum 34·0°F., maximum 84·9°F., range 50·9°F. (after E. L. Hawke, *Q. J. Roy. Met. S.*, 1944)

Great cities, as the map (p. 169) reminds us, also tend to have higher minima than the open country. Two factors are at work; first, the surface winds are lighter owing to the frictional drag imposed by the buildings. Secondly, radiation is absorbed especially in the warmer months by the roofs and sides of buildings during the day, and is given out at night; in courtyards, streets and enclosed spaces the result is that temperature falls much more slowly during the evening, and in any case does not attain the minima found outside the built-up area. Nothing is more familiar to the young London motorist in a hot summer than the evening return across the cabbage-fields of Middlesex when it becomes a question whether he and his passengers should don their jackets. Coming up the Great West Road however the air along the suburban roads at 10 p.m. is still several degrees warmer than over the fields, and in Mayfair every door and window stands open in the hope that some cooler air will arrive.

It follows that daytime maxima in summer within a city, even in parks of some size, are also in general slightly higher than in the country; given that the wind is light and the sun powerful. But if the streets and courtyards are sufficiently narrow to admit little sunshine local patches of cool air remain within them. Long ago it was observed that the average maxima at Old Street in the City tended to be slightly lower than those in the parks during summer. Compare for the hot July 1901 mean daily maxima at Old Street, 74·3 with Regent's Park, 75·8; for July 1900 76·8 and 78·1; and for the hot August of 1899 75·3 and 76·3. In the same months the mean minima were 58·2, 55·8; 59·9, 56·8; and 59·3, 57·0 respectively. At Manchester on the other hand the Oldham Road record was kept in an enclosed yard of some size (60 × 40 yards) and surrounding buildings are not high; invariably in sunny weather higher maximum temperatures were recorded than for example in Whitworth Park (16 acres) and these in turn are appreciably higher than at stations outside the city limits.

In winter radiation received by buildings is less, but a certain amount of warmth escapes from the heated interior. Outward radiation is also to some extent hindered by smoke, and slight air movements between streets and parks all tend to check the ponding and cooling of surface air by comparison with the country. Hence minima are again considerably higher in cities, especially among the buildings, and appreciable variations develop between one part of the city and another. For example in London if there is an overall slight drift from the north, the minima in the northern suburbs approximate more closely to those in the country than those in the southern suburbs.

In our excessively urbanised country the great cities undoubtedly form a very prominent element of the British scene and their climatic

---

PLATE 25
ROWSLEY, Derbyshire: heather moor, September. Fine-weather cumulus in early autumn, cloud base about 3,000 feet; cloud tending to decrease in amount eastward.

PLATE 25

Cyril Newberry

PLATE 26

B. A. Crouch

effects are considerable. The amelioration as far as frost is concerned in London is now quite appreciable. A minor effect has also become noteworthy in the last 30–40 years; fogs in the centre of our great cities are not quite so great an impediment to traffic, at street level, as they formerly were. In bad London fog visibility outside Bedford College in Regent's Park and along Kensington Gore may decrease to 10 yards; but along Oxford Street one can see 100 yards or more. There is little doubt that this effect arises from the very slight warming of the air in the streets, just sufficiently to absorb a part of the suspended moisture. The reputation for fogginess of the great suburban by-passes owes much to the fact that they often run through the less-favoured damp clays and secondly to their freedom from buildings.

The fact that the nocturnal fall of temperature is checked by the radiation from walls and buildings has long ago been recognised. The Elizabethans may have been among the first to observe the advantage when they tried their new flowers and vegetables under the warm brick walls of their houses, which in more ways than one gave scope for individual initiative. Walled gardens became the fashion; grapes, peaches and apricots were ripened on south-facing walls. Even in Teesdale there is a Jacobean walled garden where peaches are still grown with success. Eleven hundred feet above the sea at Garrigill in the Cumberland Pennines near Alston two plum trees under a south wall ripen frequently, and provide a rare example of the possibilities attached to the provision of some shelter in that otherwise cold district. Walled gardens must however be properly designed in order to prevent the accumulation of cold air within them at night (here Sir D. Brunt's papers should be consulted by those interested).

In all this we observe one of the most impressive features of the British climate. The range of temperature is such that small changes of location, aspect and shelter play a very appreciable part in the success or failure of fruit cultivation and gardening operations of all kinds. The gain resulting from a little extra care in choice of site, or in provision of protection from wind or frost, is considerable; in these and other respects intelligent forethought has for fifteen hundred years

---

PLATE 26

From RED PIKE, Buttermere looking N.E. towards Skiddaw, Cumberland: sun and shadow, early summer afternoon, June. Strato-cumulus rolls at 3,500 feet very characteristic of westerly weather in June; clear above.

or more received its reward. Even the several fields of a normal farm have each their varying characteristics arising not merely from soil but also from the significant local variations in the incidence of frost. These in turn make for forethought in regard to type of seed, date of sowing, and even more notable, the characteristics of local breeds of stock. It is difficult to resist the conclusion that while co-operation in times of calamity is desirable, individual enterprise and the sharpening of intelligence could not help but be developed in an environment offering such local differences leading to abundant opportunities for freedom of choice on every hand.

Anyone who travels about Great Britain and contemplates the varied degree of development of almost any country activity must bear in mind the climatic environment and the extent to which under differing economic circumstances any given type of cultivation or craftsmanship has been in demand. Given care, a very great variety of possibilities exists. Grain has been grown in war-time at 1,400 feet. Many old orchards in Cumberland remind us that before the days of Covent Garden's penetration to the northern counties quite a fair amount of fruit was grown, and might be grown again.

REFERENCES

Chapters 8 and 9

BRUNT, D. (1945). Some Factors in Microclimatology. *Q. J. Roy. Met. S. 71:* 1–10.

(1947). Where to Live. *J. Roy. Inst. 10:* 1–13.

DARLING, F. Fraser (1947). *Natural History in the Highlands and Islands.* London, Collins' *New Naturalist.*

DEFOE, D. (ed. Cole, 1927). *Tour through Great Britain: vol. I:* 71–2. London, Davies.

DEPARTMENT OF SCIENTIFIC AND INDUSTRIAL RESEARCH (1948). *Atmospheric Pollution in Leicester.* London, H.M.S.O.

GOLD, E. (1936). Wind in Britain. *Q. J. Roy. Met. S. 62:* 167–205.

HAWKE, E. L. (1933). Extreme diurnal ranges of temperature in the British Isles. *Q. J. Roy. Met. S. 59:* 261–65.

(1944). Thermal characteristics of a Hertfordshire Frost Hollow. *Q. J. Roy. Met. S. 70:* 23–48.

LAMB, H. H. (1938). Industrial Smoke Drift and Weather. *Q. J. Roy. Met. S. 64:* 639–43.

LEWIS, W. V. (1938). Evolution of Shoreline Curves. *Proc. Geol. Ass. 49:* 107–27.

MANLEY, G. (1944). Topographical Features and the Climate of Britain. *Geogr. J. 103:* 241–63.

(1945). The Helm Wind of Crossfell. *Q. J. Roy. Met. S. 71:* 197–219.

MINNAERT, M. (1940). *Light and Colour in the Open Air.* London, G. Bell & Sons.

PATON, James (1948). The Optical Properties of the Atmosphere. *Weather, 3:* 243–49.

RAISTRICK, A. (1943). The Pennine Walls. *Dalesman, V:* 5–13, 21–28, 45–53.

SALISBURY, E. J. (1939). Ecological Aspects of Meteorology. *Q. J. Roy. Met. S. 65:* 337–57.

SCORER, R. S. (1949). Theory of waves in the lee of mountains. *Q. J. Roy. Met. S. 75:* 41–56.

SPENCE, M. T. (1936). Temperature Changes over Short Distances in the Edinburgh District. *Q. J. Roy. Met. S. 62:* 25–31.

WHITE, Gilbert (1931). *Journals of Gilbert White.* London, Routledge, ed. Walter Johnson, 24, note for 23 February 1770.

CHAPTER 10

MOUNTAINS AND MOORLANDS: THE EFFECT OF ALTITUDE

Your snows and streams
Ungovernable and your terrifying winds
that howl so dismally for him who treads
Companionless your awful solitude

WORDSWORTH

EVERYWHERE in our highlands we are reminded that above a relatively low altitude the potential yield of the land steadily decreases. The conditions of existence of a farming population become increasingly precarious, until at length a level is reached at which farming of any kind is no longer practicable.

Over the world as a whole the limit of permanent settlement for an inland population is, in general, the cold treeline. Broadly it is found that trees in the normal sense of the term do not grow unless the mean temperature for at least two months exceeds 50°F. Even then, shelter and in higher latitudes reasonable freedom of drainage are desirable. In Iceland where the mean temperature at Reykjavik is 52° in July and 50° in August an occasional tree can be found on the coastal lowlands under the lee of a farmhouse; in sheltered valleys farther inland the average height of birches in woods may exceed twenty feet. Similar results are found on the better-drained uplands in Northern England and Scotland. Towards the coast however, winds are stronger and it becomes very difficult to establish trees at all on exposed uplands even at 1,000 feet where the mean July temperature ranges (in Scotland) from 55° to 52°. It is also necessary to recall the deleterious effects of salt spray near the coasts, which after severe gales have sometimes been observed fifty miles inland. In June, 1938, this was noticeable in Hampshire. In Wales we may compare the bare uplands of Cardiganshire with the wooded upper Towy valley. Inland, given some degree of shelter, occasional trees are found all over Britain above 1,500 feet and in the extreme instances may grow up to, or very slightly above, 2,000 feet. In the Eastern Highlands of Scotland many who know Rothiemurchus forest will be familiar with the last struggling pines and occasional birches found at this level on the way towards Cairngorm or the Lairig Ghru from Aviemore. In Cumberland a small mountain ash can be espied at this level in a gully above Ullswater; its survival owes much to the protection its position gives from the nibbling sheep. For similar reasons one or two trees nearly reach the 2,000 foot contour in Grainy Ghyll, above Seathwaite. Farther to the eastward in the high northern Pennines there is a very good example indeed of the struggles of the trees to survive. Above Ashgill, south of Alston, a plantation was established over a century ago and at 1,600 feet Scots fir can be found forty feet high. Extending up the hill-side however the trees rapidly dwindle in size. At 2,000 feet those which still survive are rarely six feet high, despite some slight shelter from an adjacent low stone wall. Moreover quite a number of the adjacent trees are dead, and others are dying. A compact summary for the meteorologist of the effect of wind on vegetation has been given by Sir E. J. Salisbury.

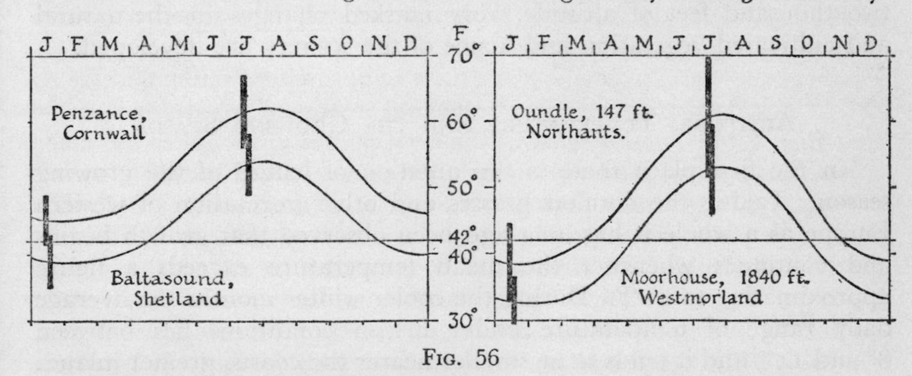

FIG. 56

The trend of monthly mean temperature at selected stations (1906–1935). The average daily maxima and minima for January and July are shown; the average length of the 'growing season' above 42° will be noted

In Northern Britain the rate of fall of mean temperature with height is about 1° for 270 feet, based on existing mountain records. Accordingly we can see that the level at which mean temperature exceeds 50° for two months is about 2,200 feet under present conditions, and hence that although a great part of our uplands are bare, trees if planted and protected might be found up to this height above the sea.

Further, in relatively recent centuries scattered trees were probably found in many of our upland valleys much more extensively than now. The decrease owes much to the combined efforts of men and sheep. One of the first essentials in re-afforestation is a sound wire fence, whose rectangular assertiveness is now too familiar in Ennerdale, at Dalwhinnie, or flanking the Plynlimon road as Midland motorists bound for Aberystwyth can see.

Under British climatic conditions we have in any case a remarkably low treeline compared with many other temperate countries. On account of the warmer summer inland we find that in Central Norway birch scrub occurs here and there above 3,500 feet; and that the pine ascends to over 2,500 feet. No recollection is more vivid in the writer's mind than that of a well-known Swiss professor of geology who, confronted at 1,680 feet—the level of Berne—with a wide stretch of the Pennines between Teesdale and Weardale, surprisingly declared "this is the tundra". Yet our climate permits, within a very few miles, the survival of delicate evergreens. We may proceed to illustrate why, within the relatively small range of mean temperatures arising from a mere two thousand feet of altitude, very marked changes in the natural and cultivated vegetation and hence of the scenery are developed.

Altitude, Temperature and the Growing Season

In the first place there is the question of length of the growing season. Taking the familiar grasses and other vegetation of western Europe as a whole it has long ago been observed that growth begins and continues whenever the mean temperature exceeds a figure approximating to 42°F. During the cooler winter months the average daily range of temperature under British conditions lies between 8° and 14°; and it tends to be smaller nearer the coasts, greater inland. Therefore, if the mean temperature is 42° it is broadly true that on most nights the minimum will remain above the freezing point.

At a representative Midland station such as Oxford the mean temperature for March is rather above 42°, which implies that, by and large, the grass begins to grow a little before the middle of the month. Much depends on whether the month is mild or severe. In the English Midlands some no doubt who take note of such things will remember mowing the lawn before the end of a mild March such as that of 1938 or 1945. Others will contrast the very slow oncoming of that inevitable suburban spring-song in the severely cold April of 1917. In some years dry clear afternoon warmth in March is accompanied by a much greater daily range of temperature than usual and hence by frosty nights, as in 1929; yet the mean remains low and grass makes little progress.

As a whole however, we can take out an approximate average date for the beginning and end of the 'growing season' from the curve showing the march of average temperatures during the year. At normal Midland stations the season of growth can be expected to end sometime in November (fig. 56, p. 179).

It will at once be evident however that somewhere to the south-westward, in lowland Devon and Cornwall for example, the mean temperature throughout the year does not fall below 42°. Hence even in mid-winter growth although very slow does not entirely cease, save for the few days of dry, cold weather that normally occur two or three times between December and February. Cornish coastal gardeners mow their lawns the whole year round; pastures in general remain green and the cattle need but little additional feeding, even by comparison with the West Midlands. Narrower belts, almost equally favourable, can be found in Pembrokeshire, and even in south-west Galloway (cf. *Scot. Geogr. Mag.* 1946); and in Southern Ireland a relatively large area offers similar opportunities.

But we may at once observe that the curve of the annual march of temperature is rather flat, that is, the rise in the mean temperature in spring is slow compared with more continental countries. With increased altitude the rate of decrease in the length of our growing seasons is accordingly very rapid. At Nenthead in Cumberland, one of the highest villages (1,500 feet) in the whole of Britain, enclosed pastures run up the adjacent hillside to 1,800 feet. Compared with the lowlands the mean temperature is lowered sufficiently to decrease the "growing" season (18 April–23 October at 1,500 feet) by about ten weeks, and at the limit of settlement (about 1,900 feet) the decrease is to 4 May–16 October. Hence, whereas on the south-west coast of Cumberland sheltered corners can be found in which the grass almost remains green throughout a normal year, forty miles away it is necessary to feed cattle, if any are kept, for six months at the very least; hay is also given to the sheep.

Lambing time is closely associated with the date when the fresh growth can be expected. At Moor House (1,840 feet) the lambs are expected about the first week in May, whereas in the lower Yorkshire dales early March is customary. Expressed otherwise; in some districts, a change of level of 1,800 feet approximately halves the growing season. Climatic data from Malham Tarn Field Centre are noteworthy (p. 196).

The rapid change in the length of the growing season means that with a small change of altitude great contrasts are found in but a short dista [illegible] gether on 28 June besic [illegible] upper Weardale, five [illegible] Cumberland. It is part [illegible] t marked for, as we shall [illegible] th height is more mar [illegible]

A [illegible] t from the curves. Wit [illegible] aracteristic of the tem [illegible] of growth depends on t [illegible] he warmest month exce [illegible] begin. With a July mea [illegible] of crops is perhaps twic [illegible] 56°. These figures are [illegible] ount and intensity of s [illegible] and the retentive-ness [illegible] ever, that a rough rela [illegible] nd the consequent yiel [illegible] e annual march of mean temperature and the line representing 42°. This explains at once why to our eyes the progress of vegetation is noticeably rapid in more continental climates such as that of Central Europe, despite the greater severity of winter frost. Further, in Britain a cold late spring is for many farmers a greater calamity than a severe winter.

Now a little thought will show that where the curve of the annual march of temperature is rather flat, for example in Scotland, a small change of altitude makes a much greater difference to the area above the 42° line (Fig. 55) than would result from a similar change in a climate such as that of the Eastern United States or Vienna. Hence we find that under British climatic conditions not only is there a sharp decrease with height in the length of the growing season; the decrease in the rate of growth and the yield of crops is again rapid. In the Northern Pennines for example the average yield per acre for oats at 1,000 feet may be put at scarcely half of that near the coast.

Further; the effect of seasonal variations is much more marked on higher ground. If in a cool summer the overall average temperature of the three summer months is 58° instead of 60°, the yield of grain, for example, can be expected to decrease. But a thousand feet higher up the mean (other factors disregarded) will be 54° instead of 56°, and without doubt a decrease of 2° from 56° will tend to be associated with a greater proportionate fall in yield than a similar decrease from 60°. Moreover, in a cool and cloudy summer with excess of rain, the amount of that excess is likely to be greater on high ground; hence quite apart from seasonal variations of temperature, the year to year variations of rainfall also render agricultural operations very much more precarious on higher ground. Thirdly, in a cool cloudy summer it is more often than not found that the aggregate deficit in mean temperature is greater on high ground than near sea level. Data for several summer months which were persistently cool and cloudy throughout are given below:—

COOL AND CLOUDY SUMMER MONTHS

Departures of the Mean Monthly Temperature below normal at high and low level stations

		July 1920	August 1912	September 1918
DEVON	Princetown (1359 ft.) °F.	53·4 : -3·1	51·2 : -5·3	50·8 : -2·9
	Teignmouth (19 ft.) °F.	60·2 : -1·9	57·4 : -4·7	57·0 : -1·7
		July 1920	July 1922	August 1922
SCOTTISH HIGHLANDS	Braemar (1114 ft.)	50·6 : -4·5	50·6 : -4·5	50·7 : -3·0
	Perth (76 ft.)	56·1 : -3·0	55·6 : -3·5	55·6 : -2·1
		July 1936	August 1943	August 1946
PENNINES	Moorhouse (1840 ft.)	51·5 : -1·3	51·7 : -0·6	49·6 : -2·7
	Penrith (559 ft.)	57·5 : -0·6	57·3 : +0·4	55·3 : -1·6

Given longer records we could with a little trouble almost express in figures the risk attached to arable farming on high ground. Shrewd men however, have long ago expressed the risk in the rent. To a certain degree the upland farmer's disappointments resulting from cool and cloudy summers are offset by the gain if the summer happens to be dry and fine. In warm dry anticyclonic weather with little wind the mean temperature on the high ground approaches much more closely than one would expect to that at low levels. This is partly because of the nocturnal tendency for the coldest air to seek the lowest level. Two or three weeks of warm dry weather at the right time are proportionately even more beneficial in the uplands than below. The effect is well shown by the following figures from upland and lowland stations, covering months in which warm anticyclonic weather prevailed for the greater part of the time.

PENNINES	June, 1940	DARTMOOR	August, 1947
Moorhouse (1840 ft.) °F.	56·1 : + 6·7	Princetown (1359 ft.)	63·7 : + 6·8
Mean max. and min.	67·0, 45·2		71·5, 55·8
Penrith (559 ft.) °F.	59·4 : + 4·7	Cullompton (202 ft.)	66·7 : + 5·6
Mean max. and min.	71·8, 46·9		78·6, 54·8

At other times of the year the incidence of calm quiet anticyclonic weather also results in relatively greater benefit to the uplands. This is shown by data for March 1933 and 1953 and December 1935—when anticyclonic weather prevailed for two or three weeks.

	Mar. 1933	Dec. 1935	Mar. 1953
Moorhouse (1,840 ft.)			
Monthly means and departure	39·2	30·9	38·5
from normal	+5·1	−3·0	+4·4
Mean max. and min. °F.	45·7, 32·7	33·9, 27·9	47·9, 29·2
Penrith (559 ft.)			
Monthly means and departure	43·4	33·4	41·9
from normal	+3·7	−4·9	+1·2
Mean max. and min. °F.	51·5, 35·3	38·2, 28·6	51·4, 32·3

The converse effect is seen when the weather is prevailingly windy and air derived from polar sources with a long sea travel predominates. The lapse-rate is thus rather high in the surface layers. Under such conditions, as we saw in an earlier chapter, everywhere on our uplands and mountains cloudiness is greater and showers are more frequent while at the same time surface winds are stronger. Hence the difference in temperature with height is more marked than usual, especially in spring when the sea surface is cool and polar air reaches us with less modification than would otherwise be found.

	Mean	*M. Max*	*M. Min.*	*Extremes*
Moorhouse (1840 ft.)				
Normal May, °F.	44·7	51·8	37·6	
Dull, wettish cool May, 1932: (2·4° below normal)	42·3	46·8	37·8	56, 19
Penrith (559 ft.)				
Normal May, °F	50·1	58·8	41·4	
Dull and rather cool May, 1932: (1·7° below normal)	48·4	55·3	41·5	67, 26

Note that the average daily maximum on the upland fell below the normal to a greater extent than in the lowland.

That such differences are not confined to the Pennines is quite evident from Dartmoor, or even the Chilterns:

Devonshire	March 1945 *Generally warm and dry*	May 1946 *Generally cool, wet with N.E. winds.*	January 1940 *Exceedingly cold; often quiet and sunny.*
Princetown (Dartmoor) 1359 ft. °F.	42·7, 3·5 above normal	2·4 below normal	6·3 below normal
3 Lowland stations	2·3 above normal	1·1 below normal	8·6 below normal
Chilterns	March 1945	May 1946	February 1947 *V. cold, dull, E. wind.*
Whipsnade 720 ft. (no normal available) °F.	44·7	48·5	25·9
Cambridge, 41 ft.	46·5, 4·2 above normal	51·8, 1·7 below normal	28·5, 11·2 below normal
Difference, Whipsnade—Cambridge	1·8 *Warm dry month*	3·3 *Cool wettish month*	2·6 *Cold and very cloudy month, occasionally clear and calm.*

From all that has been said it will be evident that the effect of increased windiness and freedom of air movement in our uplands is more pronounced with regard to the lowering of day-time maxima than as regards night minima. All walkers know this.

We may now generalise. First, the altitudinal rate of change in the potentialities of our island climatic environment is especially rapid; as a result we have an unexpectedly low climatic tree-line bearing in mind the fact that delicate evergreens can flourish in the open at many coastal resorts. The average yield of crops falls off rapidly with height, and the variability of the yield increases. Hence for all arable crops a level is soon reached at which the risk of wasted labour in cultivation is too great unless it is offset by high prices, for example in war-time.

The rate of change of climatic environment is so rapid as to call for a marked degree of adaptation in local breeds of cattle, horses, sheep and other domestic animals. These all form an essential part of the make-up of our British scene. The Eastern Aberdeenshire uplands are not complete without the presence of an extremely resistant and hardy terrier whose hairy coat defies both wet and cold. The Lakeland Herdwick sheep are adapted like no other breed to the peculiarly variable winter with its intermittent spells of excessive rain. Among sheep, the Lonk of the Lancashire fellside will stand rain, but is not so adaptable to heavy snowfalls as the Swaledale—a particularly interesting feature bearing in mind those north-eastern districts in which our heaviest snowfalls generally occur. Adaptations of farming technique are called for on every hand; and those farmers succeed best who in the long run combine sharpness of wit and readiness to adapt their actions to the weather with the needful caution before committing themselves and their dependents wholly to any given policy.

Altitude and the Incidence of Frost

In the previous chapter we discussed the effects of relief, of soil and of proximity to water-bodies with regard to the local incidence and severity of frost. Hill slopes and lower summits are freer from frost than hollows. As we ascend however we generally approach the region in which the whole air-mass is frequently below the freezing-point. To offset the diminished frequency of frost arising from upland position on those quiet clear nights when the cold air slides downhill and local inversions develop, we have an increased frequency of days when even with the wind blowing vigorously the temperature remains below the freezing-point. Consider for instance an overcast day of east wind such as 29 January 1939; this was a day of helm wind along the N. Pennines (p. 149).

		°F. *Maximum*	*Minimum*
Dun Fell	(2735 ft.)	27	23·5
Moorhouse	(1840 ft.)	31	25
Durham	(336 ft.)	37	31
Manchester	(125 ft., urban)	40	34

On balance the greatest freedom from the effects of frost at inland stations is found on slopes at intermediate levels between perhaps 200 and 500 feet above the sea. (Cf. the data for Ushaw and Malvern on pp. 166, 168). Higher up, there comes a point, roughly at the 2,000 feet level in N. England where the mean temperature of the coldest months can be expected to fall below 32° more often than not. Here is the region of scurrying low cloud and pitiless raw wind, as those who frequent our high moors in winter know. The impacted supercooled droplets from the low stratus cloud build up frost-feathers on the windward side of fence-posts and wire, walls and cairns; snow, useless for ski-ing covers the ground more often than not. The uniquely exhilarating horribleness of such weather is the more appreciated when one knows that two hours distant one finds the ham and eggs. But it becomes grimly testing when as in 1947 the persistent wind, snow and lack of food led to complete exhaustion of so many sheep, who could not rally when the thaw came.

At intermediate levels however upland valleys and basins can also be found offering with greater frequency than elsewhere the possibility of air drainage from snow-covered uplands. Probably these hollows are the site of the most severe extremes of temperature we experience. In the Highlands there is still much room for determining which upland valleys have the most severe frosts. Farther south, the former reputation of Buxton for severe frost owed much to the fact that the station was maintained in the bottom of a hollow surrounded by the bleak Derbyshire plateau, whereas the present position for the instruments is on top of a rise.

Above the Tree Line: Sub-Arctic Britain

Our knowledge of the high mountain climate of Britain owes most to the Ben Nevis Observatory built and maintained largely by the subscriptions of the Scottish Meteorological Society. Continuous observations were made from 1883 to 1904. Intermittent efforts have been made elsewhere; among the earliest were those of J. F. Miller of Whitehaven, previously mentioned as the founder of the Seathwaite rainfall record in 1845. He left minimum thermometers on Scafell Pike and other summits; but they did not give trustworthy results. From 1937-1941 I kept a record on the summit of Great Dun Fell, two miles south-east of Crossfell in the northern Pennines. There is now (1961) Lowther Hill, 2,377 feet, in Lanarkshire.

The chief characteristics of our higher mountain climate are increased wind force and increased duration of cloud cover; and all-pervading dampness is only too evident. On Dun Fell (2,780 ft.) the average wind force over the summit throughout the year appears to be about force 5 (20 m.p.h.). On Ben Nevis (4,406 ft.) the mean wind speed is upwards of force 6 (about 30 m.p.h.) and gales were extraordinarily frequent and often attained tremendous force. Above about 3,000 feet in Scotland snow is liable to fall in every month of the year, although even at 4,000 feet it does not often lie for more than a few hours between mid-June and late September.

The higher and more exposed the summits are, the less is the average difference of temperature between day and night. On Dun Fell the mean daily range is in general less than half that of the lowlands; or Ben Nevis, less than a quarter. Excessively low winter minima do not occur, as on such occasions the lowest temperatures are found in the valleys; Ben Nevis never gave a minimum lower than 1°.

It will, however, be noted that the warmest month on Ben Nevis only averaged 41°, and it is not surprising therefore to find only a little extremely hardy Arctic-alpine vegetation among the stones. Indeed with such excessive rain averaging 157 inches yearly, there is virtually no soil. On the Cairngorms at 4,000 feet (Plate 27, p. 190) the drier climate (less than half the rainfall of Ben Nevis) and greater proportion of sunshine allow more plant life. No one who wanders over our higher summits can fail to be impressed by the gallant struggle of the vegetation; it should further be noted that shelter and aspect become very important at high levels. On favourable slopes facing the sun the shallow soil may warm up tremendously during the occasional anticyclonic spell, such as that which in June 1939 gave several successive days with maxima over 70° on Dun Fell (2,735 ft.) where the normal daily maximum for June can be estimated as about 51°.

Snow and snow-cover are considered in more detail in the next chapter. The small summit plateau of Ben Nevis is generally bare of snow, apart from remnants of drifts, from early June to the beginning of October; it can be regarded as "covered" on an annual average of about 215 days. The climatological snowline probably lies at about 5,300 feet; that is, Ben Nevis would have to be about 5,500 feet high in order to harbour a small glacier to-day. Even in October and November snow-cover is rather intermittent. Warm air approaching from the south-west ahead of the normal Atlantic depressions is quite likely to give a temperature over 40° on the summit and if there is heavy rain the snow-cover is rapidly melted. After the mild and very rainy December of 1934 the summit carried very little snow even in early January.

The average annual sunshine duration on the summit of Ben Nevis is one of the lowest yet observed anywhere; about 750 hours. Brief anticyclonic spells occur from time to time and a few hours of clear sky may follow each passing depression. Nevertheless the impression gained by anyone who has to sojourn for one purpose or another on our high summits is likely to be one of pitiless and nearly incessant raw cold wind and a great deal of low cloud, with abundant rime deposit throughout the winter. Even on a lower grass-grown summit such as Dun Fell the sheep do not wander up to graze before the end of May, and the run of the mean temperatures confirms that a growing season characterised by great day-to-day variability lasts only from late May to mid-September.

The variability is the greater for plants at ground level, as they must accommodate themselves to cold wind, rain and lack of sunshine, with intervals of intense baking of the coarse, quick-draining shallow soils if sunny days occur; evaporation, too, in the exposed windy situation may be considerable for brief periods. And, while the winter temperatures are not excessively low, they are quite cold enough to inhibit any activity. Further, the snow-cover on the summits is often shallow owing to the severe drifting by wind, hence vegetation receives little protection. Elsewhere, the accumulation of drifts is such that in more sheltered locations the vegetation cannot quickly take advantage of any warmth, as the ground may be chilled by the melt-water for some time even after the snow has disappeared. In the wetter parts of the Lake District the snowline to-day would probably be found at about 5,900 feet, and at 6,300 feet on Snowdon. Tables giving a summary of high-level data will be found in the Appendix. Note that the mean temperatures are estimates based on the old observations, but brought for purposes of comparison to the period 1911–49.

Altitude and Rainfall

The increase of rainfall with altitude is a very well-known feature of our climate, arising as we all know from the additional uplift and consequent cooling of moist air-streams when they impinge against our high ground. But there are great local variations among our hills, some of which have already been mentioned. In the Lake District the convergence of the major valleys towards the central region, comprising the heads of Borrowdale, Langdale and the adjacent uplands, is a principal factor in the very high rainfall. For exactly similar reasons air-masses must often surmount the Crossfell range and the Merrick hills in Galloway, of similar height; but in both these areas the fall on the uplands is about half as much as that which is measured in the central part of the Lake District.

Farther north, a small patch of south-west Inverness-shire at the head of Glengarry has probably over 200 inches a year. Yet hills near the sea of similar height, but slightly different trend, in Sutherlandshire appear to receive little more than half this amount, although gauges in these remote highlands are few.

At altitudes up to 2,500 feet, wherever the slopes are less steep and the underlying rocks are impervious to water or nearly so, the result

---

Plate 27
Cairntoul and other summits looking S.W. from Cairngorm (4,084 feet), Aberdeenshire, Banffshire and Inverness-shire: early June. Strato-cumulus and cumulus, fresh west wind. Snow beds in corries, with remnants on plateau. Arctic vegetation. Excellent visibility characteristic of Highlands in this type of air. Temperature on summit 38°.

PLATE 27

Cyril Newberry

B. A. Crouch

Cyril Newberry

of excessive cloud and rainfall in the past is seen in the thick mantle of peat. On the flat summits of the high Lancashire moorlands underlain by resistant sandstones with nearly horizontal bedding, peat is found up to twelve feet or more in thickness in places where the present-day annual rainfall is of the order of eighty inches. This peat is the result of the decay of past vegetation under conditions of excessive rainfall and very poor drainage. Few plants can grow on such a cover, cotton grass and crowberry among them; heather prefers the patches of ground which are slightly better drained. 'Cotton-grass moor' as it is often called reminds us of the brief period in late June when even the most uncompromising Pennine moorland is dotted with the white blooms nodding in the wind. (Pl. 18b, p. 111).

The varying aspect of our high moorland owes much to climatic factors. All the roads leading to Whitby over the lower and drier moorland of North-East Yorkshire, on which in general the rainfall is of the order of 38″ are known to thousands for the magnificence of the display of heather at the end of summer. Elsewhere at intermediate levels with rather more rainfall there are wide stretches of coarse upland pasture, difficult to reclaim and improve, on peaty soils often underlain by boulder-clay. Upper Wyresdale in Lancashire with a fifty-inch rainfall is typical; very wide stretches are found throughout the uplands of Central Wales, which is also a region where the prevailing annual rainfall is of this order of magnitude.

We have already shown that the upper limit of effective settlement is determined by the tree-line. The highest inhabited house in all Britain, and the highest reclaimed pasture, lie at 1990 feet flanking the Harwood valley which is tributary to Upper Teesdale. But we can also recognise that the reclamation and improvement of moorland by drainage is again limited, not by actual altitude so much as by annual rainfall. The limits of reclaimed pasture are found on Dartmoor at 1,500–1,600 feet with about 65″ of rainfall, except for the slopes in the

---

PLATE 28

a. From GRASMOOR: Buttermere and Scawfell Pike, Cumberland: fine afternoon, September. Characteristic flattened cumulus over mountains with westerly wind. Cirro-stratus spreading above.

b. RADBOURNE, Derbyshire: unsettled weather, July. Disturbed sky in the trough of a depression, showing cirrus, alto-cumulus and nimbo-stratus to right, with fragments of strato-cumulus.

immediate neighbourhood of Princetown where over 80″ may be expected. Here additional labour has long been available. In the gritstone Pennines, the limit appears to be set by about 60″, for example north of Bolton, in Bleasdale, on Pendle Hill, above Malham, in Mallerstang and in Weardale. On the Bewcastle fells east of Carlisle, and in the Dumfriesshire valleys the profitable limit for reclamation is again of this order. Around Snowdon and within the Lake District it is broadly true that only those valleys which are exceptionally steep-sided and well drained have been settled beyond the limit of 60″. Here and there, patches of limestone offer better drainage in parts of the Pennines with a high rainfall, for example above Dent. In the Highlands of Scotland the upland limit for reclamation appears to be set by a similar figure, except in a few mild steeply-sloping areas very close to the west coast. Throughout Britain occasional examples can of course be found of settlement in areas of excessive rainfall for different reasons. Blaenau Festiniog, for example has grown up on the slate-quarrying industry, and many hundreds of Welshmen accordingly endure a 97″ rainfall.

Reclaimed pastures mean habitations; and with regard to one of the most conspicuous features of British scenery—the wide stretches of completely uninhabited moorland—we may say that the determining factor is climate. Beyond a certain limit of rainfall the endless peat-hags on our moorlands are hopelessly unreclaimable; elsewhere, few plants will stand up to acid soils, cold and wind; and below the limit, economic factors have played, and continue to play, a large part. With a fifty- to sixty-inch rainfall most of Exmoor was profitably reclaimed a century ago, but it is a little uncertain whether similar efforts would be economically sound to-day. All those who walk in our upland districts will find abounding opportunity to correlate settlement and rainfall. Many who do not know upper Teesdale and Weardale will have had the opportunity of seeing another high-lying farm in Scotland. Fourteen hundred feet up in Perthshire lies Dalnaspidal; and as the Inverness express panted up the Drumochter on an October morning, many travellers during the war years have seen a pathetic field of oats, at best languidly ripening in a place with a July mean of 53° and fifty inches of rain. At a similar elevation in Teesdale we learn that two centuries ago a miserable bread was perforce often made from the oats cut before they were ripe. Farther north, after 'the black auchty-twa'—the fearfully late cold spring and wet summer of 1782—it is recorded that the oats were carried in from the upland Banffshire fields in January. Such harsh seasons were long remembered. Urbanised though we are, dependent on the bounty of other lands for much of our food, few intelligent Englishmen or Scotsmen can fail to note something of the annual struggle of the upland farmers against those chances of variations of temperature and rainfall whose effects we have seen.

At this stage it may be asked, how far is the additional rainfall offset by better drainage, by aspect, and by the drying effects arising from increased exposure, to wind? The first question can only be answered empirically as yet. The inner Lakeland valleys owe their habitability to good drainage, as we have seen. Indeed as a result of the coarse soils and rapid run-off the effects of drought are quickly felt, and much more often than the southern visitor is prone to think. Similar effects can be found in the West Highlands. Elsewhere, the drift-covered slopes of many upland valleys are sufficiently free-draining to support quite a considerable farming population but no precise evaluation of the degree of slope or permeability of soil can as yet be given. It is significant however, that aspect plays relatively little part in our uplands, compared with the High Alps where the sites of farms and villages have been carefully chosen in the most sunny parts of favoured slopes; reference may be made to the exquisite diagrams in papers by Dr. Alice Garnett of Sheffield. In Britain the upper limit of reclaimed pasture is, in general, the same on either side of the valley; so that it would appear that in our cloudy and rainy climate free drainage is more important than direct exposure to the sun. Some studies have however been made which indicate that in a few localities such as Great Langdale the utilisation of the land by the valley farmers shows some recognition of the need to avoid as far as possible the shadows cast by neighbouring hills in winter. It has also been suggested that the successful arable farming on the Braes of Glenlivet, at 1,200 feet north of the Cairngorms, owes much to the fact that this shallow upland basin is particularly well exposed to the light; moreover recent statistics suggest that here on the lee side of the mountains the average duration of bright sunshine is about 10% greater than at Braemar on Deeside.[1] Diffused radiation from the sky is also important for plant growth in northern latitudes.

[1] I am indebted for this information to Miss M. E. Charles and Mr. I. C. Reid, during discussions in the Department of Geography at Cambridge.

With regard to wind, a good drying wind in September is certainly greatly welcomed by the upland farmer anxious to harvest his oats. But exposure to drying wind on one day implies exposure to driving rain on another. On 20 September 1942, a field of oats, on a breezy and generally favoured south-west slope near Garrigill in Cumberland, was scarcely at any point other than green. At fourteen hundred feet above the sea, most of it was already beaten down by rain and wind, and on that day an unforgettable sense of hopelessness was reinforced by the weather; a gloomy sky with driving rain and a hard south wind. Too often in the past have such Septembers wrecked any prospect for the bold farmer who, encouraged by high prices and by no mean sense of patriotism however little expressed in words, has essayed the cultivation of even the most promising-looking breezy and sunny slopes at fourteen hundred feet—with a fifty-seven inch rainfall and scarcely twelve hundred sunny hours in a year, in the instance mentioned. On the Braes of Glenlivet the rainfall, after all, averages thirty-five inches and the limy drift soils add a material advantage.

For upland position not only adds to the quantity of rainfall; the frequency and duration are also increased, though it must be emphasised that this increase is not in the same ratio as the quantity. Seathwaite receives five times as much rain as London, falling on about seven days for every five in the south. More precisely, measurable rain falls on 236 days at the head of Borrowdale, 169 in London. Other representative totals include 207 at Falmouth, 196 at Cardiff, 194 at Manchester, 217 at Greenock, 240 at Fort William and 263 at Stornoway in the outer Hebrides. Upland position means that there are a good many days in the year with one or two showers which the lowlands escape. Many good meteorological texts give diagrams to show that there will be occasions when the additional ascent of moist air over hills leads to instability and the growth of towering cumulus to a greater height than in the lowlands, hence a greater likelihood of showers. To such days must be added many others when the familiar

---

PLATE XIII*a*: Colliery smoke spreading and rising in turbulent daytime wind.

b: November 1947: a rainy day in Manchester suburbs. The stand pipe had just been erected as a result of the prolonged late summer drought, when domestic supplies throughout the city were about to be cut off owing to severe shortage of water.

PLATE XIII*a*. *By permission of the National Smoke Abatement Society*

b. *The Manchester Guardian*

PLATE XIV*a*. *C. H. Wood*

b. *C. H. Wood*

persistent wetting drizzle associated with the low cloud and strong wind helps to swell the annual total.

At a few stations automatic gauges record the number of hours during which rain of measurable intensity falls. Above Kinlochleven, in the Argyllshire highlands near Fort William, the annual total of 84 inches falls with measurable intensity on about 240 days and during 1,550 hours, whereas in London the annual total of 24 inches falls on 163 days during 437 hours. At an intermediate station (Eskdalemuir, 800 feet, in Dumfriesshire) the annual total of 56 inches falls during 1,217 hours; at Southport 32 inches fall on 189 days during 685 hours, but at Armagh the same amount falls on 215 days during 848 hours. It will be seen that the total amount, the number of days, and the number of hours do not lie in the same proportion. Most of the additional rainfall in our mountain districts results from more intense and prolonged rainfall on the wet days.

With altitude, on the whole, low cloud, rainfall and wind force increase; range of temperature tends to decrease; average day-time maximum temperatures in particular decrease with height more rapidly than average minima. The amount of bright sunshine decreases; and the length of the growing season, and rate of maturing of crops, decrease very noticeably.

Some slight advantage is gained, it would appear, with regard to thunderstorms. Over most of our highland districts they appear to be rather less frequent than in the lowlands, although at intervals they may produce considerable devastation due to the rapidity of the run-off.

Here and there our northern hillsides show the scars of long past torrents due to sudden violent downpours. One of the most appalling visitations of this type befell on Stainmore in 1930, and is described in *British Rainfall* for that year. Brast Clough, on the western flank of Pendle Hill in Lancashire, is said to have been largely eroded by the deluge in a violent thunderstorm in the sixteenth century.

Snow, again, is an element whose frequency increases very rapidly with altitude. So large a part in our impression of the scenery of

---

PLATE XIV*a*: A characteristic cloud-sheet over Lancashire in a humid air stream from S.W. round the margin of an anticyclone. Smoky patches are visible in the cloud. Warm subsiding air above, hence little or no upward growth of cumulus.

b: Liverpool and its smoke seen through a gap in the cloud-sheet.

Britain is played by snowfall in some parts, and by lack of it in others, that the subject deserves a chapter to itself.

REFERENCES

Chapter 10

BUCHAN, A. and others (1905–10). The Ben Nevis Observations. *Trans. Roy. Soc. Edinb. Vols. 42–44* (also many articles in *J. Scot. Met. Soc.*)

GARNETT, Alice (1935). Insolation, Topography and Settlement in the Alps. *Amer. Geogr. Rev. 25:* 601–17.

(1939). Diffused light and sunlight in relation to relief and settlement in high latitudes. *Scot. Geogr. Mag. 55:* 271–84.

MANLEY, G. (1942). Meteorological Observations on Dun Fell. *Q. J. Roy. Met. S. 68:* 151–65.

(1943). Further Climatological Averages for the Northern Pennines. *Q. J. Roy. Met. S. 69:* 257–60.

(1945). The Effective Rate of Altitudinal Change in Temperate Atlantic Climates. *Geogr. Rev. 35:* 408–17.

PEARSALL, W. H. (1950). Mountains and Moorlands. London, Collins' *New Naturalist.*

SALISBURY, E. J. (1939). Ecological Aspects of Meteorology. *Q. J. Roy. Met. S. 65:* 337–57.

TANSLEY, A. G. (1939). *The British Islands and their Vegetation.* Cambridge, University Press; indispensable; gives abundant references for further study.

NOTE

Since 1950, in addition to the station at Moor House (1,840 ft.) now under the Nature Conservancy, upland observations have become available at Malham Tarn Field Centre (1,300 ft.), at Onecote (1,350 ft.) in Staffordshire, at Bwlchgwyn (1,297 ft.), Alwen (1,100 ft.) and Llyn Stwlan (1,656 ft.) in North Wales. Observations near the summit of Ben Machui were also kept for a few months by Dr. P. Baird, then at the University of Aberdeen.

CHAPTER II

SNOWFALL AND SNOW-COVER

The belching winter wind, the missile rain
the rare and welcome silence of the snows
R. L. STEVENSON: *To My Old Familiars*

. . . the uniformly scandalous
Condition of the snow!
CLIMBER'S DITTY

OVER A large part of the British Isles the more impressive extremes of our winter weather occur with somewhat dangerous rarity. Further, their effects on our consciousness tend to be mitigated in towns and cities. To this we must add the fact that a rapid improvement in the amenities and ease of transport and in the efficient heating of public buildings has taken place during four decades, 1898–1939, characterised by a predominance of mild winters.

Eighty per cent. of our people live in towns; hence many of our people are inclined to forget that snow is still a very important factor in our northern uplands, perhaps more so on account of the very wide variations in its occurrence, whether in amount, persistence, or season of year. Snow came back noticeably into the Londoner's consciousness, for the first time for several years, in December 1938. Since then the severe winters of 1940–41–42, the cold January of 1945, the very severe February of 1947 and the cold December of 1950 have affected all Britain. February 1954 and 1955 were snowy; February 1956 severely cold.

It remains to be seen whether, in 1980, we shall continue to describe winter extremes as dangerously rare. Such a phrase may yet be advisable when we consider how little provision is made. The stupendous absurdity of our plumbing arrangements, dating from a time when the revenue of private water companies was considered to be more important than the comfort of householders, could only be tolerated in a climate which for several years allowed the drivers of the early locomotives to stand, like coachmen, inconsiderately exposed to the weather. The awful January of 1838 called attention to the need for improvement. Our tendency to hope for the best with regard to frost and snow may owe a good deal to that element in us which for centuries endured windowless huts and a smoky fire—such as we still see in the west of Ireland. The Iron Age camp on the summit of Ingleborough testifies to a curiously able people (compare the Desborough mirror) with an astonishing endurance of cold and damp, who carried their scorn of Continental heating arrangements right down to the eighteenth century. Some of those who had gone to America were then gradually seduced from their old allegiance; and to meet the harder winters of a new continent Benjamin Franklin devised his stove. At last the New Englander was able to resist the charms of bundling.

But their cousins retain much of their ancestral endurance, having less need for technical improvements unless the British climate takes a markedly colder turn. With regard to frost, snowfall and transport problems February 1947 probably did good, in reminding us that tolerance of out-moded institutions may be carried too far.

Average temperatures being what they are (January, from 45° to 39° at sea level) every part of the British Isles must expect an occasional winter day with sleet or snow rather than rain. In an earlier chapter we saw that in the cooler months air-masses either of polar maritime or polar continental origin may reach us with a surface temperature even on the coasts below 40°; but as they travel over an open sea it is rarely that the wind from seaward gives a surface temperature below freezing-point at coastal resorts. Hence unless a snow-cover is very deep and the wind persistently off the land the ground adjacent to the coast will rarely remain covered for more than a few hours. Even the Scilly Isles however are occasionally snow-covered, when an exceptionally severe outbreak of cold air from the Continent reaches the islands as a north-easterly gale from the direction of the nearest land thirty miles away. Perhaps the most unpleasant experience those normally mild islands have had for fifty years past was the gale of 29 January 1947, when the maximum temperature at Plymouth was 23° and even at the Scilly Isles only 28°, in spite of their being surrounded by sea-water with a surface temperature of nearly fifty degrees. February 1955 again gave a heavy drifting snowfall in West Cornwall.

For snowflakes to be observed it appears that the freezing level should not be more than about 2200 feet above the observer. If the British Isles were entirely flat we might expect the frequency of days with snow or sleet falling to increase steadily northward and eastward, as a result of the greater proximity and frequency of arrival of air-masses sufficiently cold at the surface to allow snow to reach the ground. More often than not it can be observed that on a day with continuous precipitation, i.e. saturated air, from low cloud in winter, rain only will fall if the air temperature is 39° or above. At 38° an alert motorist may see an occasional blob of sleet among the raindrops on his windscreen as he drives; at 36° a few flakes falling in the chilly rain will be seen by walkers; and at 34° largish half-melted snowflakes will predominate. Note that if the surface air is not saturated, snowflakes in a passing shower might thus reach the ground with an air temperature as high as 43°, for example on cold, windy but sunny days in late April; this comforms with our experience.

But if the temperature falls to the freezing-point, fine dry snowflakes will drift before the wind, and the lower the temperature, the finer the snow as a general rule.

Eastbound drivers leaving Exeter on a sleety winter morning soon appreciate these facts as they approach the Somerset uplands. Even in outer London, the climb from Croydon to Woldingham provides abundant experience.

The rapid diminution with temperature in the characteristic size of snowflakes leads to an increasing tendency to drift, and to penetrate into houses. Hence we find that drifting snowfalls occur quite rarely on low ground near the sea, but that they are a regular feature of our higher mountains. Further, from what has just been said with regard to rain, sleet and wet snow at temperatures above freezing-point it will at once be evident that on many days when nothing more than chilly rain is observed in our lowland cities, heavy snow will be falling at the 2,000 feet level.

Altitude then combines with latitude and proximity to colder air supplies to increase the frequency with which snow falls. Moreover, just as in the last chapter we saw that the presence of hills and mountains tends to increase the frequency of rain as well as the amount, we can now see that precisely similar results will apply with regard to snow. Snow showers are very commonly a consequence of instability in Arctic or Continental air reaching our shores. Cold dry Continental air leaves Denmark with a temperature of perhaps 20°; after crossing the open North Sea its surface temperatures at Tynemouth or Spurn Head may be 34°. But the result of the warming, in combination with the increased humidity, is not only an extraordinarily unpleasant rawness; the surface air is thoroughly unstable, and on meeting any obstacle giving a slight upward push it will rise, forming cloud and vigorous showers. At the temperatures named these are sometimes merely sleet and hail on the coast, but a few miles inland on high ground they are of snow.

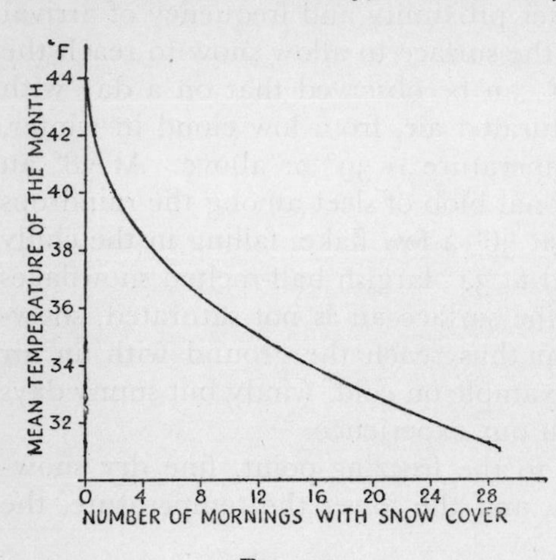

Fig. 57
Average frequency of snow cover in relation to the mean temperature of winter months: based on nine upland stations

Nothing is more characteristic of our north-east facing coasts than the driving snow-showers accompanying the onset of Arctic air throughout the winter months. These are particularly noticeable where the coast rises steeply, as in Cleveland, or where high hills rise but a few miles inland, for example, on the Lammermuirs. Dwellers in Norfolk will recall how often they break on the morainic hills behind Cromer and Sheringham, and farther south, the relatively high frequency of snow on the Downs in East Kent, by comparison with the hills nearer London, derives largely from the same cause. Instability snow showers due to the arrival of cold air over a relatively warm sea also occur from time to time elsewhere, in Eastern Ireland and, with a north-wester, in such places as the North Wales coast.

We can see the combined result of these factors in the map (Fig. 58), which shows for stations *at low levels* the average frequency of days per annum on which snow or sleet has been observed to fall, taken over the period 1912–38. In the Scillies and in the far south-west of Ireland the average is three or four days only. In the Outer Hebrides, although the average temperatures are but little lower, proximity to

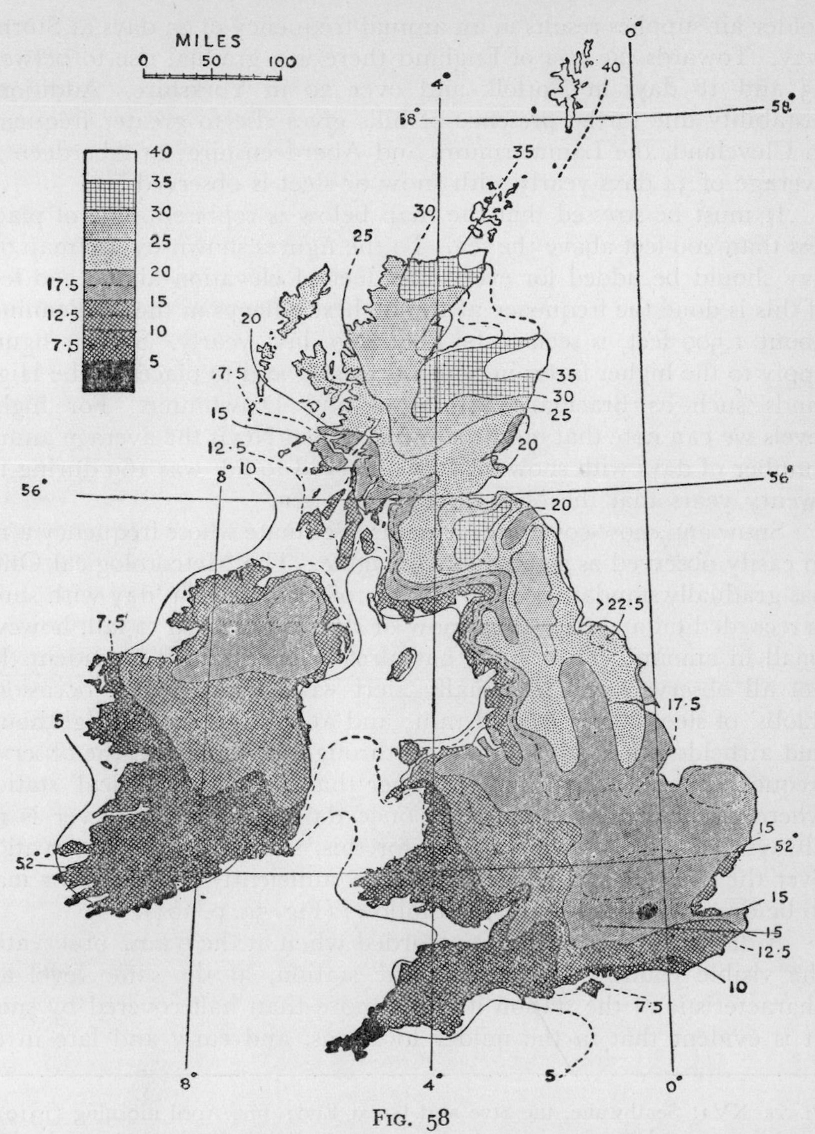

FIG. 58

Average annual number of days with snow or sleet observed to fall (after the author's paper in Meteorological Magazine, 1940, by permission of Her Majesty's Stationery Office). The data refer to ground below 200 feet. For higher levels add to the figure shown by the map one day for each 50 feet above 200

colder air supplies results in an annual frequency of 25 days at Stornoway. Towards the east of England there is a gradual rise to between 15 and 18 days in Norfolk and over 20 in Yorkshire. Additional instability due to the presence of hills gives rise to greater frequency in Cleveland, the Lammermuirs and Aberdeenshire; at Aberdeen an average of 34 days yearly with snow or sleet is observed.

It must be stressed that the map below is representative of places less than 200 feet above the sea. To the figures shown by the map one day should be added for every fifty feet of elevation above 200 feet. If this is done the frequency at the highest villages in the N. Pennines, about 1,500 feet, is seen to be about 50 days yearly. Similar figures apply to the higher farms in South Scotland and to places in the Highlands such as Braemar, Tomintoul and Dalwhinnie. For higher levels we can note that on the summit of Ben Nevis the average annual number of days with snow or sleet observed to fall was 169 during the twenty years that the observatory was open.

Snow and snow-cover are elements of climate whose frequency is not so easily observed as some might imagine. The Meteorological Office has gradually standardised the manner of recording; a 'day with snow' is recorded on any day when snow or sleet is observed to fall, however small in amount. From what has already been said it is evident that not all observers will be equally alert with regard to the occasional 'blobs' of sleet falling in cold rain; and at stations such as lighthouses and airfields where a watch is kept throughout the night, the observed frequencies tend to be slightly higher than at 'climatological' stations where the instruments are read once daily and the observer is not always present. Allowing a little for this, since 1912 the observations over the country as a whole have been sufficiently consistent for maps to be drawn, such as that which follows (Fig. 59, p. 204).

A day with 'snow-lying' is recorded when at the 9 a.m. observation the visible country surrounding the station, at the same level and characteristic of the station itself, is more than half covered by snow. It is evident that in the milder locations, and early and late in the

---

Plate XV*a*: Seathwaite, the Stye and Great End; fine April morning (1910), a few winter snowdrifts above 2000 ft. The wettest place in England; annual rainfall 122 inches by the farm, 185 inches locally on the fell beyond.

b: Ski-ing on Ben Lawers. Typical fine windy winter day; snow slightly drifted, grasses showing.

PLATE XV*a*. *G. P. Abraham, Ltd.*

b. *The Times*

PLATE XVI.*a.* *The Times*

b. *C. H. Wood*

season, such a definition does not imply a 'day's duration of snow-cover'.

Broadly however, it can be said that the number of 'days' duration' is about the same as the 'frequency of snow-lying at 9 a.m.' whenever the mean temperature for a month is less than 37°. In what follows it will be convenient to refer to the annual frequency of days with snow-cover; but it should be remembered that as observations are at 9 a.m. an annual average of 1·3 at Torquay, or 2·1 at Eastbourne does not necessarily mean that snow lies throughout one or two whole days. But in the Scottish Highlands at Braemar where the annual average (1913–1949) at the 9 a.m. observation is 66·4, the number of days during which to a casual observer most of the ground by the village is covered throughout is probably very close to this figure.

The number of days of snow-cover depends largely on the mean temperature, which is chiefly a question of altitude; and partly on the frequency with which snow falls, partly on the quantity. We have already seen that frequency is greater, on account of relief, in certain areas towards the north-east coasts. Correspondingly, frequency of snow-fall is less on the lee-side of those hill ranges over which northerly and easterly winds must descend. One of the most snow-free areas of the Scottish mainland lies at Prestwick; in England, the head of Morecambe Bay is similarly favoured. Further, as most of our heavy snowfalls arrive on winds from an easterly point, when depressions are centred to the southward, we find that there is a very marked orographic influence. Just as the south-western flanks of our mountain districts lead to rapid increase in rainfall, the easterly and north-easterly slopes are almost always associated with rapidly increasing depth of snow whenever there is a major fall. Exceptions occur now and then, for example when the dominant wind direction is southerly and a stationary or slow-moving front lies north to south in the Irish Sea. Such a situation in January 1940, January 1947 and again in March 1947 gave a particularly heavy snowfall in the Lake District. Occasionally too the rather rare development of what is called a polar-air depression gives a heavy snowfall attaining its greatest depth in somewhat unexpected places, for example north-west Ireland in April 1917

PLATE XVI*a*: Snow on the beach at Perranporth in Cornwall, December 1938.

b: Floods in Wharfedale. Winter floods following heavy rain are often reported from the middle courses of many of the Pennine rivers.

FIG. 59

Number of mornings with snow lying, annual average, 1912–38 (after the author's paper in Meteorological Magazine, 1947, by permission of Her Majesty's Stationery Office)

and the Lancashire Pennines in May 1935. (See also Fig. 36, p. 113, 10 May 1943).

In general however the regions of heaviest snowfall are well marked. Throughout the eastern highlands of Scotland snowfalls tend to be heavier than farther west; Deeside for example is more snowy than Speyside. In England snowfalls are most common in the eastern Pennines, and at high levels they can be very persistent. Sheffield's high-lying suburbs are well-known in this respect; also the efficiency of its transport system. Huddersfield is more snowy than Oldham; Consett in Durham and Queensbury in the West-Riding of Yorkshire are probably the snowiest urban communities in Britain. Nearer the coast, the wide windswept Yorkshire Wolds repeatedly figure in our newspapers; again, the amount falling is generally considerably greater than in the plain of York to the leeward, or in low-lying Hull. The Lincolnshire Wolds are prominent for the same reason. The snowy sections of the Great North Road are well-known. The Stamford-Grantham section is a mere four hundred feet above the sea, but, rising like the Wolds, from the plain to eastward, it has a heavier snowfall. In 1947 the hold-up of road transport here was remarkable; similarly, in 1708 Ralph Thoresby was forced to spend five days at Stamford on account of deep snow. Other uplands characterised by heavy snowfall against which easterly winds are forced to rise include the hills of north-east Wales, East Kent, and Dartmoor. Dartmoor snowfalls are occasionally extremely heavy and are the more noticeable

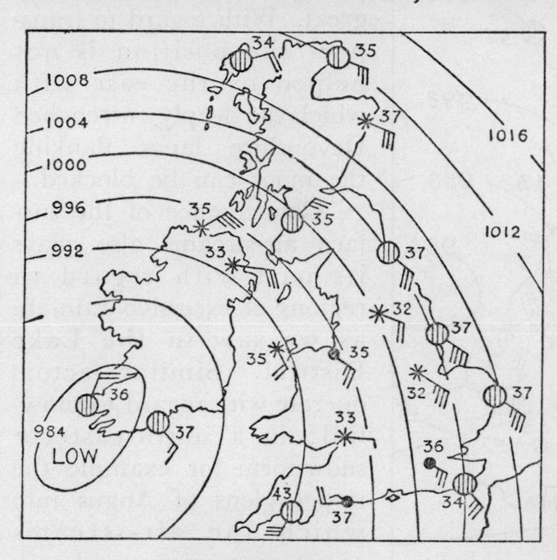

FIG. 60

0700 hrs., 25 February 1933. Very heavy orographic snowfall in N.E. and N. England and in N.E. Ireland associated with a low which moved westward; snow 6 inches deep at Durham but 30 inches deep in Teesdale at 800 ft. (For notation, see p. 5)

in the relatively mild south-west; again we can see how on steeply-rising moorlands facing the damp unstable east wind as it sweeps down-channel from the cold Continent not only the quantity of snow, but the associated drifting at higher levels can be very great. With regard to transport the position is not helped by the ease with which the deeply entrenched Devonshire lanes flanking the moor can be blocked.

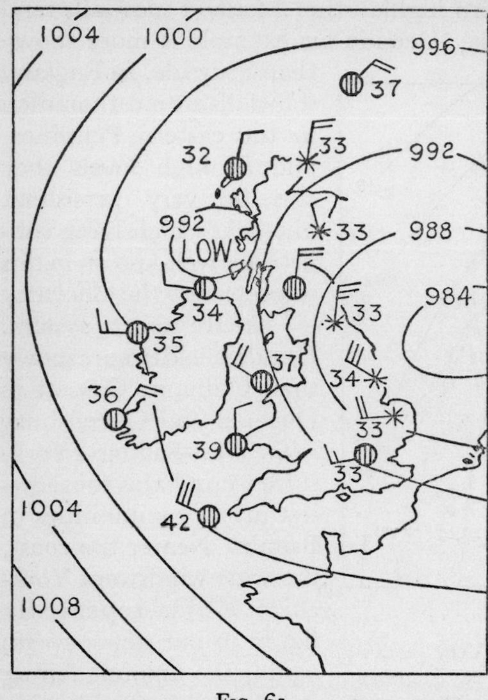

FIG. 61

1800 hrs., 20 February 1941. Towards the end of the great N.E. coastal snow storm, depth 30 inches in Newcastle and 42 inches at Durham, 18-20 February (see p. 5)

Convergence of the surface air-streams also plays its part with regard to regions of excessive rainfall, as we saw in the Lake District. Similar factors operate with regard to snowfall; in a south-easterly snowstorm for example the upper glens of Angus into which the air-streams converge, receive very heavy falls. Convergent air-streams also affect Dartmoor snowfalls; with an east wind, very exceptional depths were reported near Holne in 1929. Upland snowfalls may occur during many months of the year. During the past half-century the earliest noteworthy fall was that which covered the Pennines above about 800 feet on 20 September 1919; while an exceptionally late fall, nearly a foot in depth in some upland Lancashire towns, occurred on 16 May 1935. Some

---

PLATE 29

HARTINGTON, Derbyshire: brilliant noon in the Peak District, March. Snow slightly glazing on surface, air temperature 28°. Cold anticyclonic day with gentle north wind, slight growth of cumulus over the distant plain.

PLATE 29

Cyril Newberry

PLATE 30

Cyril Newberry

representative figures with regard to frequency of snow and snow-cover are given below, together with maps; see also Table VIII in the appendix.

Estimates for higher levels of the average monthly and annual number of days with snow-cover prevalent during the last three decades.

	J	F	M	A	M	June to Sept.	O	N	D	*Year's total*
N. ENGLAND Moorhouse (1, 840ft.)	19	17	17	6	1	0·1	2	6	12	80
SCOTLAND Ben Nevis (4,406 ft.)	31	28	31	30	20	9	14	22	30	215

Taking all the available evidence it appears that everywhere in Britain at *higher* levels the overall increase with altitude in the average annual frequency of snow-cover is nearly linear. In Scotland it can be put at about 1 day for 22–25 feet. Further south data are more scanty but the increase is probably not so rapid; about 1 day for 30 feet is a fair estimate for South Wales. It appears that at levels up to 800 feet the range of variation between very mild and very severe winters lies from about one-quarter of the average figure (zero where the average is less than 10) up to about fifty days above the average. Even from a small scale map it will be seen that in spite of the appreciable chance of snow in September and May virtually no part of the British Isles below 2,000 feet has on the average as many as 100 days with snow-cover. Throughout the winter warm rain, such as accompanies the onset of maritime tropical air (or even, at times, returning *mP*), is capable of removing practically all traces of snow.

From this it is evident that at any inhabited level snow-cover is neither regular nor continuous in its occurrence. We can judge this very well from the figures for Braemar (1,120 ft.) where snow can be expected to lie for about half January and February. During the past

PLATE 30

THE CAIRNGORMS and the Lairig Ghru from Rothiemurchus, Inverness-shire: early June. Anticyclonic weather with clear skies above; strong southerly wind giving cloud on the mountains and clearer skies on the descent (partial "föhn"). The lowest visible snow patches are at about 2,700 feet.

37 years with an average of 66, the range of variation in the number of days with snow-cover has lain from 34 (in 1938) to over 122 (in 1919). Nearby at Balmoral (930 ft.) the figures are 61, 27 and 116. At Braemar there were twelve winter months during which snow has lain throughout the month in the village; these include 1 January, 8 Februaries, and 3 Marches. By the time we reach levels at which an average of 100 days with snow-cover can be expected, we can be pretty certain that in an average year ground will be found covered from mid-December to some time in March, except for brief periods.

Hence it might be argued that ski-ing accommodation might profitably be established at this level (2,000 to 2,200 ft.) where indeed the chances of snow-cover in Scotland are about equal to those at the lower Alpine resorts. Unfortunately, the conditions for ski-ing are rendered the more difficult, not only by the weather; terrain plays its part. Damp wind, mist and succeeding frost create appalling surfaces; as a result of fusion they are commonly made up of large and thoroughly uncomfortable crystals, and may be so hardened on the surface that control is difficult. Drifting moreover occurs to such an extent at higher levels that the depth of snow is extremely variable. (Pl. XVb, p. 202). Icy patches interspersed with projecting boulders, followed by irregular drifts hardened by the incessant wind often provide challenging difficulties even on a sunny day, which are hardly conducive to the relaxed enjoyment offered by a high-altitude resort in the central Alps within the European winter axis of high pressure. And unless one lives in Edinburgh or Glasgow the cost of the journey is still considerable. Nevertheless many of those who live near our higher hills know how to take advantage during winter week-ends of the fine interval following a fresh powdery snowfall. Nothing is more welcome than such a day of opportunity seized and held within the Scottish winter, when under the low January sun in a wedge of high pressure, the opalescence of the polar air lights up the slopes of the Central Highlands above Dalwhinnie or Loch Tay with astonishingly beautiful pearly colouring. Opalescence is probably a by-product of salt particles suspended in the lower atmosphere after stormy weather; it appears to belong only to high latitudes, for example Iceland. Further south, there are the rare occasions when the snow-covered Kinderscout peat-hags dazzle the Manchester man's unaccustomed eye, and only the dreadful brown smoke-haze penned within the surface inversion-layer over the plains to westward remains as a reminder of the murky gloom in the city. Such a day was 5 January 1941; in Westmorland minima in the Eden valley fell to zero, and in the clear air above the valley inversion the rosy flush of dawn on snowy Crossfell was unforgettable. On the same morning the minimum at Barton, outside Manchester was also 0°, but at Oakes on the ridge above Huddersfield (760 feet) 19°; compare Ushaw and Houghall in County Durham, page 167.

Indeed the significance of snowfall in Britain lies in the further variety of scene it provides, often at most unexpected times and places. Occasionally among our hills a fiercely brilliant May sun from a cloudless sky blazes in Alpine fashion on a deep snow-cover, such as covered the Isle of Skye almost to sea-level on 2 May 1923. (Cf. Pl. 29, p. 206). At other times, large corniced drifts have been seen flanking the Great North Road within a few miles of London. Cuttings half a mile long in the snow at the head of Weardale were eight to twelve feet deep when, after a seven weeks' blockage, the first car passed over Killhope summit on 2 April 1937. In December 1938, the *Times* published a photograph of the snow-covered sands of Cornwall terminated only by the Atlantic (Pl. XVIa, p. 203).

Snow is capable of giving a remarkable amount of trouble in this normally mild country. Even the innocent monotony of the outskirts of Cambridge has provided its story. In February 1799 Elizabeth Woodcock floundered into a drift on her way home from Cambridge market to Impington three miles away, and was buried for eight days before she was rescued—still alive—after hearing the Histon church bells twice. Farther north each snowy winter is generally marked by tragedy when keen but unprepared walkers lose themselves in snowstorms on the moors. When an exceptionally snowy season befalls, such as February 1947, catastrophic losses of sheep must often be faced.

Taken over the country as a whole, February 1947 was probably unmatched by any other month since January 1814 for persistence of cold, dullness and repeated heavy snowfalls. In the London area, statistics of occurrence of snow can be assembled since 1668.

Permanent Snow

We have already seen that, on high ground above 1,000 feet or so, drifting is characteristic of the majority of heavy snowfalls. Roads, especially where they run between walls, are only too easily blocked, and if the resultant depth of snow is greater than about eighteen inches they must be dug out. Sometimes weeks of labour are needed, as in Teesdale in March 1937. Drifts in the lee of walls and other sheltered places may persist long after the country as a whole is free from snow.

As a result it is common to see occasional drifts here and there among our mountains, especially in gullies facing north and east and thus sheltered from sun (and driving rain; continuous warm rain is perhaps the most effective agency for removing a snowfall). In a normal year the last drifts at the 2,000 feet level in the N. Pennines linger till about the middle of May. In the northern gullies of Helvellyn (2,800–3,000 feet), towards Red Tarn and Brown Cove, some small patches of snow may be found till early June; and the same is true of the Great End Gullies above Borrowdale, and the northerly crags of Carnedd Llewelyn in North Wales.

Rarely there have been reports of strangely persistent drifts in exceptional places at much lower levels. In 1947 in a shady location, remnants of a drift in the Northern Cotswolds were visible until mid-July, and Bonacina has reported that in the same district in 1634 a snowdrift is said to have survived until August. Nearby, the village of Snowshill, eight hundred feet up with a northerly aspect, is likely to owe its name to past experience, reminiscent of Cold Ashby in Northamptonshire.

Persistent snowbeds occasionally linger in the full sunshine; following a north-easterly blizzard in March 1937, snow accumulated just under the summit plateau of Crossfell on the south-west side. In spite of a sunny May, the drift remained visible until 10 June. After the terribly severe March of 1785 followed by a warm May, one of our first Lake District tourists commented on the large snowdrifts on the Crossfell range still visible from Penrith on 18 June.

There is no doubt that the lingering snowdrifts of winter add their quota to the charm of the mountains in spring. In April in the Lake District the fresh green of the larches around Grasmere (however much they were detested by Wordsworth) is the more noticed when the northern slopes of Helvellyn or Bowfell still carry much snow. Even more is this true of the higher Scottish mountains. Above 3,000 feet, on the Cairngorms and Ben Nevis, on Mam Soul, Ben Wyvis and other high summits large snowdrifts linger until July, and in some recesses through August. In two exceptional locations they normally just last through the year, the critical period being the last half of September.

These semi-permanent snowbeds both lie at about the same level, 3,750-3,800 feet, on the north-east flank of Ben Nevis and on the same side of Braeriach, in the sheltered and deeply shaded gullies descending steeply below the summit ridge (Pl. XVIIIb, p. 227). Forty years ago these beds were judged to be permanent; and it is on record that some considered that the degree of consolidation of the snowbed on Ben Nevis almost justified the view that it was an incipient glacier. But alas for the enthusiasts; the amelioration of the North-west European climate in the last four decades has been reflected, in Scotland, in the total disappearance for a brief period in certain years (1933, 1935, 1938, 1945, 1953 and 1958) of these drifts, towards the end of September. It is interesting to note that in each year in which the beds failed to survive, a warm and rainy autumn and a winter of little accumulation was followed by a warmish summer, rainy in the latter part (*i.e.* August).

Eighteenth-century travellers in the Highlands espied the snow-drifts on the Cairngorms (August 1727) and on Ben Nevis (September 1787), while Pennant in 1771 believed those of Ben Wyvis to be permanent. For Helvellyn in the Lake District there is some interesting evidence that in the Napoleonic era early last century snowdrifts were more persistent than we should nowadays expect. John Dalton was a man of very regular habits, and year by year made the ascent about the end of the first week in July, at the beginning of his annual holiday from Manchester. He was fond of determining the humidity of the air, and to do this he cooled a small vessel if possible to the dew-point. For this purpose melting snow was convenient. In several years between 1805 and 1823 he records that he found snow remaining "in the usual place, about a quarter of a mile north of the summit," that is overlooking Brown Cove. To find snow remaining there nowadays so late in the year would be most unusual; but in 1812 a large drift was still to be found on 18 July (1951, 10 July). Late in June 1817 Dorothy Wordsworth noted that snowdrifts remained above Red Tarn; no doubt her brother's line about the recess "that keeps till June October's snow" derives from this excursion. Protracted cold springs were rather frequent during the years in question. In mid-July 1843 Miller noted a remnant near the summit of Scawfell Pike; in August 1818 Keats crossed snowbeds as he walked up the *west* side of Ben Nevis.

Studies of the extent to which our average temperatures of summer have varied in the past 200 years suggest that after exceptionally cold springs and summers it is just possible that a rare snowdrift might survive through the year in the N. Pennines and the Lake District, and perhaps even in North Wales. Odd stories to the effect that 'some say the snow never melts up there' have been heard in the north of England, and were probably handed down from the occasional cold year in the past such as 1695. That such stories were current long ago is shown by an account of an ascent of Crossfell in 1747 when the writer emphasised that "on August 13 in spite of careful inspection of all the likely places, no trace of snow could be found."

Such stories indeed remind us that there is always a tendency to romanticise with regard to such extremes as snow. Nevertheless, it does not do to neglect the possibilities of snow, whether as an element in scenery, an impediment to comfort, or a dangerous menace to survival. The experience of 1947 is too fresh in memory. In the last decade the winter of 1950-51 was marked by exceptional persistence of mountain snow-cover above 2,000 feet; and for the first time for many years snowdrifts in the Lake District and North Wales lasted far into July. February 1954 and 1955 also gave heavy snowfalls.

Calculations based on the average temperature of the summer months lead to the conclusion that in the regions of heavy precipitation the present day snowline would be found at about 5,300 feet in the Ben Nevis region, 5,900 feet in the Lake District and 6,300 feet in the Snowdon District. That is, if the mountain summits exceeded this height small glaciers would probably develop under present climatic conditions.

REFERENCES

Chapter 11

ASHMORE, S. E. (1952). Records of snowfall in Britain. *Q. J. Roy. Met. S. 78*: 629-32.

BONACINA, L. C. W. (1927). Snowfall in the British Isles, 1876–1924. *British Rainfall, 67*: 260.

(1936). Snowfall in the British Isles, 1925–1936. *British Rainfall, 76*: 272.

MANLEY, G. (1939). On the Occurrence of Snow-cover in Great Britain. *Q. J. Roy. Met. S. 65*: 2–26.

(1940). Snowfall in Britain. *Met. Mag. 75*: 41–48.

(1947). Snow-cover in the British Isles. *Met. Mag. 76*: 1–8.

(1949). The snowline in Britain. *Geogr. Annaler, 31*: 179–93.

CHAPTER 12

SECULAR VARIATIONS OF THE BRITISH CLIMATE

The cold earth slept below
Above the cold sky shone
And all round with a chilling sound
From caves of ice and fields of snow
The breath of night like death did flow

SHELLEY

MANY readers of this book will not lack acquaintance with the story of the rocks and the climatic conditions under which they were laid down. For an introduction to this subject others may well be commended to the companion work in this series on *Britain's Structure and Scenery* by Professor Dudley Stamp. Therein they will be reminded of the evidence that two hundred million years ago, and to a less extent thirty million years ago, the climate of this country over endless ages by our present time-scale was warm and humid to a degree which we should nowadays associate with lower latitudes. Such climatic variations have indeed affected the character of the rocks we see; but are more properly considered in a work on geology.

We should rather concern ourselves with the more recent climatic variations that have affected the surface aspect of the country, the vegetation and through them the life of animals and men. The period of man's existence and development covers the past million years. During that span of time those parts of the earth in which man has most notably evolved have been subject to the exceptional climatic vicissitudes culminating in the several phases of the last Ice Age. In the British Isles the effects of the glaciation on our landscape are so widespread and conspicuous that it would be convenient to begin any discussion of secular variations with an account of the factors governing the greatest climatic variation of all, which took place but yesterday by comparison with the geologist's time. Indeed for the past ten thousand years the country as we now see it has been slowly recovering from the effects of the last phase of the Ice Age. Even in comparison with the million years of man's evolution this period is very short, but in it all the developments of the civilisations we know have been comprised.

We must first try to envisage the state of affairs before this series of wide climatic oscillations began. In the earlier Tertiary period it appears that the world was nearly, if not entirely, free from ice. Geological evidence from deposits of that period in Spitsbergen and the Antarctic indicates that the seas were open while on land, even in high polar latitudes, considerable vegetation was then to be found which we should now regard as typical of more temperate countries. Weather was stormy and disturbed; and the vigorous air movement now characteristic of the Icelandic seas prevailed in winter throughout the Polar Basin.

Later in the Tertiary very extensive mountain-building movements took place, culminating in Europe in the uplift of the greater part of the Alps and associated southern European ranges. Towards the Poles extensive uplift of existing land masses such as Greenland and Scandinavia was associated with a correspondingly widespread degree of foundering in the North Atlantic, and with the great volcanic outbursts of which Iceland is the principal remnant.

It has been shown that even with an ice-free Polar Sea the winter temperatures, though very similar at the North Pole to those we experience in Britain now, were such that snow would intermittently fall on the neighbouring land. The uplift of high mountain ranges adjacent to the Polar Seas, therefore, would not only lead to an increase in precipitation; at higher levels much would fall as snow, and gradually accumulate. Evidence is still rather controversial with regard to the exact period at which sufficient accumulation of snow on a high land-mass such as Greenland existed to initiate an ice cap. Undoubtedly, however, there should be a gradual spread of melt-water, and floating ice, from Greenland's marginal glaciers; this, being less dense would form a sea-surface layer above the denser salt water—as it still tends to do at the present day. Further, in winter such a surface layer would freeze again much more easily than if it were salt.

Gradually then we observe the chain of events; high land in high latitudes; ice caps within the marginal mountain ranges from which glaciers descend towards the sea; melt-water spreading over the sea, favouring as a whole the slow increase of sea-ice winter by winter. Once the sea-surface is extensively frozen, it assumes to a large extent the properties of a snow-covered land surface. Radiation on clear winter nights leads to cooling of the adjacent surface air, and the slow building up of the resistant "cushions" of cold air that we recognise as a characteristic of the present-day Siberian winter. In an earlier chapter we named them cold anticyclones, and we noted their propensity to fend off the surface incursions of warmer air.

In his standard work *Climate through the Ages*, Dr. C. E. P. Brooks has shown by ingenious argument that whereas over long periods of the earth's history the polar seas were just warm enough to be open, if such an open polar sea began to freeze, it is probable that over a period of years the ice would gradually increase in area until its edge lay about 10° from the Pole. Consequent upon the existence of such an extensive icy plain winter temperatures in the surface air would fall to far lower values than formerly prevailed. Cold winds spreading outward from this cap of Polar air would be likely to engender much more vigorous depressions along the Polar Front between them and the warmer maritime air to southward. In turn, however, the resulting stronger winds near the margin of the Arctic would tend to break up the floating ice, so that the sea-ice would not continue to spread indefinitely towards the equator.

Meanwhile we must recall that in the Pliocene or late Tertiary Scandinavia also stood higher than at present; with the wider spread of cold air and the increased storminess the accumulation of snow on the Norwegian and Swedish highlands began. Ice-covered highlands, like ice-covered sea, create their own cold. The reflexion of solar radiation from a snow surface is so great, amounting to about 80% of that which arrives, that it is not difficult to envisage the steady development of an ice cap from a stage where not merely scattered snowdrifts, but many square miles of ice, resisted the sun, warm wind, and rain of summer.

All these processes owe something to the changes in the elevation of land in high latitudes relative to the sea. It is probable that the continents were also more extensive. It can be shown that in higher latitudes an increase in the degree of continentality, that is, of the area of land surrounding a given station, will lead to a greater fall in the winter temperatures than can be compensated by the rise in summer. If for example, the present-day Baltic became dry land the net result would be a fall in the overall average temperature of Sweden. Without doubt we can build up a tolerably satisfactory explanation of the establishment of vast ice-caps in Greenland, over Scandinavia and over eastern Canada based on the greater extent and elevation of land in high latitudes towards the end of the Tertiary. Fringing the great Scandinavian ice sheet the mountains and highlands of Britain were also heavily glaciated and from them as most of us know, the ice flowed southward towards and across the English lowlands, almost as far south as the Thames valley.

For the detailed results with regard to scenery reference is better made to Professor Stamp's companion volume. It will serve here to remind the reader of the widespread mantle of glacial drift; the diversion and subsequent modification of many of the rivers, here rejuvenated, elsewhere incised in attractive gorges such as that at Durham, at Ironbridge, or above Llangollen. Our mountains are of relatively low altitude but they are nevertheless true mountains; in this respect they surprise many visitors from abroad. Peaks and ridges, screes and crags, corries and their tarns all testify to the effects of extensive and severe glaciation. So much so, that it is at times difficult for the climber to believe that once above the rocky lip of the corries in Skye there will not after all be a small remnant of the ice, however convinced he is from his reading that such remnants were last visible about 8300 B.C.

But the Ice Age, represented for us by a vast development of ice over Scandinavia and by extensive glaciation of our own islands, was not a single episode. Within the last million years or so it is probable that the great ice-caps of Scandinavia and Eastern Canada waxed and waned in at least four major phases, accompanied to a varying degree by corresponding advances and retreats of the mountain glaciers of the Alps and other ranges, and by minor fluctuations of their peripheral extent. Between these major phases of the Ice Age, interglacial periods of considerable length occurred and in some districts there is evidence that the climate became milder than at the present day. Deposits of interglacial age have been identified in the Alps in which the plant remains indicate a more southern type of vegetation than is permitted by the present temperatures.

This problem of multiple glaciation is one of the world's most fascinating puzzles. Those who would be prepared to accept an explanation based entirely on the results of the elevation of land in high latitudes find it very difficult to conceive that Scandinavia could bound up and down, so to speak, several times in such a short period having regard to the earth's history as a whole. Other suggestions have been made which postulate considerable variations in the intensity of solar radiation. It has been demonstrated by Sir George Simpson that a small increase in the power of the sun would ultimately give rise to increased cloud and precipitation in highland regions towards the poles; assuming that the land was already sufficiently elevated, the resultant increased cloudiness and snowfall would gradually give rise to an ice cap. He points out the importance of the fact that a widespread cloud sheet, once formed, reflects a great deal of the radiation falling upon it. The elegance with which his theory can be extended to explain the occurrence of cooler and warmer interglacials is attractive; it was published in the *Quarterly Journal of the Royal Meteorological Society* for 1934, with some revision in 1957. But unfortunately, sufficient geological evidence is not forthcoming with regard to the relative coolness or warmness of the several interglacials which Simpson's theory would require; interglacial deposits are rare, as they are generally removed by the succeeding glaciation. For this reason the elucidation of the full story of the British glaciations is tardy. Moreover full agreement has not been reached with regard to the number and extent of the several glaciations in other parts of the world.

Other theories with regard to the cause of widespread refrigeration, as some have called it, rest on small variations in the intensity of solar radiation arising from slight changes in the tilt of the earth's axis and the eccentricity of the orbit. The effects arising from such causes are of disputable magnitude. Then, apart from variations in the behaviour of the sun itself, the amount of solar radiation transmitted through the atmosphere to the earth's surface may have varied on account of changes in the gaseous constitution of the atmosphere. Such changes at low levels appear rather unlikely; we are as yet uncertain of the possibilities higher up. Another suggestion arises from our knowledge that terrestrial volcanic eruptions result from time to time in such an explosive outburst of fine dust, that the climate of the whole world is affected due to the slight curtailment of solar radiation reaching the earth's surface. But although certain historic cold years are known to have been so caused (1784, 1816, 1912 for example) the finest volcanic ash does after all sink back to earth within two years or so of the eruption. To explain an ice age lasting a hundred thousand years would demand eruptions on a scale which again appears incredible for other reasons. It cannot be denied however, that vast eruptive disturbances in combination with several other factors might, so to speak, have pulled the trigger, had they occurred at the right moment. But until we have satisfied ourselves that the glaciations in various parts of the world were contemporaneous we cannot answer this and many other problems.

Puzzling features of the glaciation problems keep appearing. For many years it was thought that we were definitely living in the waning of the last Ice Age; barring some slight evidence of a slight recession towards greater cold about 500 B.C. Since about 1920 it has been recognised that the climate of much of the north temperate zone took a turn towards greater cold within the last seven hundred years. As part of this comes within the period of organised science and instrumental recording, we will discuss the details later in this chapter. We must leave the questions of why such a significant climatic fluctuation has occurred as one part of the greater question—that of multiple glaciation, for whose solution the combined efforts of workers in a variety of sciences will for long be required.

Let us rather endeavour to reconstruct the climate of the Ice Age in Britain at its maximum. Evidence for such a reconstruction is forthcoming from a variety of sources. First, we know that there must have been floating ice adjacent to our S.W. coasts, and that many of the icebergs melted but a short distance from land. This is shown by the many erratic boulders dredged from the sea bed. Banks of shelly drift near the East Anglian coast testify to an earlier stage in which, while the North Sea was still unfrozen, very strong and persistent easterly winds prevailed. Thirdly, floating ice debouched to the north-west of Scotland, and we must therefore allow for a broad ocean to the west of Britain. For very long periods however most of the North Sea was a vast plain covered by slowly moving ice from the Scandinavian Highlands, indicated by the many erratics of Norwegian origin found on and near our east coasts. Similarly, the shallow Irish Sea, the North Channel and the channels among the Hebrides were completely covered by the great tongues of ice, or piedmont glaciers, spreading from the adjacent highlands. For convenience the map from *Britain's Structure and Scenery* is reproduced on the next page.

We have to devise a scheme which allowed the survival of forest trees in Northern France, although Southern England was a bare windswept tundra snow-covered for a great part of the year. The English Channel was merely a wide gulf extending eastward from the Atlantic. We have to devise a pressure distribution to allow for certain features elsewhere; in Bermuda the sand deposits indicate an excess of strong south-westerly winds in glacial times. Further, if there were so great an amount of floating ice in latitudes south of Iceland and west of Britain, we may assume that the surface waters of the North Atlantic owed much to the melting of the ice and hence that they were much colder and presumably less saline than now.

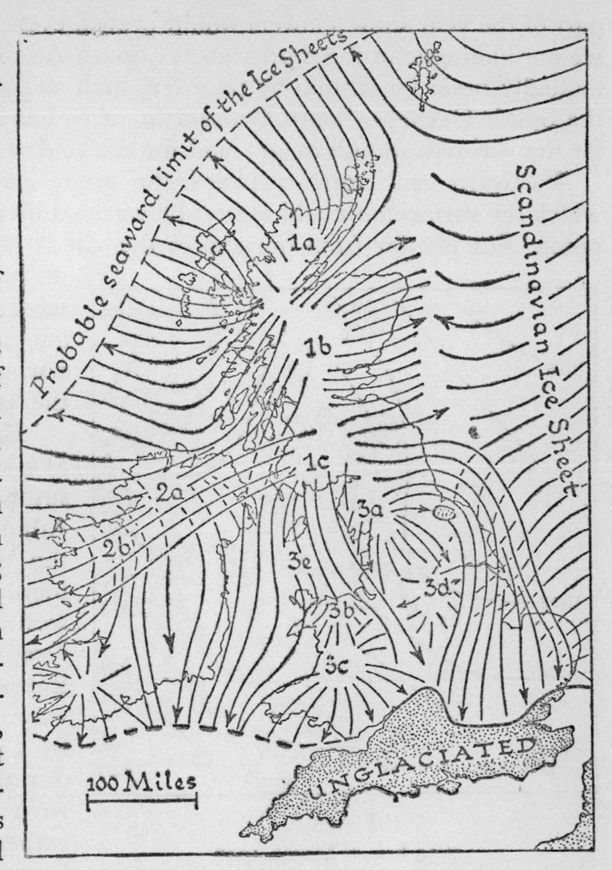

FIG. 62
The geography of the period of the maximum glaciation in the British Isles. This map shows the approximate position of ice sheets at the second glaciation. The home ice-caps—centres of ice accumulation and dispersal—are numbered; lines and arrows show direction of ice flow; where lines are broken the earlier directions were superseded by those shown by solid lines (by courtesy of L. D. Stamp)

Polar air from W. to N.W. reaching Britain would do so over a very much colder sea than at present. In such air, through the greater part of the year snow showers would prevail even at sea level. We can see our analogue in those present-day South-Atlantic islands which are annually beset for a time by pack-ice, such as South Georgia or even the South Orkneys. South Georgia in 54°S. has a mean temperature for the warmest month of 43°, and for the coldest 29°.

Eastward and north-eastward the great ice-cap prevailed over which the surface air could at best be warmed little above the freezing-point. But not far from our shores the salt Atlantic water remained unfrozen and indeed it seems likely that off Portugal its temperature was not much below that of the present day.

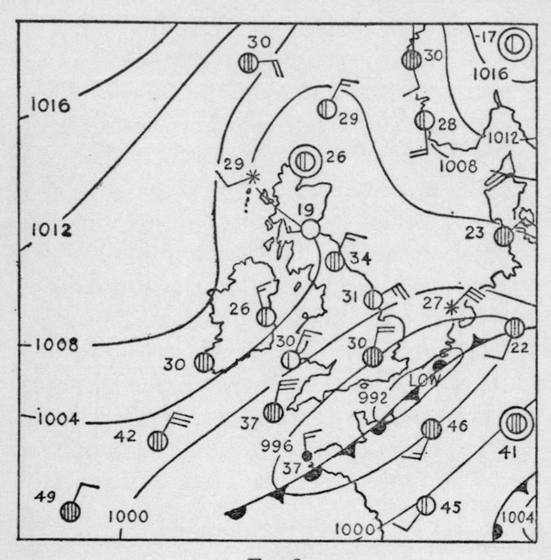

Fig. 63
0600 hrs., 6 March 1947

Hence we can arrive at a scheme of winter and summer weather. Milder Atlantic air would meet the polar air very commonly just to the south-west and south of Ireland. Deep depressions following this polar front would give very heavy precipitation in the south-west, but during much of the year rain would often fall rather than snow. Even in winter, with a warm open sea but a few miles to the southward, rain would occur now and then sufficient to wash away much of the snow on low ground; the like occurs nowadays in extreme S.E. Greenland. We might thus hazard a climatic explanation for the remarkable survival of certain relatively delicate plants in S.W. Ireland, while not far away behind the Killarney mountains to the north-eastward much of the country lay under ice. For the mountainous south-western promontories would be almost constantly under heavy cloud; and in clearer weather, almost incessant stormy and damp winds would hinder the ponding of the cold air close to the coast. At the present day, by the gloomy fjords of western Tierra del Fuego, stunted evergreens survived in an appallingly cloudy, wet and chilly region in which it is nevertheless rare to record more than ten degrees of frost. A hundred miles to the eastward, in the lee of the mountains, the climate is far drier and much colder in winter; persistent lowland snow-cover and temperatures around zero are quite common.

Similar results would be found in Cornwall and Devon. The onset of milder air from the not-distant Bay of Biscay was evidently frequent enough to prevent the winter accumulation of snow on Dartmoor from persisting sufficiently to establish glaciers except perhaps on a very small scale. Yet, a hundred miles further north, the glaciers from the Brecon Beacons were sufficiently powerful to reach the Bristol Channel. The mild damp air from the seaward, flowing over the fringe of icy waters near our coasts and then across the barren cold land would give heavy low cloud and drizzle for much of the summer in Cornwall, Devon and along the south coast. In winter, the very sharp contrast with the cold air inland would be marked by vigorous depressions in which heavy cold rain, sleet or snow would fall depending on the distance from the warm sea water. An almost exact working model of the conditions of the Ice Age can be seen in the events of February 1947, in south and south-west England. (Cf. Fig. 63, p. 220). In the events which culminated in the great Midland snowstorms of 5 March, in place of the expected thaw whose northward travel did not extend beyond the Thames, we can see why the limit of the ice sheet in one glaciation almost reached the Thames, while in later advance it scarcely spread south of Yorkshire. For thousands of years the battle between tropical and polar air was repeatedly fought out over Southern England.

Farther north almost endless snowstorms would be expected in winter along the western fringe of Britain associated with depressions making their way up the west coast along the chilly half-open seas. So marked, however, were the maritime effects that even in the Outer Hebrides the slopes of the higher hills appear to have been free from ice; and to have allowed the survival of a number of hardy plants. Farther to the eastward these coastal snowstorms would be less intense, and with distance from the open sea the weather throughout the year would become appreciably drier and more settled, with some sunshine, especially in spring and early summer. As a result the quantity of snow farther to the eastward was not sufficient to cover so much of the high ground throughout the year. In summer, parts of the Peak and North-East Yorkshire stood above the ice as broad, barren, frost-shattered uplands with occasional exceptionally hardy Arctic plants. Farther south, although the Downs were free from glaciation there is still some reason to suppose that in the shady hollows facing north-east persistent snowdrifts remained throughout most years. Some consider that such snowdrifts played a part in sculpturing the hollows by 'nivation', not only there but along the Marlborough Downs and perhaps elsewhere. Such coombs are clearly to be seen along the scarp of the South Downs near Eastbourne. Not all authorities, however, are agreed on this point.

Other evidence agrees with the meteorological reconstruction, according to which few depressions would penetrate to north-central Europe with sufficient energy to give heavy rain. Throughout Central Germany and even in North-East France the widespread 'loess' soils derive from the immense clouds of fine, dusty soil raised in a dry climate by the cold north-easter flowing outward from the ice-cap, especially in spring. Just as in the Fenland to this day the chilly north-east wind of spring, flowing over the warm black soil on a sunny morning, becomes turbulent as a result of surface heating and raises clouds of the fine soil, so in the past; for most of the phenomena of the Ice Age we have to-day admirable working models. One difference between conditions then, and the initial stages reproduced in February 1947, lies in the much greater snowfall of the eastern flanks of our uplands in the north compared with the west. This characteristic feature of modern snowy winters arises from the presence of the broad North Sea; in glacial times such a sea was not in existence. The greatest snowfalls probably occurred throughout our western highlands in a manner reproduced for us by the events of 29 January 1940 and 12–13 March 1947. On the former date the snowfall along the western flanks of the Pennines and the Lake District was so heavy that the southbound L.M.S. express was held up for 36 hours a few miles south of Preston. In mid-March, 1947 a rather similar "south-wind snowstorm" affected much of north-west Britain, especially on the uplands. It may be added that a minor event of the same type befell

---

PLATE 31
LOCH ACHTRIOCHTAN, Glencoe, Argyll: August evening. Intermittent sunshine with patches of characteristic rather disturbed strato-cumulus and cumulus at 3,000–4,000 feet among mountains.

PLATE 31

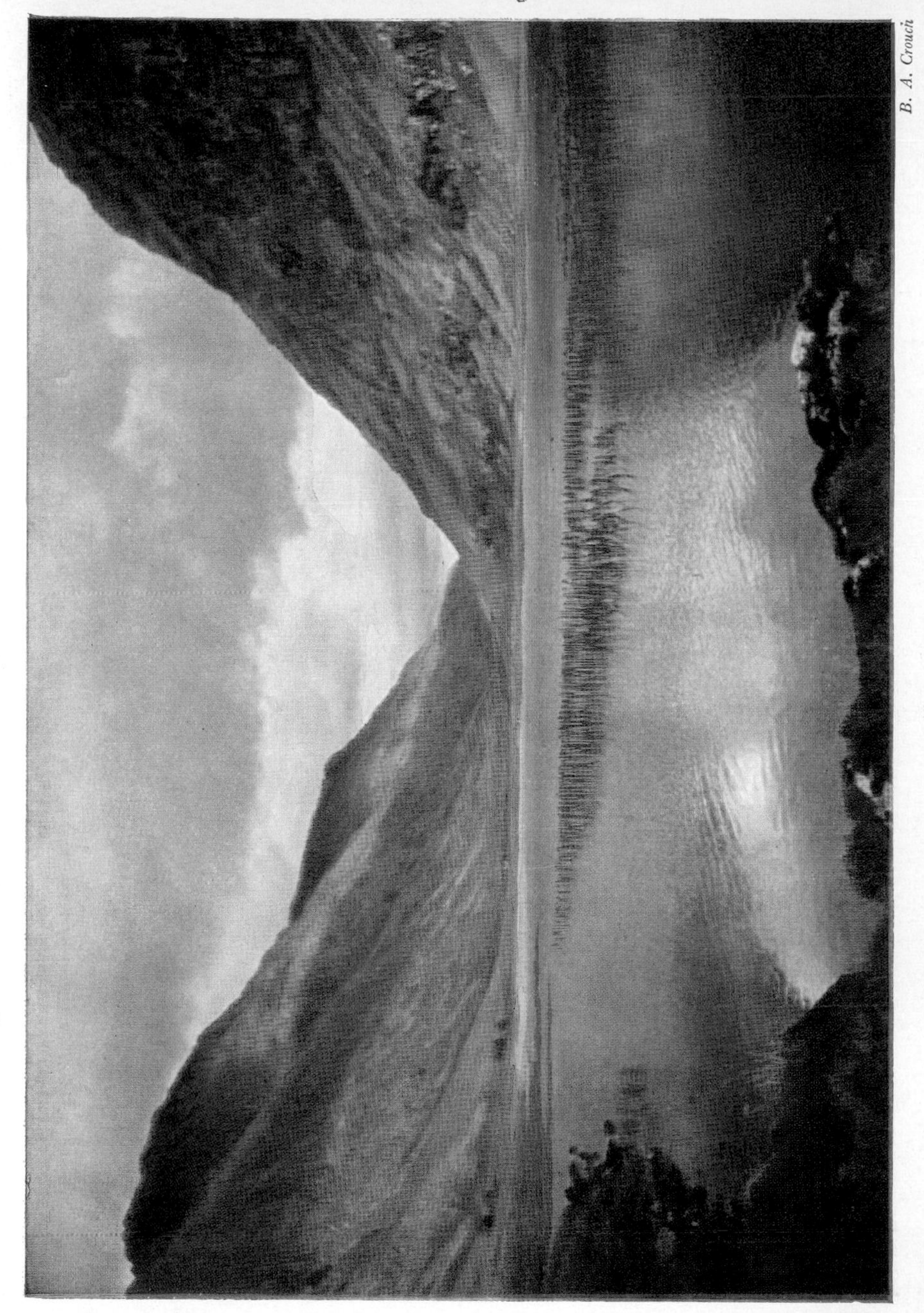

B. A. Crouch

PLATE 32

Anne Jackson

on 6–7 January 1947; with a nearly stationary front running N–S, through the Irish Sea, the south wind on its eastern side gave about nine inches of snow in Borrowdale, but very little east of Penrith, or on the Pennines beyond.

Estimates have been made of the extent to which temperature in England at the height of the Ice Age was lowered compared with the present day. In the extreme south the July average was probably in the middle forties. Farther north the presence of such a large ice sheet would imply that at sea level in Eastern Scotland the summer averages were a few degrees above freezing-point, while in winter they may have lain between 10° and 15°F. At the latter season severely cold clear weather and stormy cloudy weather with a much higher temperature alternated; the increased amount of cloud and wind towards the west of Scotland would there lead to milder winters, just as at present. It seems likely that the difference in average winter temperatures between, say, Edinburgh and the Hebrides was of the order of three times that which prevails to-day. Yet even with regard to Eastern Scotland it must not be forgotten that a largely open and very stormy sea lay at no greater distance and separated by a smaller mountain barrier than that lying between Ostersund in central Sweden and the Norwegian coast, across which the present-day January temperatures differ by 12° or thereabouts.

From the meteorological standpoint there would also be a considerable improvement in the weather, away from the Atlantic. Deep depressions would be found travelling eastward towards the Bay of Biscay, giving in Northern France heavy precipitation, often as rain. The existence of a massive surface layer of cold air over most of the British Isles would ensure that many warm fronts in winter would remain nearly stationary near the south-west coasts. The mild Atlantic air ascending over the cold surface air would give snowfall much more frequently than rain in districts such as the English Midlands.

But at intervals other depressions would pass northward along the partly open sea west of Scotland. The air in their warm sector would

---

PLATE 32

WESTMORLAND: June afternoon. Normal fair-weather strato-cumulus and cumulus of turbulent humid westerly current in early summer, ascending over hills. Kent estuary in foreground.

however be considerably chilled by its northward journey over the cold icy waters, or up the ice-covered plain where we now find the Irish Sea. Probably many of these depressions would rapidly decline in vigour, although the frequent sweep of the cold moist south wind along the western isles of Scotland and beyond to the coasts of Norway would still, as we have seen, give frequent and heavy snowstorms and very extensive low cloud. Following the northward passage of such depressions cold west to north-west winds would prevail with frequent snow showers on our mountainous coasts; a situation resembling that at the present day, save that the air was probably ten to twenty degrees colder. We can imagine ourselves sitting on the Donegal mountains in June. For miles around we should see the gently-sweeping contours of the ice, covered by melting snow at the higher levels, grey and rather dirty-looking below. Overhead, a broken sky of ragged snow-laden cumulus with hard blue gaps; here and there in the distance the curtains of blinding snow showers passing over the higher hills, but falling in wet and heavy flakes on the greyish bare ice around Londonderry. Towards the Atlantic the dark water sky from which the raw west wind blew would be seen; eastward, the white glare on the horizon would remind us of the clearer skies and sunshine extending over hundreds of miles of lifeless waste, only broken by the dark streaks of rock on the sunny slopes of the Isle of Man, a few struggling lichens perhaps on Crossfell, and beyond the bare stony warmth of the hills of North-East Yorkshire from which even the gleaming Cheviots might often be seen refracted upward on days of calm clear air. The photograph (Pl. XVIIIa, p. 227) of a retreating glacier on a cloudy day in Central Norway has been chosen as we may expect that something very similar would be found occupying the upper Pennine valleys during the waning of the last phase of the Ice Age, about 15,000 years ago, and rather later in the valleys of the Scottish Highlands. There would also be many days of widespread low cloud, such as we often now find covering the gloomy basaltic fjords of North-west Iceland.

Nothing indeed fires the imagination more than the evidence of the immense range of variation in the effective climate of these islands over a relatively short space of time. In countries such as Norway small glaciers still exist and the ice age does not appear so remote. In the lowlands of France there was no glaciation at all, and although pine and birch replaced the deciduous forest of the present day, there was still abundant life. But in Britain the change to a temperate climate to which an astonishing range of plants can be adapted from one in which almost the entire realm was lifeless, is the more startling. Yet it agrees well with the evidence of our eyes; that under British climatic conditions a very short journey is needed to produce a remarkably effective change in the environment. No definition can be given of such a climate; no precise classification can apply. Suffice it to remind ourselves that even to-day, although many exotics flourish in mild coastal locations, our tree-line lies at a mere two thousand feet. To find a similar contrast it is necessary to climb six thousand feet or more in the Alps or the Rockies.

Post-Glacial Changes

Having reviewed the circumstances of the greatest of the climatic vicissitudes of which we have evidence we might now consider the minor, but none the less significant variations under which the evolution of the present inhabitants of the British Isles has proceeded. Primitive varieties of the present human stock certainly existed in England in the interglacial period between the last major advance of the ice and the last but one; they may have been here in the previous interglacial period, retreating to warmer lands during the tens of thousands of years through which each ice-cap persisted. But from many points of view we can consider that the slate was virtually wiped clean when, about 20,000 years ago, the final retreat of the last great phase began. Southern England and the Midlands were then and for long remained a tundra; ice had not spread quite so far as in previous glaciations, but except perhaps in shelter in the extreme south the climate was cold enough to prevent the growth of trees other than the dwarfed Arctic birch and willow creeping here and there along the ground. Hence practically all the vegetation and the animals we now know (including the more evolved varieties of man) have made their way into Britain within the last 15,000 years.

Within the great barren tundra-heath lying to the south of the retreating ice-sheet drainage was often poor and immense hollows existed in which the slow growth and decay of boggy vegetation led to the formation of peat. As the climate gradually became warmer trees spread. First there came the hardy tree-birch followed by the resistant pine, both of which we still find advancing up the valleys towards the highest levels at which trees will grow in Norway, and northward towards the shores of Lapland by the Arctic sea. The pollen from the trees growing in the neighbourhood of bogs and of small lakes was carried over them by the wind, and some deposited. Now, by microscopic analysis, the number of grains of pollen representative of several different kinds of tree at any given depth in the bog, or lake-bottom deposits, can be counted, hence the proportion of different trees in the surrounding woodlands at various times during the past ten thousand years or so can be estimated. This ingenious technique, largely developed since 1921, has given us for most of Northern Europe with the British Isles the character of the plant succession and hence the general nature of the climate. For a predominance of birch and pine indicates that while the winters were still cold, summers were relatively warm and dry, such as we now find at 2,000 feet in eastern Norway. At slightly higher levels towards the surface of the bogs the proportion of pine increases further, then it decreases and the proportion of alder pollen becomes large, together with hard woods such as oak and elm. Without doubt this indicates a change to a considerably wetter climate with milder winters.

In some parts of Denmark (Allerød) and locally in South Britain at least as far north as Berwickshire the late-glacial and post-glacial deposits reveal a narrow layer containing a little tree pollen between wide layers which were laid down where the landscape was devoid of trees. This warmer phase is often known as the "Allerød oscillation" and appears to have lasted for upwards of a thousand years. From this and other evidence we are led to assume that the amelioration of climate following Glacial times was not at all regular. We can imagine the tide of vegetation advancing northward as the summers grew warmer, but then being thrust back for some hundreds of years before the advance was resumed. Recent work by Winifred Pennington on the sediments of L. Windermere has led to an approximate dating in agreement with the Swedish geochronology; roughly 10,000 to 8,850 B.C. for the warmer phase and 8,850-8,200 B.C. or thereabouts for the reversion to a colder climate. It appears probable that this was the period when the last small glaciers descended into the heads of the Lake District valleys, leaving their moraines for example near Stockley Bridge on the path from Seathwaite to the Stye.

To a certain extent those climatic fluctuations can be correlated with other events. The increased dampness accompanied by much

C. H. Wood

PLATE XVII: Snow in Victorian Bradford, February evening.

Plate XVIII: *a.* Norwegian glacier in retreat, Jotunheim, September 1949 (Styggebre above Spiterstulen). The same appearance would be presented by the retreating glaciers in the upper Yorkshire Dales possibly about 12000 B.C. (*Gordon Manley*).

b. The Ben Nevis Snowbed in the Observatory Gully at minimum, September 22, 1948, with with the first autumn powdering of snow. The bed is about 60 yds. in length along the gully and 40-50 yds. across; estimated depth in centre is 25 ft. (*Gordon Manley*).

milder winters, shown by the rapid and widespread increase of alder, oak and elm, accompanied the changes in the extent of land and sea. A general rise of sea level was associated with the opening of the Baltic to Atlantic water and a considerable enlargement not only of the Baltic but also of the early North Sea. These events are placed by recent workers about 6,000–6,200 B.C. and readers of Professor Stamp's work in this series will recall his suggestion that the rapid amelioration of climate owed much to the fact that the circulation of salt ocean waters round Britain became freer. For at some time during this period the evidence indicates that the Straits of Dover were also finally opened. But as the general rise of sea-level was largely the result of the world-wide melting of the great ice-caps it is not easy to disentangle the primary cause of such extensive and rapid melting, remembering how long the ice had been in existence and the intermittent character of the subsequent climatic recovery to which the earlier Allerød oscillation testifies.

A rough time scale is given in the table.

SUMMARY TABLE

VARIATIONS OF CLIMATE IN BRITAIN DURING THE PAST 15,000 YEARS

Before 14,000 B.C.	Ice Sheet slowly retreating over Scandinavia and N. Britain with halts and brief readvances at times. Tundra and heath throughout S. Britain, birches slowly advancing in S.E. Climate cloudy, windy and raw, with minor oscillations.
Before 11,000 B.C.	Spring and early summer relatively dry. (July mean temperature about 45°, Midlands). Windermere becomes a lake after retreat of ice.
After 10,000 B.C.	'Allerød phase'. Climate appreciably warmer. Tree birches reach Berwickshire, Lake District and Central Ireland. (July in the Midlands about 54°). Occasional permanent snowdrifts among mountains, but probably no glaciers S. of Scotland.
About 8,850 B.C.	Deterioration. Birches retreat S.E., tundra in N. England (July—N. Midlands—about 48°–50°). Re-establishment of small glaciers in Lake District, Wicklow and N. Wales; extension in Scotland (the 'Highland Readvance') lasting about five centuries.

Following 8,200 B.C.	Improvement to 'Boreal'; birches spread N. followed by pine; summers steadily warmer; sea becoming warmer off British coasts. N. Sea still largely dry land, winter severe at first, milder later, but still drier and colder than today. Weather probably more anticyclonic than now. Elm and oak spread later in period: summers quite warm at end. Final disappearance of remaining mountain glaciers.
About 6,000 B.C. and subsequent millenia	'Atlantic' phase; sea level rises considerably, North Sea broadens; possibly final opening of Straits of Dover. Summers remain warm but ground generally more moist, winters milder than to-day. Rapid and extensive spread of oak, alder and also lime replacing pine and birch. Climate gradually attains a degree of breezy warmth comparable with warm, moist and cloudy years (1834, 1852); July perhaps 65° in N. Midlands and January in lower forties. Mountains decidedly wet and windy; trees in Highlands not found quite so high as in subsequent drier phase.
About 3,000–2,500 B.C. and onward	'Sub-Boreal': in the north, drier and less windy. Winters drier with some frost, summers rather warmer than at present, equally rainy in S. Birch and pine prominent again on drier areas, also up mountains to about 3,000 feet in Central Scotland, but not so high nearer coasts. Upland settlement widespread, glaciers in Central Norway probably disappear. Decided fluctuations at intervals but not apparently of great length until :—
About 500 B.C.	Climate again much damper with considerably cooler and more cloudy summers, less evaporation, more wind and rainfall. Rapid growth of peat over previously forested uplands especially where less well drained. Tree-line lowered by perhaps 1,000 feet. Birch increases in lowlands and in damp sites oak, alder and willow especially prominent. Summers perhaps 4° cooler than previous phase, winters still rather mild due to much wind and cloud ('early Iron Age').

Historic. Possibly minor amelioration and recession in Roman times; improvement about 7th and 11th century, wetter around 1100, again more disturbed after 1300. Minor fluctuations with tendency for colder winters after 1550; tendencies probably more or less similar to those shown by Fig. 64. Minor drier and wetter groups of years in S.E., but uncertainty how far these are applicable in N. and W. Prevalence of colder winters in later 17th century, and recurrence 1740 onward; groups of generally warm summers, e.g. 1772-83; and cool, 1692-1700, 1809-18. Tendency in direction of milder winters since 1850 or earlier but not uninterrupted. Appreciable increase of average temperature in spring, summer and autumn since 1930. Despite 1959, the peak may have been passed.

Post-glacial forest history in Britain pursues a broadly parallel, but by no means identical course in different areas. The general picture however, is clear; that of the march to and fro of different types of forest as a direct response to changes of climate.

A climatic optimum was reached upwards of 2,500 years B.C. since when in spite of partial recoveries there have been several small but significant steps towards greater coolness in N.W. Europe. Perhaps the most notable of these steps is represented by the deterioration which set in about 500 B.C. The intervening slight ameliorations of climate in the direction of greater warmth and dryness have been interrupted by further slight recessions and if we accept the Scandinavian dating as being broadly applicable to much of Britain we may put the recessions at about 400 and 1300 A.D. It is however probable that the effects on temperature and rainfall in these last fluctuations were not greater than those we now experience between groups of warmer and drier years, and later groups tending to more unsettled and windy summers with more cloud and less evaporation. Minor fluctuations of this type lasting for a decade or so can be recognised even in the past two centuries of instrumental recording.

These very significant variations within the period since Neolithic times are important. On what do they depend? That we cannot say as yet; perhaps a clue to the problem will eventually be forthcoming when the smaller fluctuations within the 'instrumental' period have been fully analysed. For although the variations have not been large, even within the last 250 years short spells have occurred which, if they persisted, would be likely to affect the natural vegetation and it is a matter of deciding how such spells might become more lasting.

We can summarise the effect of these climatic fluctuations on the habitability of the British Isles. At a time when the earliest civilisations were well developed in lower Mesopotamia and Egypt, between 4,000 and 3,000 B.C., we can imagine a decidedly damp and mild Britain heavily forested except in limited hilly areas where exposure to wind, better drainage and the character of the soil encouraged a more open vegetation. Mesolithic cultures may still have been represented by scattered groups living here and there towards the sea-coast, for example, on the limestone uplands of East Durham. Neolithic peoples were about to arrive, possibly across a narrow strait of Dover, or making their way along the North Sea and Atlantic coasts. Among other things they brought with them was the art of pottery; and the distribution of the fragments they left gives us a good indication of the routes they followed towards the more favourable areas for settlement.

Slowly the climate became drier, probably not without interruption; and the dense tangled woodlands throughout the clayey valleys for long remained an obstacle. Fifteen hundred years later we find the evolved men who built Stonehenge, who quarried for flint so successfully in sandy, dry West Norfolk, who for some time had cultivated grain and were beginning to know of the use of metal, still living on the uplands and on the naturally-drained areas where trees were more scattered. In Northern Britain the evidence indicates that about 2,000–1,500 B.C. the climate was drier than at present. Examination of the peats indicates that again the pine and hazel were spreading; and other evidence which it is tempting to relate to climatic factors can be found. In Dorsetshire the recent excavations at the great ramparts and earthworks of Maiden Castle show that for several centuries it was occupied as a pre-historic fortress-site by a relatively accomplished people until a period indicated by the finds of bronze weapons as somewhere about 1,500 B.C., perhaps two centuries or so after the erection of Stonehenge. Then it was suddenly abandoned, and left unused and untenanted for twelve hundred years, until about 300 B.C. the obvious advantages of such a defensible site again appealed to later Celtic-speaking tribesmen of the Iron Age. Now it is tempting to think on the dry chalk downs the abandonment of the site was enforced by the lack of sufficient water; as the winter climate became drier, the water-table in the underlying chalk would fall below that to which wells could be dug. That the summer temperature and rainfall was not seriously changed in these extreme southern parts of England is however suggested by other evidence. During the Maiden Castle excavations under Dr. Mortimer Wheeler large numbers of charcoal specimens were examined dating from Neolithic times, the Iron Age and Roman times. The width of the annual growth rings does not differ significantly from that of present-day representatives of the same species.

Yet, at about the same period, the presence of Irish gold ornaments in Scandinavia is attested, and points to the development of a good deal of traffic across the North Sea. It is argued that such navigation in the very primitive vessels of the time would only be practicable if the weather was quieter than we should now expect. The meteorology of the period might plausibly be explained by analogy with the situation during occasional dry years such as 1887, 1911, 1921, or better still, August 1947. The Azores anticyclone, well to the north of its normal position and extending far to the north-east, would give dry quiet weather with light variable winds throughout Northern Britain and across the North Sea to Norway. But in Central Europe and also in Southern Britain thundery summer rains would still contribute, as they do to-day, an appreciable proportion of the annual rainfall.

As the vegetation of Central Europe depends largely on the summer rainfall we should not therefore expect such marked evidence of drought, even in S. England, as might for example appear in north-west Scotland.

If during the winter months the Azores High still tended to occupy a much more northerly position than at present, Southern England would again be decidedly drier at that season than now. We may seek an analogy in the phenomenally dry, mild February of 1932. For the replenishment of the chalk in Britain it is the winter rainfall that is necessary; and the phenomena of the Bronze Age in Britain and Scandinavia might reasonably be explained on the assumption that the Azores High spread farther north both in winter and summer than at present and was more persistent. Icelandic depressions probably were less vigorous, and also moved on tracks well to the northward. Such a situation might well arise if there had been a considerable decrease in the amount of ice in the Arctic Seas. This again would favour Western

Scandinavia in that winters would be less stormy; and the greater dryness would undoubtedly be of benefit to Ireland. Further the dryness of Ireland at this period is supported by the evidence; in a Donegal peat bog twenty-six feet beneath the surface, relics of a well-built log hut dating from Late Neolithic times, were discovered some years ago. Elsewhere, the growth of good timber in areas now completely covered by peat supports the view that the Bronze Age climate was not only drier, but less windy than now. Within the peat flanking the summit of Crossfell, stumps are found indicating that at some time trees grew up to 2,500 feet above the sea; and in Central Scotland trees are known to have reached 3,000 feet. Yet from the nature of the vegetation growing in Bronze Age times at lower levels it is clear that summer temperatures as a whole were at best but little higher than now. We have however seen (p. 184) that if in our uplands summer weather is predominantly anticyclonic the mean temperature deviates above the normal to a greater extent than in the lowlands. For example if in a dry sunny summer month the mean in the lowlands is, say, 2° above normal, that on the high Pennines may be 3° above normal. Such an excess of temperature relative to the plains below would be just enough to raise the climatic tree line by five hundred feet or so compared with the present. Without doubt the balance of the evidence favours the view that Bronze Age summers were over long periods considerably drier, calmer, and probably rather warmer than at the present day, particularly in Ireland, Northern England and Scotland. In the south of England, at least near the Channel, the changes were less marked. It is interesting to observe how in the fine dry August of 1947, the lack of rainfall was much more marked in N.W. Scotland than in the extreme south.

Evidence is as yet somewhat uncertain with regard to the happenings in Britain towards the end of the Bronze Age, say between 1,000 and 500 B.C. Following minor fluctuations, at the end of this period it is evident that especially towards the north and west the climate again became considerably wetter, with more wind and cooler summers, and greatly diminished evaporation. There is, literally, overwhelming testimony in the massive mantle of peat which now covers all our high summit plateaux unless they are for one reason or another especially permeable or well-drained. Everywhere in our wetter districts more or less peat can be found; and in the Irish plain it becomes particularly prominent in the ill-drained lowlands. It can be seen at its grimmest and most intractable on Rannoch Moor or, farther south, covering smaller areas on Fairsnape Fell and the Kinderscout plateau on the gritstone Pennines. The balance of opinion favours the view that practically the whole of our hill-peat developed in a few centuries about 500 B.C. In many places it is now being eroded probably by the action of the weather, more quickly than it is being renewed by the further decay of vegetation; but our upland plant-cover is a complex matter on which reference should be made to Professor Pearsall's *Mountains and Moorlands* in this series.

With regard to the 'Sub-Atlantic' the weight of the evidence suggests that in England the mean temperature of a summer month such as July was about 4°F. cooler than in the preceding optimum periods. This is not greater than the difference between a dull cool July nowadays, and a fine month. Compare July 1888 or 1922 (56° in the N. Midlands) with July 1887 or 1921 (62°–63°).

Exactly what caused this effective increase of wetness is still an unsolved problem. That it had many repercussions with regard to the movement of peoples is probable. Some have gone so far as to suggest that the great influx of northern peoples towards the Mediterranean, shown in the early invasions of Greece, can be ascribed to climatic causes. This is very uncertain; many factors besides climate have influenced Greek politics throughout all ages. In North-West Europe it used to be supposed that the invasions of Britain of the early Iron Age could be attributed to the more severe climate and particularly the colder winters about this time. Unfortunately more recent evidence of the spread of the beech supports the view that the winters were if anything a little milder than at present; but the summers were definitely cooler and more cloudy. The greater wetness and storminess is plausibly reflected in a marked decline in the energy and vigour of Scandinavia as shown by archaeological finds. There is however some danger in drawing too many inferences with regard to climate from the cultural characteristics of peoples. It will be interesting to know if our descendants in two thousand years' time will be persuaded that the arrival of an Iron Age in New Zealand about 1830 was due to starving refugees fleeing from the fearful winters of Napoleon's day.

Opinion in recent years has tended towards the view that for some reason as yet unknown, which may for example be connected with minor variations in the intensity of the sun's radiation reaching the earth, operating over periods of time which may amount to centuries, the vigour of the circulation of the earth's atmosphere varies considerably, at least around the North Atlantic and possibly more widely.

If the circulation is feeble, the great cold anticyclones over the continents and the polar regions in winter build up more strongly and are more persistent. Less frequently than usual are they whittled away by the surging streams of air on their flanks or over the top, whose energy depends at least in part on those differences of temperature over the earth's surface which in turn ultimately derive from the sun through the variable medium of the atmosphere. Correspondingly in summer the great anticyclones over the oceans will tend to spread their influence over a larger area, and farther towards the Poles. At the present time we can observe such tendencies at work in individual seasons. In February, 1947, extremely persistent high pressure developed over the lands and seas of the Arctic. In the droughty hot summer of 1921 high pressure extended with extraordinary obstinacy from the Azores to the Baltic; this was in part repeated in 1949 and 1959. Possibly such tendencies developed over longer periods of years, arising out of slight variations in the general vigour with which the surface circulation was maintained; the dry quiet summer warmth of the Bronze Age might thus be explained. Earlier still, the amelioration of temperatures indicated by the spread of tree-birches during the "Allerød oscillation" is again of the same order of magnitude as that which we find to-day between quiet anticyclonic summers and cool windy summers when the atmospheric circulation is more lively (*Geographical Journal*, March 1951).

But referring for the moment to the Northern Hemisphere a second factor in the argument arises from the great variations in the extent of the Arctic sea ice. If a period of feeble circulation happens to coincide with a very icy Arctic we might expect a group of very cold winters; but if there was very little ice, the consequence might be a 'Bronze Age' of dry warmth in the British Isles. For it is tempting to deduce from the evidence forthcoming from the North Pacific that in a region where the surface supply of cold Arctic water is impeded, due to the narrow Behring Strait, the oceanic anticyclone takes up a much

---

PLATE XIX*a*: Chalk country: cumulus cloud over the Sussex Downs. Compare also Plate IV*a*, p. 47.

b: Snow-laden winter sky over the High Chilterns, January 1936, near Ivinghoe Beacon. Trees show the effect of prevailing westerly wind.

PLATE XIX*a*. *The Times*

b. *E. L. Hawke*

H. Jenkins, Ltd.

PLATE XX: Trawlers going out at sunrise off the Norfolk coast, showing N.W. wind in late September with characteristic low cloud, breaking a little over the sea.

more northerly position in summer compared with the corresponding Azores High in the Atlantic. One result is that southern Vancouver Island now enjoys English temperatures in summer but with less rainfall and with 25% more sunshine.

It certainly appears that the vigour of the circulation, represented by the frequency with which the contending air-masses surged over the British Isles and North-West Europe, was much increased five centuries or so B.C. That the increased circulation of the Sub-Atlantic followed a phase of increased spread of the Arctic Ice and cooling of the N. Atlantic surface waters is a hypothesis awaiting further results of ocean research (Ovey, p. 250).

In Roman times the evidence favours the view that the climate of Southern England was still rather damper than at present; but it is extremely difficult to say how much of this was attributable to the extensive damp uncleared and undrained forest and marsh. Little of the country was yet cultivated, although clearance of forest had probably begun. We obtain from Tacitus and other writers a contemporary indication of temperature conditions little different from the present. The frosts of winter were in general neither so severe nor so prolonged as on the adjacent continent, and no mention is made of frozen rivers or harbours. Attempts to cultivate the vine met with much trouble and little success. As for Tacitus's mists and fogs, many a peninsular Italian would say very much the same thing to-day; and it may be added that Prussians are prone to remark on the rarity with which the cloudless blue summer skies of Pomerania are to be seen, even in East Anglia. With regard to the effect of lack of drainage, Brunt has pointed out that over heavy soils the retention of water close to the surface will favour a decreased diurnal variation of temperature. But the evaporation of water from the ground surface would produce a superimposed lowering of air temperature by day and night. Forests would tend to check the development over wide areas of very low winter minima, but again the warming of the ground in spring and summer would be hindered. Nevertheless it must be remembered that there was undoubtedly a good deal of cleared land and open heath even in Anglo-Saxon times, on which conditions might be much as at present.

In the Dark Ages following the Anglo-Saxon invasions there are few written records and the climate is as difficult to elucidate as any other aspect of that period. The weight of the evidence however favours the view that the period centred about the sixth–seventh century A.D. was again considerably drier and less stormy in N.W. Europe. The amelioration of the Scandinavian climate led to better and safer harvests and to an enterprising, vigorous and energetic population, increasing rapidly enough for many to be willing to seek their future overseas, and for others to display the splendid craftsmanship and incipient literary ability of which we have abundant evidence. Deplorable though this hearty accomplishment may have appeared to those who preferred the reminiscent ruins of the Mediterranean, it lies behind much of the present-day Britain. Many will reflect not merely on the natural landscape but also on the works of man. Lindisfarne and Whitby, the ancient trading streets of York and the fair-skinned matrons of quick-dealing pugnacious Leicester owe much to the spring-time north-easter round the Scandinavian anticyclone of twelve hundred years ago. Even the Cornish saints of the seventh century show signs of having profitably occupied the more permanent sources of water, in a period when the dry east wind blew more often than now. And in an Icelandic saga (Laxdæla) we read how Unn had ships built in Caithness where it would now be very difficult to find any trees well enough grown for the purpose. Pre-Conquest vineyards existed in England.

On rather scanty evidence it seems that the climate tended to become wetter soon after the Norman Conquest. Much of our evidence depends on entries in monastic chronicles; and an entry from a single abbey in Kent of "a bitter winter with deep snow" might not in any way represent the experience of a monk belonging to a less ascetic order in Devonshire. In the cold February of 1942, for example, the impression of wintriness was very different in the Plynlimon moorlands where there was but little snow, from that in Cambridgeshire. Careful investigations of all available medieval material have also been compiled elsewhere in Europe. To mention two; one inquiry concerns the frequency with which the Danish entrances to the Baltic were blocked by ice, while others have studied the frequency with which rivers such as the Seine at Paris were frozen, or gave rise to excessive floods at unusual seasons. The dates of harvest and of the vintage have also been studied.

Further, reports of the success or failure of the harvest from S.E. England would not necessarily reflect the course of events in, say Galloway. A comparative investigation of the annual rainfall at southern English stations with that in the mountainous West Highlands has shown that for the last fifty years or so there has been a decided though not uniform tendency for a wet year at Oxford to be normal or even rather dry in the Argyllshire mountains and vice versa, that is, there is no significant correlation between the fluctuations in these two regions. Reference should be here made by those interested to papers by Dr. Glasspoole in *British Rainfall*, 1925. No year showed this more remarkably than 1937, which ranked as a very wet year indeed in parts of Sussex (over 150% of normal) while as the maps in *British Rainfall* (1937) show, at Ullapool on the north-west coast of Scotland the year was phenomenally dry, with only 61% of normal. A similar distribution prevailed in 1958. We have elsewhere seen that for many places an annual rainfall of about 65% of normal is likely to represent the "driest year on record" over periods up to a century or so in length. In general therefore we must exercise much care in the interpretation of historic records of success or failure of crops. Further, with regard to the connection between the yield of crops and weather, temperature, rainfall and sunshine all play their part; a complex problem attacked by Hooker.

All manner of fragments of information have been pieced together. A study of the distribution of medieval water-mills in Kent indicates that for a considerable period about 1275 the flow of the streams was deficient. It is extremely interesting to observe that about this time catastrophic droughts beset the western United States in the dust-bowl region with which the events of the past decade have made us familiar. Later, the frequency of reports of climatic catastrophes of one sort of another becomes so marked as to suggest that the fourteenth and early fifteenth centuries were a period of great variability. For example, we hear of great dearth in 1315-16. It has recently been shown that great inundations arising from tidal surges in the Southern North Sea are associated with strong north-westerly winds accompanying the arrival of particularly deep depressions on the west coast of Norway. One of the most recent of such surges gave rise to the Chelsea floods in London in 1928. Many of the severe inundations of the later medieval period may have been associated with exceptionally deep Atlantic depressions of the same type. There also befell an extraordinarily severe winter in 1323, and a tremendous gale in January 1362 —this last may have been the greatest gale ever known in Southern England, save the 'Great Storm' of November 1703. Our first British weather record, indicating the frequency of wind and rain at Oxford between 1337–44, nevertheless points to climatic conditions very like the present, perhaps with slightly milder winters. But the evidence from Scandinavia and Iceland points to a decline towards less favourable conditions from about 1300. It is possible that some great gales have left their mark on the scenery to this day. The great breach of the dykes along the Cardigan Bay coastal marsh land may have taken place in the sixth century. Fourteenth-century inundations are recorded from the Solway as well as the east coast; great damage was done by inundations and blowing sand in Lancashire in the sixteenth century and round the Bristol Channel in 1607. Some of these events and their effects are noted by Professor Steers (1946); and we may note the possible effects at Dungeness on the building of shingle-ridges.

Later in the fifteenth century there is a little evidence of a return towards quieter and warmer conditions; among other things, the cultivation of the cherry spread northward, and it seems to have been known in county Durham in the sixteenth century, even at 800 feet. There is too, a long break in the reports of the freezing of the Thames at London—from 1434 to 1540. In Elizabethan days however, the climate as a whole appears to have turned a little cooler. Some mention should be made of the freezing of the Thames as no British river is better documented. The last year in which it was definitely passable on foot within London was 1814. But in 1820 old London Bridge was pulled down and later the Lambeth marshes were embanked; and as a consequence the tidal scour increased. There is little doubt that in any case with the same degree of cold the river is less liable to freeze over; the effect of old London Bridge for example was to jam the ice floes coming down the river. This is known to have occurred in 1740 and 1814. (See Pl. XXII, p. 271). Hence caution is necessary in drawing conclusions from these old accounts. In 1895 the river was so nearly unnavigable that without doubt it would have become completely covered under the earlier circumstances.

Mention of American data in a preceding paragraph will serve to remind many that the vicissitudes of annual rainfall, especially in the arid S. Western States, have been closely correlated with the successive growth-rings of trees. A chronology of the great droughts has been constructed from the evidence not merely of trees of exceptional age, but also from the beams used in the constructions of pre-Columbian buildings. Efforts have been made to extend this in more temperate climates both here and in America but the rate of growth of trees is influenced by too many factors to justify some of the earlier conclusions.

Explanations of the entire course of European historic evolution from the indications of drought or additional rainfall given by a very small number of Californian trees have long been condemned as too bold. Nothing in the diversity of Western Europe operating on a variety of peoples through five thousand years of history, lends itself very happily to the solutions appropriate to a continent of greater simplicity both in space and time; not even the weather. Yet with the slow assemblage of data from various parts of both hemispheres it appears that whereas some climatic fluctuations may appertain largely to the Atlantic margins others have affected the whole Northern Hemisphere and even the whole earth. For example the present retreat of glaciers is widespread. But with regard to climatic fluctuations we are still largely engaged in collecting and assessing the observational material. Some time must elapse before we can decide how far they are to be attributed to one or other of the several causes which have been brought forward in the past, such as changes in the amount of solar radiation reaching the earth's surface, or in the extent and elevation of the earth's land masses.

Careful studies have been made in recent years of the recorded movements of Alpine and Scandinavian glaciers. From these, the somewhat surprising conclusion has been reached that, since the final retreat of the last great glaciation, the greatest advance of the mountain glaciers of the Alps, in Norway and in Iceland has taken place within the past 400 years. The advance and retreat of glaciers has been shown to be influenced in the first place by variations of temperature. At intermediate levels in Norway, for example, a slight overall fall of temperature means that a greater proportion of the precipitation falls as snow rather than rain, and through the summer months there is a shorter period during which melting can occur to offset the greater accumulation of the past winter. A slight overall rise of temperature is best effected in a country such as Norway by an increase in the frequency of south-westerly winds from the warm Atlantic. This in turn commonly leads to increased precipitation, but often in the form of rain rather than snow; further for a great part of the year rapid melting is associated with the onset of the moist air, so that in the aggregate the glaciers retreat. Dr. Mannerfelt of Stockholm considers that the great retreat of the Ice Age was conditioned above all by a greater frequency of warmer winds from the open seas to the south-west.

In South-East Iceland, where the Vatnajökull ice-cap has been extensively studied, it has been shown that although over the last two or three decades the annual precipitation has increased by about 20%, the rise in the mean annual temperature of about 2°F. has been sufficient to lead to rapid retreat of the glaciers debouching from the ice-cap above. This ice-cap, the size of Yorkshire, lies close to an open sea; many of our conclusions with regard to the character of Ice Age weather in Britain are derived from knowledge of what happens on these smaller ice-caps farther north. Recent investigations support the view that many if not all the smaller glaciers in Iceland and Norway are not to be regarded as remnants from the Ice Age; they may well be the result of the deterioration which set in about 500 B.C. The semi-permanent Scottish snowbeds—our nearest approach to glaciation—appear to have responded in similar fashion.

The greatest advances of the Icelandic and Scandinavian glaciers are now dated; taking the overall view, it appears that they reached their maxima about 1745–1750; a second maximum was reached about 1850. Since 1890 retreat has predominated, but it has not been regular; there was a slight tendency to advance about 1920, since when the retreat has become very marked.

The course of events in Britain since Elizabeth's day can be roughly sketched. The material is surprisingly fragmentary, but it appears that her reign was characterised by a number of severe winters; the coolness or otherwise of the spring months remains to be examined. In 1564, the Thames was frozen; in January 1570 after the 'Rising in the North' had petered out the rebels were last heard of in the impassable snowbound Cheviots. In 1571 and again in 1574–5 Derwentwater was frozen, and the newly arrived Tyrolese miners had a lot of trouble with iced-up machinery. Stow mentions the extremely heavy snowfall of February 1579 in London, and there was another heavy fall late in April of the same year. The spring of 1587 was very cold; another very cold April befell in 1595. Summers were variable, and for example, 1594 was excessively wet; but the Armada summer of 1588 was so magnificently English in all respects that it cannot be wondered that the nation was thankful for its weather.

Many hints as to the weather come from the accounts of voyages. The cheerful departure from Rotherhithe of the Muscovy Company's first trading expedition on 10 May 1553 gives us one of the most delightful pictures from Hakluyt; but as nearly three weeks then elapsed before they left Harwich we may well presume that in that year the Scandinavian anticyclone with its easterly winds was not only late, but very persistent.

John Davis's quick passage to Greenland in June 1585 also hints at favourable winds; but as the scope of this book is confined to Britain we cannot go into the fascinating story of the Elizabethan seamen in the Arctic ice save to say that it was found in much the same place by the Victorians.

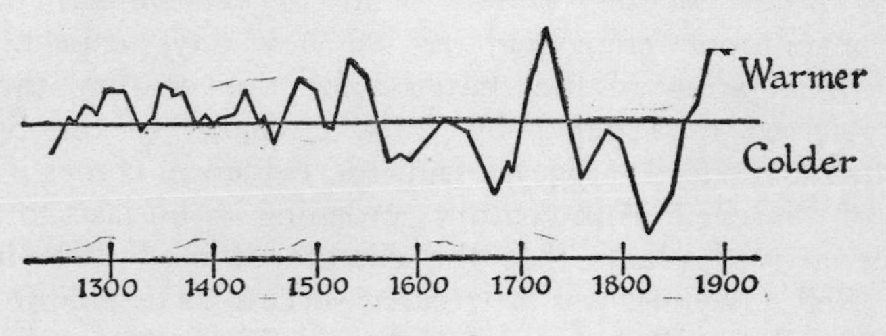

Fig. 64

Fluctuations in the mean winter temperature in N. Europe since 1250. Decline set in and prevailed about 1550–1680; short but marked amelioration about 1720: minimum early 19th century (based on Easton and Wagner)

A better testimony to the general character of the late sixteenth century comes rather from Denmark, when Tycho Brahe's careful observations included the direction of wind. The frequency in winter of winds from points east of south indicates that in 1582–1597 cold continental air then spread over Denmark more often than at present. His collected material has been considered to indicate a mean temperature for February and March about 2° or 3°F. colder than that of our recent decades, but these figures must be treated with a good deal of reserve. Nevertheless we may presume that a parallel effect, although probably less marked, would be found in England not dissimilar to that during the Napoleonic era, when a similar tendency prevailed, possibly a shade more extreme. Between 1579 and 1607 we have no reports of the Thames at London being completely frozen, as it was in 1795 and 1814. From the Alps there is evidence that the glaciers advanced considerably around 1600. It has even been suggested that the profusion of thick clothing characteristic of this era in

England and Holland owed much to the climatic conditions; but something must surely be allowed for a demonstrative age in which the technique of heating dwellings was still very imperfect.

Throughout the seventeenth century it appears that the fluctuations of climate were very similar to those we have experienced in our own lives. There were intermittent severe winters (1607, 1616, 1632, 1658, 1676, 1683-4, 1694-5, are noteworthy) and the concensus of opinion favours the view that following a poor wet year in 1648, Cromwell's Protectorate was favoured by several fine dry summers. That of 1652 was particularly fine in Scotland. Pepys' diary, and the many others of the period point to very much the same amount of outdoor enjoyment as we nowadays expect. For a detailed investigation of the material for this century the reader may be referred to a paper by J. N. L. Baker of the School of Geography at Oxford (1932). Indirect evidence comes from the attempts of the seventeenth century gardening enthusiasts to acclimatise newly arrived plants that the climate differed little from the present. Nell Gwynne was a greater success as a saleswoman of oranges than John Evelyn as a producer; this feature arising from our climate has never been overlooked by Londoners, from Covent Garden wholesalers to barrow-boys. We may continue to expect every effort to be made to persuade us that Italian peaches are to be set above Scottish raspberries.

At last we are coming to the period of instrumental recording already mentioned in the second chapter, and here we shall interpose diagrams to illustrate the trend of temperature and rainfall during the past three centuries. To these we may add the results of interpretation of other early data. It must be remembered nevertheless that before about 1800 all early instrumental records must be treated with the utmost care. The variety of errors to which meteorological observations are liable, when made with doubtful instruments by pioneer enthusiasts in questionable exposures is terrifyingly large.

From these diagrams and those for rainfall on pp. 266–268 the nature of the variations to which our climate has been subject will emerge. If with regard to temperature for example we smooth the curves by taking ten-year 'running means', we can get a better picture of the overall trend. It becomes evident at once that the temperature of the mid-winter months tended to be lower than at present over most of the early nineteenth century, whereas, spring, summer and autumn

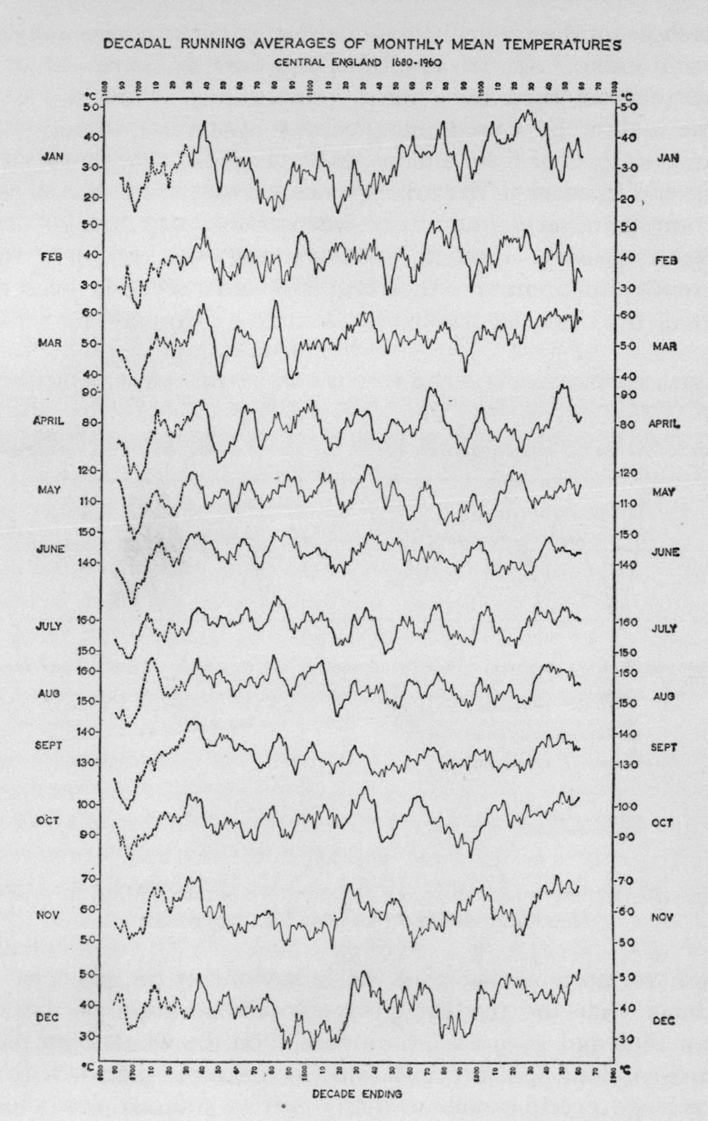

FIG. 65

Ten-year running averages of the monthly mean temperature for "Central England" from 1680-1960. These are broadly representative of the West Midlands; and before 1730 are much less certain as they are extrapolated from a distance.

months show on the whole little difference from the average experience of recent decades. At the same time, however, it appears that 'severe winters' (characterised by a mean temperature below say 34°, and therefore marked by a predominance of snow instead of rain, together with a good deal of frost and probably considerable skating on still waters) tend to occur in irregular groups. Using this criterion we shall find a number of severe months between 1688-1702, 1740-48, 1760-68, 1776-89, 1795-1803, 1808-20, 1826-30, 1837-55, 1878-97, 1940-47. The trends apparent in the English and Scottish records are supported by those in Holland, Denmark, Norway and Sweden.

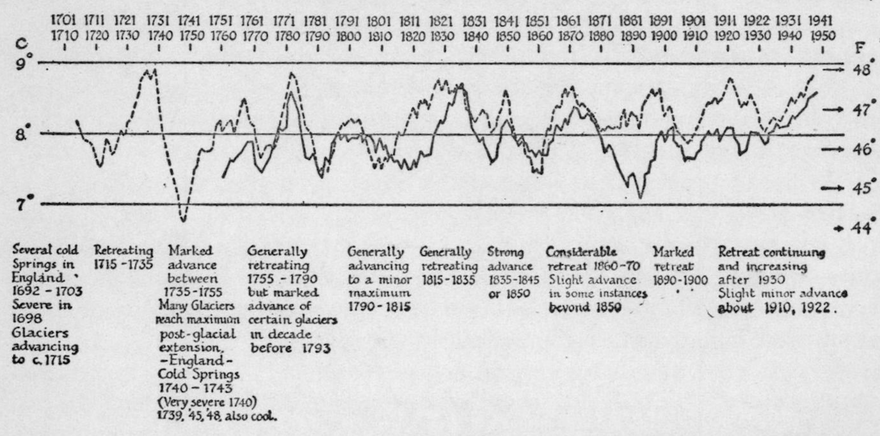

Fig. 66

Ten-year running means of spring temperature (March–May) in N. England and Holland (see also Fig. 73, p. 281)

Further, we must take note of the behaviour of the glaciers. There is evidence that the tendency towards amelioration of the winters between 1898 and 1939 was accompanied on the whole by a tendency to increased cloud and precipitation; for example, between 1911 and 1930 average precipitation at many British stations was about 5% higher than over the period 1881–1910 and the sunshine duration was slightly decreased. Since 1930 there has been an appreciable rise in the average temperature of spring, summer and autumn; but it would be unwise to assume a continuance.

Studies of the frequency of wind from various directions have shown that the 'resultant' in N.W. Europe lies for 1901–1930 slightly more to the westward of south than formerly. In Scandinavia these facts are interpreted as the accompaniment of a renewed 'rise in the vigour of the circulation'—more frequent, and probably deeper depressions finding their way along the immemorial track from south of Iceland into the Norwegian Sea and onward towards Spitsbergen or, sometimes, the Baltic; leading in turn to more frequent sweeps of the Atlantic air over that region, especially effective in raising the temperature in winter. These no doubt would affect Britain, especially in Scotland; we might appropriately say, the climate became more maritime during at least the three earlier decades of this century with a distinct reversal since 1930. In this period was a minor fluctuation represented by a slight advance of the Norwegian glaciers, within the general retreat, and in England by the rather cold winters and springs of 1917, 1919, and 1923 and the cool summers of 1919, 1920 and 1922. But the lack of consistency even in a brief spell of this kind is shown by the fact that in Southern England 1921 was particularly warm and dry.

While it appears that fluctuations of some magnitude develop, yet as far as we can see they are not necessarily periodic, continuous as regards any given characteristic or of uniform length. The picture is only slowly emerging. If in the decades when Stonehenge was new there were eight fine summers and but two were wet, we recall within our own experience the general opinion that in the wet 1920's and 1950's there were two very fine summers and eight wet. Yet rarely have the extremes been so great that there was even widespread failure of crops. Over four thousand years the amplitude of variation has not been large.

One of the most significant fluctuations which can just be recognised in the instrumental period has been mentioned, namely, the decided swing towards greater dryness, in Southern England at least, through the earlier part of the eighteenth century. Continuous series of rainfall measurements are available for Lancashire (1677–1704) and in Essex (1697–1716) but then there is a gap until 1727. From 1727 the overlap of each record can be compared with others in a continuous succession down to the present day. This is necessary in order to eliminate several possibilities of error in measurement. Although for many years there were but two gauges, they were so placed as to enable us to judge with considerable accuracy the extent of the variations over England as a whole. For Scotland however, we cannot begin a series until 1785, by which time quite a number of records were becoming available farther south.

All the indications show that the period 1700-1750, possibly 1680-1750, gave at least in southern and eastern England averages of rainfall distinctly below the 1881-1915 normal. Since 1727 the driest decades in England have been 1740–49 (86%) and 1850–59 (93%); compare the diagram on page 266, based on Dr. Glasspoole's deductions. Four years in succession, 1740–43 were phenomenally dry; fortunately, the rainfall occurred just sufficiently and at the right time for the grain crops. In the interval for which measurements are lacking contemporary writers name 1716–19 as very dry; and in S.E. England 1714 seems to have been one of the driest years ever known, comparable with 1788 and 1921.

It is very tempting to relate all this to the English scene. How much do we owe to the great agricultural improvements of that age when harvest after harvest served to encourage the enterprising farmer? The era of comely brick in the Norfolk market towns; the rising revenues of College estates; the speculative fever of the South Sea Bubble may owe more to climate than we think. Farther north, the widespread rebuilding of Lake District farmsteads from 1680 onward has drawn attention. Domestic buildings in stone had been erected for at least a hundred and fifty years; but all over the wetter Pennine dales signs of a quiet prosperity are to be found which together with miles of stone wall indicate that there was energy to spare. In the rise and spread of Quakerism; in the establishment of country banks; in the spare energy available leading to the widespread cultivation of remarkable observational talents by North and West country medical men climate may have played its part.

The drainage of fields and clearing of woodland no doubt assisted the generation of farmers who had to face the wetter years of the 1760's and early 1770's, the cold seasons of 1782, 1784, and 1799. Rising prices too made it here and there profitable to attempt the reclamation of land far up the hill sides in the Napoleonic Wars, in spite of the setbacks which must have beset the upland farmer in cold summers such as that of 1816. Eventually, however, the conflict of agricultural and industrial interests was resolved; the years from 1820 to 1845 were more often wet than dry and the succession of poor harvest and dearth did much to foster the repeal of the Corn Laws. In the north the rising industrial towns were largely recruited from the country population. The immense profusion of present-day surnames of territorial origin in Lancashire and Yorkshire industry cannot fail to be noticed by those who contemplate the names of the fellside farms on the one-inch Ordnance Survey.

The fluctuations of early industry were not unaffected by the vicissitudes of the weather. Shortage of water for mill-lodges and canals during the intensely hot dry summer of 1826 played a big part in the distress of that year, partially relieved by widespread road-building in Lancashire. As a further consequence we may note a considerable increase in the number of rain gauges kept on either side of the Pennines; incidentally, the predominance of dry summers in the 1850's led to a further interest in rainfall measurement and was followed by the establishment of the British Rainfall Organisation in 1860. Our earlier industrialists were dependent on weather in many ways; in 1838 a prolonged east-wind spell led to much delay and difficulty for the ships bringing the cargoes of cotton into the Mersey.

The diagrams will serve to illustrate the extent to which temperature has varied. Apart from the wide swings shown by the predominance of colder winters in the Napoleonic, the early Victorian, and the late Victorian eras, occasional exceptionally cold summers owe their character, dismal from the farmers' point of view, to volcanic eruptions in distant lands. Such were 1784, 1816, possibly 1845, 1860 and 1885; also 1902 and 1912. Preceding each of these years a violent explosive eruption of ash, characterised by the spread at high levels of minutely fine dust, resulted in a large part of the earth's surface being to some extent screened. It is in summer, rather than winter, that the effects of such screening of the incoming radiation become most prominent with regard to temperature.

After premonitory rumblings, the great Icelandic eruption of Laki (Skaptárjökull) early in June 1783 led not only to prolonged darkening of the sky but to a rain of ash which damaged the pastures; the series of catastrophes following this eruption was perhaps the worst in Icelandic history. About two weeks later observers in England, among them Gilbert White[1] and Parson Woodforde, began to

[1] Gilbert White's version is: "By my journal I find that I had noticed this strange occurrence from June 23rd to July 20th (1783) inclusive, during which period the wind varied to every quarter without any alteration in the air. The sun, at noon, looked as black as a clouded moon, and shed a rust-coloured ferruginous light on the ground, and floors of rooms; but was particularly lurid and blood-coloured at rising and setting." Letter LXV to the Hon. Daines Barrington.

comment on the hazy sky, so marked that "the sun appeared red until it was twenty degrees above the horizon". Ash also fell in Norway and the north of Scotland. July was extremely hot and sultry, but of the next twenty-one months in the north of England all but three were colder than normal; the effects of the cold summer of 1784 were very evident in Scotland, although slightly redeemed after September which appears to have been the only month with a mean temperature above normal during that chilly year. The chilly year 1816 followed a series of violent East Indian eruptions. The most recent, 1912, was marked by the coldest August on record at many English stations; summer throughout was also exceptionally cloudy, and when it was clear many observers commented on the remarkably pale blue colour of the sky. A violent eruption of Katmai in Alaska, in April, appears to have been the cause.

He who reflects upon any view of British countryside cannot but be struck by the extent to which it is man-made, except when one crosses the fifteen-hundred-foot level or takes refuge in some carefully restricted remnant of marsh, heath or dune. Man's effect on the landscape has however been influenced throughout history by climatic factors. There have been times when nature and man have been in harmony. The age of reason may also be the age of reasonable weather. In other periods the crops have only been brought home with great difficulty or anxiety; many have been led through desperation to seek a living elsewhere if no better solution presented itself at home. There have been sheltered corners where a secure livelihood with little effort could for long be counted on. Everywhere something of both clime and time can be seen in buildings. To this day the universal semi-detached villa still shows in the varied exotics of its garden the Englishman's conscious desire to improve and adorn his own bit of the island, while the poorly fitted windows and plumbing reveal the ever-present temptation to forget that extremes occur, begotten of the maritime climatic phase of the early twentieth century and the desire for quick returns. It was when the climate was most maritime between the wars that we were most tempted to disregard the continental menace. Who knows that there is not a link between our collective attitude of mind and the airs that blow upon us, bringing with them a reminder that somewhere beyond the wind are other men?

REFERENCES

Chap 12

AHLMANN, H. W. (1948). Glaciological Research on the North Atlantic Coasts. *R.G.S. Research Series, 1.* London.

(1949). The Present Climatic Fluctuation. *Geogr. J. 112:* 165-93: gives many further references.

ANGSTROM, A. (1939). The Change of the Temperature Climate in Present Time. *Geogr. Annaler Stockholm, 21:* 119-31.

(1946). *Sveriges Klimat.* Stockholm. Generalstabens Litografiska Anstalt.

BAKER, J. N. L. (1932). Climate of England in the Seventeenth Century. *Q. J. Roy, Met. S. 58:* 421-38.

BROOKS, C. E. P. (1930). The Climate of the first half of the Eighteenth Century. *Q. J. Roy, Met. S. 56:* 389-402.

(1949). *Climate through the Ages.* London, Benn, 2nd ed.: the standard work, first publ. 1926.

and GLASSPOOLE, J. (1928). *British Floods and Droughts.* London, Benn.

and HUNT, T. W. (1933). Variations of Wind Direction in the British Isles since 1341. *Q. J. Roy, Met. S. 59:* 375-88.

BRUNT, D. (1945). Some Factors in Microclimatology. *Q. J. Roy. Met. S. 71:* 1-10.

CHARLESWORTH, J. K. (1957). The Quaternary Era. London, Arnold.

FLOHN, H. (1954). Witterung und Klima in Deutschland. Berlin.

GLASSPOOLE, J. (1928). Two Centuries of Rainfall. *Met. Mag. 63:* 1.

(1931). General Monthly Rainfall over England and Wales, 1727-1931. *British Rainfall, 71:* 299.

(1937). Rainfall over the British Isles, 1901-30. *British Rainfall, 77:* 264.

GODWIN, H. (1941). Pollen Analysis and Quaternary Geology. *Proc. Geol. Assoc. 52:* 328-47.

(1947). The Late Glacial Period. *Science Progress, 35:* 185-92.

HARRISON, J. W. H. (1948). The Passing of the Ice Age. *New Nat. J. 1:* 83-90.

HOOKER, R. H. (1921). Forecasting the Crops from the Weather. *Q. J. Roy. Met. S. 47:* 75-99.

(1922). The Weather and the Crops in S. England, 1885-1921. *Q. J. Roy. Met. S. 48:* 115-38.

LABRIJN, A. (1946). Climate of Holland in the past two and a half centuries. *Med. Verh. K. Ned. Met. Inst. Utrecht, 49:* (No. 102).

LAMB, H. H. (1959). Our changing climate, past and present. *Weather 14:* 229-318.

MANLEY, G. (1946). Temperature trend in Lancashire, 1753-1945. *Q. J. Roy. Met. S. 72:* 1-31: gives many references to older data.
(1949). The Snowline in Britain. *Geogr. Annaler, 31:* 179-97.
(1915). The range of variation of the British climate. *Geogr. J. 117:* 43-68.
(1953). The mean temperature of Central England. *Q. J. Roy, Met. S. 79:* 242-261.
(1959). The late-glacial climate of N.W. England. Liv. Manch. Geol. Journ., *2*, 188-215.

MEYER, G. M. (1927). Early water-mills . . . E. Kent. *Q. J. Roy. Met. S. 53:* 407-429.

OVEY, C. D. (1950). On the interpretation of climatic variations . . . Atlantic deep-sea core. *Centenary Proceedings, Roy. Met. S.*, 211-215.

PEARSALL, W. H. and PENNINGTON, W. (1947). Ecological History of the Lake District. *J. Ecol. 34:* 137.

PENNINGTON, Winifred (1947). Lake Sediments of Windermere. *Philos. Trans. Roy. Soc. London, 233B:* 137-75.

SCHOVE, D. J. (1949). *In* discussion on post-glacial climatic change. *Q. J. Roy. Met. S. 75:* 175-79.

SIMPSON, Sir G. C. (1934). World Climate in the Quarternary. *Q. J. Roy. Met. S. 60:* 425-78. Also ibid. *83*, 459-80 (1957).

STAMP, L. Dudley (1946). *Britain's Structure and Scenery.* London, Collins' *New Naturalist.*

STEERS, J. A. (1946). *The Coastline of England and Wales.* Cambridge, University Press.

STOW, J. (1631, ed. Howes). *Annales: or a general chronicle of England . . .* London, 685.

WHEELER, R. E. Mortimer (1943). Maiden Castle. *Research Series XII, Society of Antiquaries, London.*

WHITE, Gilbert (1789). *The Natural History and Antiquities of Selborne . . .* London.

WRIGHT, W. B. (1936). *The Quaternary Ice Age.* London, Macmillan.

ZEUNER, F. E. (1945). *The Pleistocene Period.* London (Ray Society).
(1949). *Dating the Past.* London, Methuen, 2nd ed.

NOTE ON THE DIAGRAM ON P. 243

The recent discovery of meteorological journals kept since the late 17th and early 18th century has made it possible to estimate the mean monthly temperatures characteristic of the West Midlands from 1680 onwards. From 1680-1730, however, estimates are based on rather distant stations with very imperfect instruments, so that the early part of their answers should be viewed with caution.

CHAPTER 13

INSTRUMENTAL RECORDS: THE RANGE OF CLIMATIC BEHAVIOUR

I'll show thee
All the qualities o' the isle . . . . .

SHAKESPEARE: *Caliban, in the Tempest*

FOR THE benefit of those whose curiosity is better satisfied by figures there follow some comments on the extremes shown in our statistical tables. The upper limit of possibility of overall warmth in January seems to have been approximately demonstrated in 1796, 1834 and 1916. In each of those years mean temperature for Midland stations lay in the region of 45° (6° or more above normal) and the thermometer at many inland places scarcely sank to freezing-point, if at all. At a great many stations not even sleet fell.

Such warm winter months are easily explained; provided that a continuous series of depressions passes on a track from off the west to the north of Scotland we remain throughout in the milder type of maritime polar air with intervals of maritime tropical. Not only are the skies cloudy or overcast for long periods; the air has practically no opportunity of stagnating for more than a very few hours.

It is interesting to observe that if such conditions were to persist through the winter months the consequences would appear to resemble to a marked degree those of the Atlantic phase (cf. Chap. 12, p. 238) of warmth and moisture about 4,000–5,000 B.C. In summer, warm humid weather with a predominance of moist air from southerly points is represented by such a month as July 1852, or less markedly by July 1928. The July of 1852 was one of the hottest on record in Scotland and N. England; at the same time it was particularly humid with a good deal of rain. Mean temperatures at sea level were of the order of 64° in the wetter north-west; at London, nearer the continent and frequently in a drier and less cloudy air-mass, the mean exceeded 68°.

In other words if the continental and Arctic sources of cold air were absent or not effective, and yet the Atlantic air circulation remained lively, we might expect a mild moist climatic phase rather like the wetter parts of New Zealand, and probably rather more cloudy. There would be occasional slight frost; it would not be as warm as the Azores, but many delicate evergreen shrubs might be expected to flourish everywhere.

Under such conditions the orographic component in the rainfall of the western parts of Britain would become even more marked. To illustrate this we may consider the data for March 1921 and November 1938, both being south-westerly months almost throughout.

March 1921 Pressure 5 mb. above normal in Kent, 4 mb. below normal in Hebrides.
Mean temperature 2°–4° above normal everywhere, S.W. dominant.
Rainfall (England S.E., E. and N.E.) 47% *below* normal. (England S.W., N.W. and Scotland W.) 31% *above* normal.

November 1938 Pressure mainly normal in Kent, 11 mb. below normal in Hebrides.
Mean temperature 4°–5° above normal everywhere, S.W. dominant.
Rainfall (England S.E., E. and N.E.) 13% *above* normal. (England S.W., N.W. and Scotland, W.) 73% *above* normal.

In each of these months, as the diagrams, data and maps in the Monthly Weather Report show, winds from between south and west predominated to such an extent that temperature was well above normal. It is not difficult to imagine our climate changing slightly in this direction. If such a change were to occur with rather more emphasis over a longer period we should find that over a period of years the rainfall of all our western hill districts would be increased much more markedly than that of the lowlands farther to the south and east. Evaporation too would diminish; and assuming that the country was no longer artificially drained the result would be increased water-logging of the soil and eventually the formation of extensive blanket-bog and peat. This has been and remains characteristic of the wetter phases of the British climate.

Now let us look at the worst. If continental-Arctic air prevailed in winter, gradually turning to maritime-polar and maritime-arctic in spring, summer, and autumn, the consequences would be revealed in a series of monthly averages, for lowland stations inland, somewhat as shown below. (For convenience the actual months based on available long averages in which such extremes were recorded are also given.) It will be seen that the resultant would be very similar to the present climate of Iceland in winter, and a little warmer in summer. It might be still possible to grow oats or rye in the Midlands. Apples would be rather doubtful, and wheat almost impossible. Permanent snow beds would be found on Snowdon and in the Lake District; on Ben Nevis and in the Cairngorms there would after some years be small glaciers. There would still be abundant pasture and rather stunted trees at lower levels in England.

Never in the past two hundred years have such conditions, either of warmth or of cold, persisted through a whole year. In the warm years 1834 and 1949 they were nearly approached. On the cold side, the wet, snowy and chilly year 1695 appears to have been one of the worst we know of, while 1740 was dry and cold almost throughout. The period 1692-1701 was particularly unfortunate in Scotland, with its succession of cold late springs and inclement wet summers. In Scotland, the same combination of cold spring and wet cool summer rendered 1782 exceptionally bad; in this respect 1799 too, was bad. Indeed it is generally true that cloudy and wet cold years are the worst of all, such as 1879 in southern England—the year which, coming after a series of wettish summers, was 'the ruin of English agriculture'. In Southern England 1958 and 1960 have been wet, but not cold.

Anticyclonic weather with its accompanying warmth and dryness, rather than warmth and humidity give south-east England the highest summer average temperatures. To illustrate the possibilities of dry warmth we may name 1733, and the three successive hot summers of 1779, 1780 and 1781; subsequently 1826, 1846, 1868 and at least in the south 1911, 1921, 1933 and 1947 were outstanding. To these we may now add 1949, 1955 and 1959.

THE RANGE OF MEAN MONTHLY TEMPERATURE:
Warmest and Coldest Months on Record (1751 to date) °F.
(Lancashire Plain)

	J	F	M	A	M	J	J	A	S	O	N	D	*Year*
Warmest	45·3 1916	46·5 1779	48·3 1957	50·5 sev-eral	58·5 1833	64·0 1846	64·8 1808 1852	64·8 1947	61·0 1865	55·4 1959	48·5 1818 1938	46·4 1934	50·6 1959
Coldest	25·6 1814	28·9 1855	33·5 1785	39·9 1837	46·5 1782	52·9 1823	55·3 1816	54·5 1912	48·7 1807	42·6 1817	35·0 1782	30·0 1878	(44·5*) 1740
Average 1901–30	39·5	39·6	41·8	45·5	51·6	56·0	59·4	58·6	55·1	49·3	42·3	39·8	48·2
1921–50	39·0	39·5	42·5	46·5	51·7	57·0	60·3	59·7	55·9	49·8	43·3	40·0	48·8
Overall 1751–1950	34·4	39·0	41·2	45·9	51·6	57·1	59·9	59·2	55·2	48·7	42·2	38·9	48·0

*Estimated. Next coldest, 1879. Back to 1680, none of the above records appears to have been seriously surpassed.

MEAN TEMPERATURES FOR EACH MONTH:
Exceptional Years (Lancashire Plain)

	J	F	M	A	M	J	J	A	S	O	N	D	*Year*	
1834	44·0	42·1	44·4	45·8	54·8	58·6	62·5	60·5	55·7	50·8	44·1	42·0	50·4	Warm and rather wet
1921	44·7	41·1	44·5	46·1	51·5	57·4	63·0	58·5	56·5	55·3	40·5	43·6	50·2	Warm and rather dry
1959	34·4	40·8	45·1	48·2	55·3	58·2	61·6	61·9	58·3	55·4	45·1	42·7	50·6	
Normal 1921–50	39·0	39·5	42·5	46·5	51·7	57·0	60·3	59·7	55·9	49·8	43·3	40·0	48·8	
1799	34·8	36·4	37·6	40·8	48·4	57·5	58·5	57·3	55·0	46·5	40·9	34·3	45·7	Cold and rather wet
1879	30·5	36·8	40·3	42·1	47·9	54·9	56·2	57·8	54·0	47·9	40·1	33·3	45·1	Cold; wet summer

It appears that once in a century or so the mean July temperature within London reaches 70°; and between 69° and 70° at south-country stations in lowland valleys.

So long as Britain is an island surrounded by unfrozen seas the possible range of climatic conditions for individual months lies very nearly within the extremes illustrated by the last two centuries. For it is evident that with an unfrozen sea the lowest possible mean temperature in an island of our size can scarcely fall below about 24° at sea-level. This figure was nearly touched in January 1814; reasonably well-exposed thermometers at Perth and Carlisle gave means just below 25°.

Within this frame we may now look at the incidence of extremes with regard to maxima. The highest outdoor temperature we are ever likely to get in the hottest parts of S. England lies very close to 100° as in July 1808, 1868; August 1911, 1932. It must be remembered that there is always some doubt attached to extremes before 1880 or so as they were not always kept under conditions we should nowadays accept.

In Scotland, 90° has been very rarely reported (Leith 1876, Prestwick 1948) but so many stations in later years have had values of 87° to 89° (Paisley, Elgin, Kilmarnock, Liberton, Perth) that it seems likely that some part of Scotland should attain 90° perhaps once in twenty years, while in a city such as Manchester 95° might be attained.

The lowest values ever likely to be reached in exceptional locations under severe conditions are indicated by the records kept in certain well-defined frost-hollows when a deep fresh snow-cover was also present; and even in the London suburbs occasional minima below zero have been known, the first we hear of being in January 1684.

Many places in frost-hollows from Scotland to Sussex have in recent years recorded between −3° and −6°; but the further drop below −10° seems to be very rare indeed in S. England and rare in the Midlands and Scotland even in the most favourable sites; Braemar recorded −13° in February 1955. Indeed unless frost-hollows are chosen, temperatures below zero are rather rare and in the earlier nineteenth century grave doubts were cast on anyone who claimed that such readings had been obtained. In eighty years at Cambridge zero has been touched once and +1° twice (in February 1947); in a hundred years at Durham the minimum has fallen once only (1895) to 1° below zero; while at Stonyhurst, up on a slope above the Ribble valley, nothing below 4° has been observed since 1848. Cold air at this temperature may, however, spread over snow-covered open plains and down the broader valleys to the coast; Blackpool has recorded −1° (in 1881), Swansea +2° (in 1945), Southport +2° (in 1881), Barnstaple +3° (in 1945), Aberdeen +4° (in 1895).

It must be remembered that such extremes only affect a narrow surface layer of air in exceptionally exposed open valleys and plains. On upland ridges a mere three hundred feet above, quite normal minima may occur on the same night. At Little Rissington, 730 feet up on the Cotswolds, the lowest minimum for any of the exceptionally severe nights of early March, 1947 was 21°, whereas in the Severn valley figures round zero were recorded.

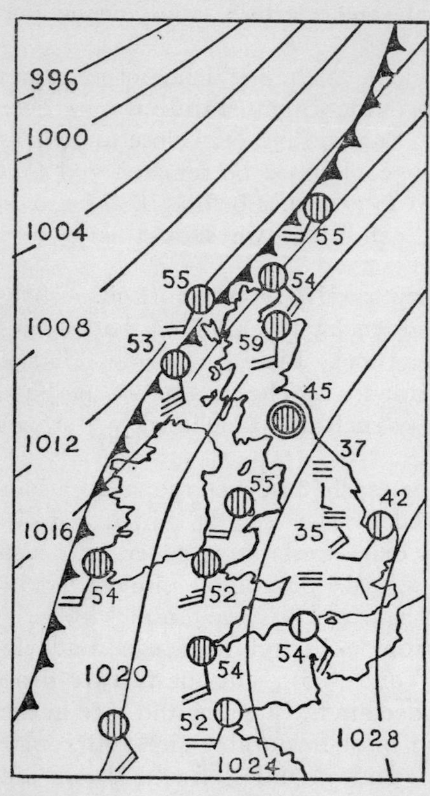

Fig. 67

Exceptional warmth with a south wind on the Moray Firth, 12 h., 27 November 1948 (59° at Lossiemouth); dense fog and low temperatures throughout S.E. and Midlands to N. E. England. Partial föhn in lee of mountains (Notation, p. 5)

More remarkable, because more dangerous, are the very severe frosts which can occur in spring; and in the other direction, some very surprising high temperatures have been known in quiet anticyclonic weather as early as the end of February. Maxima of 75° near Harrow on 9 March 1948 (map on page 88) and 77° at Wakefield on 28 March 1929 were as outstanding as a minimum of 5° near Penrith on 2 April 1919. Going back into early records, 67° was recorded at noon on the north wall of Gordon Castle as early as 10 March 1826. Without doubt this implies the possibility before the middle of March of a shade maximum over 70° in many parts of the far north of Scotland. If such things can happen 70° will occasionally be reached in February in southern England; perhaps once in 200 years. In the autumn, Cambridge had a maximum of 70° on 5 November 1938. As this exceeded by three degrees any previous November maximum anywhere in England since 1880, there is little doubt that such warmth in November is probably almost as rare as a similar occurrence would be at the end of February. For the highest mid-winter temperatures we must look to the possibilities of subsiding air combined with föhn as we have already seen (p. 130).

Some extremes for each month are summarised in the table below. Although some of the records are unofficial, a good general idea of the variability likely to be found at inland stations will be obtained.

LOW TEMPERATURES: RECORDED EXTREMES (°F.)

Month	Temperature, place	Year
Jan.	* −20, Berwickshire	1881
	−16, Kelso	1881
	* −14, Walton-on-Thames	1838
Feb.	−17, Braemar	1895
	* −13, Braemar	1955
	* −20, Aberdeenshire	1855
Mar.	−6, Houghall (Durham)	1947
	−9, Logie Coldstone	1958
April	5, Penrith	1917
May	15, nr. Thetford (Norfolk)	1941
	18, Braemar	1938
	19, Appleby	1891
June	22, Dalwhinnie	1955
July	28, West Linton (Peebles)	1926
	29, Scotland, numerous	1919–58
	30, Santon Downham, Norfolk	1960
Aug.	27, nr. Alston (Cumberland)	1885
	28, Glenlivet	1954
Sept.	20, Dalwhinnie	1942
	21, Braes of Glenlivet, Banffshire	1948
Oct.	11, Braemar	1880
	Dalwhinnie	1948
	13, S. Farnborough	1926
Nov.	−10, Braemar	1919
Dec.	* −20, Berwickshire	1879
	* −15, Cheadle (Staffs.)	1860
	* −11, Maidstone (Kent)	1796

(*approximate, reasonably authentic)

HIGH TEMPERATURES: SOME EXTREMES ON RECORD (°F.)
(EXCLUDING LONDON)

Month	Temperature, place	Year
Jan.	63, Aber., 1929; Rhyl	1916
	62, Durham	1888
	61, Wrexham	1944
Feb.	67, Cambridge	
	67, Barnstaple	1891
Mar.	75, Wealdstone	1948
	77, Wakefield	1929
	74, Haydon Bridge	1957
April	83, Cambridge	1893
	81, Peterborough	1945

HIGH TEMPERATURES: SOME EXTREMES ON RECORD (°F.)
(EXCLUDING LONDON)—*continued*

Month	Temperature, place	Year
May	91, Tunbridge Wells, etc.	1944
	89, Mildenhall (Suffolk)	1947
June	95, Rickmansworth	1947
	94, Waddington (Lincs.)	1947
	95, Northolt	1957
July	96, Cambridge	1911
	95, Reading	1923
	(>95) S.E. England	1868, 1825, 1808, 1757
	94, Cromer, etc.	1959
Aug.	97, Halstead (Essex)	1932
	96, Cambridge	1911
	96, Norwich	1932
Sept.	94, Raunds (Northants)	1911
Oct.	83, Reading, etc.	1921
	83, Rugby, etc.	1959
Nov.	71, Prestatyn	1946
	70, Cambridge	1938
Dec.	65, Achnashellach (N.W. Scotland)	1948

The range of possibilities with regard to rainfall and snowfall can be briefly summarised. Rainfall statistics over long periods have been closely analysed; and over the country as a whole it appears that at most stations the driest year can be expected to have about 60% of the normal rainfall, the "wettest" 150% taken over periods of the order of a hundred years. With regard to the months, many places in the British Isles have recorded a whole month without measurable rain; even in the very wettest districts this has been known to occur in months with persistent wind from unusual quarters. For when north and east winds prevail our normally wet highland districts lie on the lee side of the country. Hence in February 1947 no precipitation whatever was measured at Glenquoich, in the wettest area of the Highlands; and in February 1932 the like occurred at the head of Borrowdale. In August 1947 no rain whatever was recorded in an extensive area towards the west of Scotland. Months in which anticyclonic conditions have so dominated the country that a wide area recorded no rain include September 1865, February 1891, June 1925, February 1932, April 1938, September 1959. Many other spells, however, have obviously occurred in which absolute drought prevailed for a period upwards of thirty days though not coincident with calendar months. One of the most severe droughts on record lasted through the greater part of April-May 1893. Studies of the incidence of dry spells have been made which indicate as we should expect that the greatest tendency for such spells is in the spring and early summer months; there is a minor peak in September. The least chance is in the late autumn, but no month is entirely devoid of dry spells. The chance that a spring month such as May will have no rain can, however, be estimated as several times greater than a rainless August; while the phenomenal dryness of August 1947 is fresh in memory it is worth reiterating that dry Augusts are rare; August 1955 was notable.

Snowfall is an immensely variable element as we have seen. If we regard the snowiness of a month as represented by frequency and aggregate depth of snow, probably the most snowy months in the past two centuries over the country as a whole were February 1947 and January 1814. Locally snowfalls have been heavier; for example the great Lancashire snowstorm of January 1940 and the phenomenal fall in Northumberland and Durham in February 1941. In the south, the Christmas snowstorm of 1927 gave remarkable depths in Kent, Sussex and Hampshire; the classical blizzards of December 1836 and January 1881 affected the whole of the south and Midlands, while that of March 1891 was very heavy in the west country and classical on Dartmoor. In Eastern Scotland where for orographic reasons heavy falls are fairly frequent, March 1886 was very exceptional. Farther back there have been some remarkably deep out-of-season falls; in September 1673 and October 1836 in the county of Durham, in May 1782 and 1923 in Aberdeenshire, in June 1749 and 1809 in parts of Scotland, in May 1838 in the Cotswolds, in April 1908 and 1919 in Suffolk. All through our annals miscellaneous accounts of heavy falls can be found; in January 1751 a prevailing depth of 27 inches was noted at Richmond in Yorkshire, just as it was in February 1933. In February 1567 and January 1615 very great losses of sheep followed heavy snow in the same area. The same climatic factors have long been in operation, so that in many parts of Great Britain the results of a given combination of falling barometer, north-easterly wind and upland exposure are long remembered.

Snow has fallen even on low ground in June; in the northern counties of England and Scotland it may perhaps be reported here and there as sleet once in thirty years. It is stated to have fallen on 11 July 1888 at several places from Cumberland to Kent, although when we look up the meteorological events of the day in question it may be suggested that while the day was undoubtedly very chilly the distinction between sleet and soft hail was not at that time everywhere clearly drawn. Heavy falls of hail in summer have been claimed as snow in quite recent years, when in 1930 some keen Yorkshiremen ran charabanc excursions "to see the snow in August" after a violent thunderstorm on the Yorkshire Wolds. Sleet has been reported in August from one or two high-lying Scottish stations, and, doubtfully, from Gordon Castle on 17 August 1784; also from Dartmoor in 1879. But Defoe's dating of a late August snowfall on the moors between Rochdale and Halifax (in 1705?) must be doubted. Even on our lower mountains however snow has occasionally fallen to some depth in June. In July and August it is a little doubtful whether, since 1900 and south of the Border, a lasting cover below 3,000 feet has been observed at all. Generally speaking, if a keen watch is kept, Scottish summits above 3,000 feet will be found to be freshly covered for a short time on about one occasion in June, and for the first time in autumn, during the latter half of September.

Bearing in mind that the water equivalent of a foot of snow is generally taken to be an inch of rain, it is evident that for a single fall to exceed ten inches in depth is very rare on low ground inasmuch as falls exceeding an inch of rain in a day are themselves very few unless in occasional severe thunderstorms. Depths on the level, apart from drifts, have been known to exceed four feet, for example in upland Denbighshire and Upper Teesdale in February–March 1947, the result of accumulation of several heavy falls in those districts in which as we have already seen snowfalls are greater in amount on account of relief and easterly aspect.

From County Durham after a single prolonged fall four feet was reported from Consett (800 feet) in 1941, with 42 inches at Durham Observatory (336 feet); snow is said to have fallen continuously for between 50 and 60 hours. The effects even of low hills are shown by the falls of February 1947 in the east Midlands. At the time when about nine inches was the prevailing depth in Cambridge, at three hundred and fifty feet above sea level in west Bedfordshire depths of over twice this amount prevailed. Similar depths were attained on the low Lincolnshire heights and on the rising ground in North Norfolk. Many northern motorists will remember how often the difference can be seen between the depth of snowfall at Catterick Bridge and that at Scotch Corner, five miles to the northward and 300 feet higher. It is here that the Glasgow-bound driver begins to wonder what is in store for him on Stainmore, that summit which his ancestors have known for so many centuries. Bishop Nicolson of Carlisle, on his way to London in the cold January of 1709, commented with relief on his passage of Stainmore "as the snow had been deeply tracked". It is not without significance that the Romans sited this road in such a manner that to this day the road remains fairly free from snowdrifts. Easily the worst blockage by drifts occurs where the modern road-maker's gently graded diversion leaves the steeper but better-exposed Roman track, just on the Westmorland side of the summit.

The long traditions of the winter traveller through the eastern Pennines are shown again west of Penistone on the Derbyshire border. Here a rough track diverges from the main Manchester road and rejoins it after an exposed hilltop course half a mile beyond; and the Ordnance Survey marks it as 'the snow road'. It was probably used as such in the eighteenth century and earlier; that this route by Lady Cross began to be used by wheeled traffic is evident from the account of the energetic Lady Anne Pembroke who in March 1656 crossed it with her coach "where never coach before". Scotsmen can find much of interest in the annals of the adventures of the Edinburgh mail from the south; the great roads over the Lammermuirs have their stories of classic snowfalls (December 1836 notably) and twenty miles from Edinburgh the cutting at the top of Soutra Hill is a place where again the modern lorry-driver finds much the same troubles as his forebears. Snow-screens have helped recently.

Gale, fog, torrential rain, hail and thunderstorms add their quota of incident to the annals of many a parish; and even tornadoes of some violence are not unknown inland. Statistics of the possibilities are again better sought elsewhere, but a brief note may be added here as from time to time these more violent manifestations of weather also play their part in moulding the scene as well as our impression of it.

All around our coasts winds of gale force (exceeding Beaufort force 8 i.e. maintaining for an hour or more an average speed upwards of 38 m.p.h.) blow with some frequency, although as most gales blow from between S. and W. the frequency of such excessive winds is greatest on our N.W. coasts in Scotland and Ireland and least in N.E. England, where the wind speed from any westerly point has been checked over the land. Inland however winds attaining gale force for an hour or more are in many places rare; although on 29 November 1938 Cardington in Bedfordshire recorded a mean speed of 59 m.p.h. over an hour, with a maximum gust of 86 m.p.h. The gale of 16 March 1947 which gave a gust of 98 m.p.h. at Mildenhall seems to have been even more noteworthy as far as damage was concerned; for it is the gusts which damage structures and bring down trees.

The effect of friction is to diminish the *average* speed of the wind; but the fluctuations between gusts and lulls are much more marked. If the wind *averages* force 6 (25 m.p.h.) inland it is quite probable that many gusts will exceed gale force, and these are just as liable to cause structural damage as if the location were on the coast. An example is shown by the anemogram for South Kensington (Fig. 49, p. 153).

Suffice it to say that, by and large, the windiness of the British climate increases north-westward. It is not easy to give precise statistics for frequency of days with gales for any given location. The difficulties of comparison even on the coasts are considerable; over such periods of years as are available for the three Hebridean stations Stornoway gives 24 days yearly with gale force recorded at some time during the day, Castlebay 11 and Tiree 33. The reasons for these differences are apparent when local exposure is taken into account; Castlebay provides a relatively well-sheltered harbour, whereas Tiree is an island of very low relief.

Up to 1954 the highest gusts on record ranged from over 70 to 125 m.p.h.; highest hourly wind speeds from values of the order of 35 m.p.h. (force 7) to nearly 80 m.p.h. The latter speed is occasionally reached at stations such as Scilly, Pendennis Head by Falmouth, St. Ann's Head in Pembrokeshire, Southport and Fleetwood, the north-west coasts of Ireland and Scotland, and also Bell Rock off the Firth of Tay, Orkney and Shetland. But many windy coasts lack an anemograph; inland, too, the variations in average wind speed shown by open airfields in comparison with farmed and wooded areas are very evident. Little information is as yet available for our uplands. The exceptional winter gales of 1952-3 are noted on pp. 271-2.

Summer thunderstorms occur more frequently in the East Midlands than elsewhere, and in the middle Trent valley thunder is heard on an average of about 20 days yearly. Torrential rain such as that which flooded Louth in 1920, and more rarely, catastrophic hail are often associated with thundery conditions. Very extensive damage to crops, greenhouses and the like by hail was reported from Kent in 1916 and from West Suffolk in 1946; other instances have already been mentioned. Summer thunderstorms are also rather frequent in some river valleys west of the Pennine watershed, where Stonyhurst averages about 18 days yearly with 'thunder heard'. (Cf. Marshall, p. 272.)

Small tornadoes have been reported from time to time from many localities; they also are a concomitant of violent thunderstorms. The best described disturbance of this type preceded the passage of a cold front across South Wales in October 1913. In June 1937 a small tornado crossed the southern outskirts of Birmingham, blowing out doors and windows; on December 8, 1954, another crossed West London and wrecked Gunnersbury station. Occasionally small disturbances of great localised violence, probably of tornado character, have also been reported for example in Nottinghamshire, Staffordshire, Derbyshire and South Lancashire as well as several East Midland counties and East Anglia. A very interesting early account of a tornado which damaged Widecombe Church on Dartmoor in 1638 has recently been given by L. C. W. Bonacina (1946). Exceptionally unstable conditions, with cold air above over running warm moist surface air, appear to be necessary; such conditions are probably more likely to develop in spring and summer, but are not of necessity confined to that season. Indeed, one of the best defined tornadoes was that at Nottingham on 1 November 1785 (*Gentleman's Magazine*, 1786).

The incidence of violent thundery rains and damaging hail storms, and even perhaps the chance of an occasional small tornado, are probably associated with the regions of most frequent thunder. Small scale maps indicating frequency over a considerable period have been given by W. A. L. Marshall (1934).

With further study of the observational material, among which one may mention the collected data of the Thunderstorm Census Organisation under the directorship of Mr. S. Morris Bower of Huddersfield, it seems probable that in time more detailed maps can be drawn. There is a good deal of evidence for the view that the frequency of overhead thunderstorms (and presumably their consequences in the form of damaging hail) shows appreciable local variation; but precise observations are not easily contrived. Mr. A. B. Tinn of Nottingham has analysed the incidence of thundery rains in the neighbourhood of that city and shows that they tend to be more intense in certain areas, notably just west of the city where the valleys of the tributary rivers Erewash and Leen join that of the Trent (Tinn, 1939). Studies on similar lines with regard to the validity of local proverbs such as "if it thunders at all, it will thunder at Thirsk"—at the foot of the

Hambleton Hills on the east side of the Vale of York—might well be useful. Some will be tempted to suspect from Mr. Morris Bower's maps that the broad belt of oolitic limestone running across England and including the Cotswolds, may be considerably less liable to hear thunder than the valleys on either hand; but it would be wise to await further observations. It is interesting to correlate with this possibility an observation by Mrs. Ann Douglas, the well-known glider pilot, that at times sailplane pilots have found difficulty in this area from an unexpected lack of vigorous thermals. The matter, like others, needs further investigation.

The incidence of long-continued rains, and of rainfalls of exceptional intensity, have been the subject of many studies in *British Rainfall.* It is noteworthy that two of the greatest observed amounts of rainfall in a day have both occurred in the same region towards the foot of the Mendip Hills in the Somersetshire levels; namely, 9·56 inches at Bruton (28 June 1917) and 9·4 inches at Cannington (18 August 1924). None of our mountain ranges has quite attained this amount in a day, although a number of 'orographic' falls exceeding eight inches in the day have been recorded. (Lynmouth 1952, see p. 272).

Thunderstorm rains on the Pennines (compare Chapter 3, p. 52) as well as on some of our other hill ranges which are less visited, have undoubtedly given very heavy localised falls in the past. In addition to the Stainmore cloud-burst in 1930 previously mentioned, records of a flood near Todmorden in July 1870 undoubtedly point to a local fall of upwards of nine inches of rain. More recently, severe damage was caused at Holmfirth by a similar storm in 1944, causing a sudden spate in the streams of the West Riding of Yorkshire. Daily rainfalls in excess of four inches have at some time been recorded from most of the English, Welsh, and Scottish counties; but such torrential downpours naturally play more part in moulding the scenery in the uplands where the run-off is rapid. Professor Austin Miller has commented on the results of such a downpour in Wales. (Lynmouth, see p. 272.)

Spells of excessive rain of several days' duration are more usually associated with slow-moving depressions in which the fronts are nearly stationary, especially in the summer months when the water-vapour content of the air is greater. In 1930 from 20-23 July nearly twelve inches of rain fell at Castleton in the North York Moors, and the consequent flooding in the Whitby Esk was very great. Other instances include the "Moray floods" of 1829 and the extremely heavy and persistent rain of 11–12 August 1948 which did so much damage to bridges, roads and railway in S.E. Scotland; these have already been

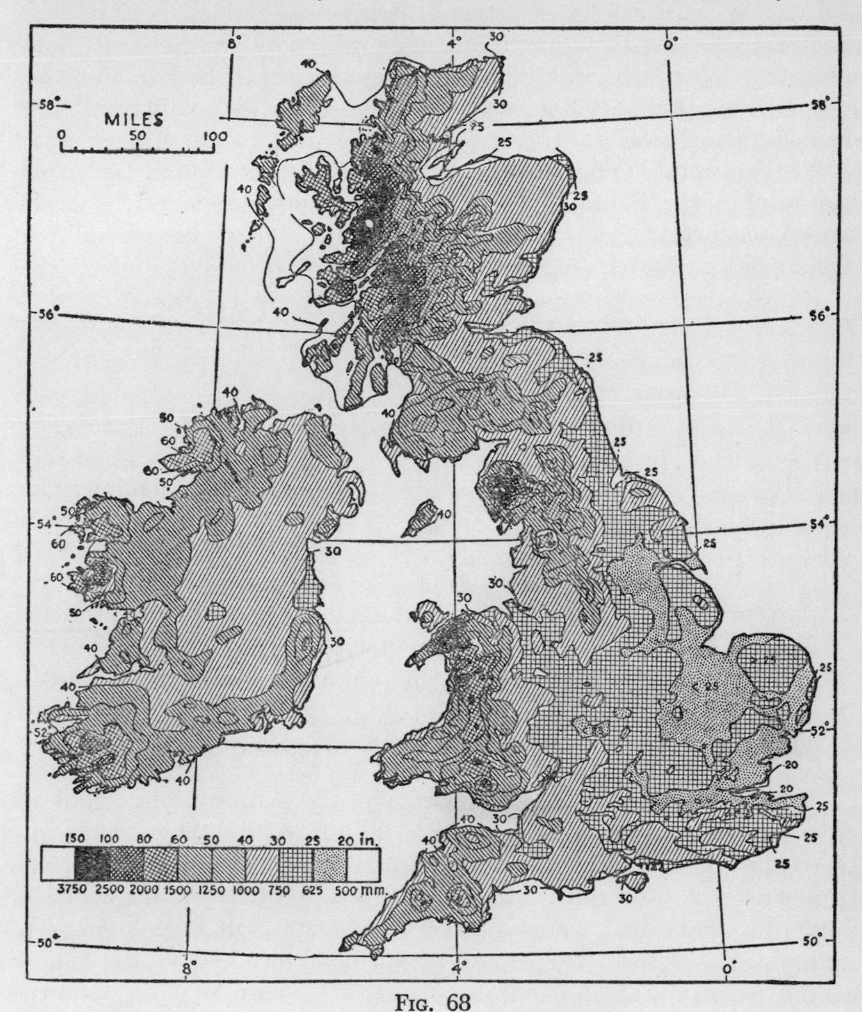

Fig. 68
Annual rainfall over the British Isles

mentioned (p. 123). From every part of Britain however there are accounts of historic floods; some of the worst have arisen not as a result of long-continued summer rains, but from the combination of heavy warm rain falling on a deep snow-cover. The great March floods in the Fenland in 1947, like those of the Severn were largely due to this cause, as an exceptional depth of snow had accumulated on all the surrounding uplands and the underlying ground was already saturated. Very extensive flooding in both areas befell in February 1795 as a result of a sudden thaw with rain. Farther north, apart from such Highland rivers as the Spey, the great and rather sudden floods of the Tees and Tyne, draining a large upland on which both snow and rainfall are liable to be heavy, are often mentioned; here the historic calamity which washed away almost every bridge on both rivers befell in November 1771.

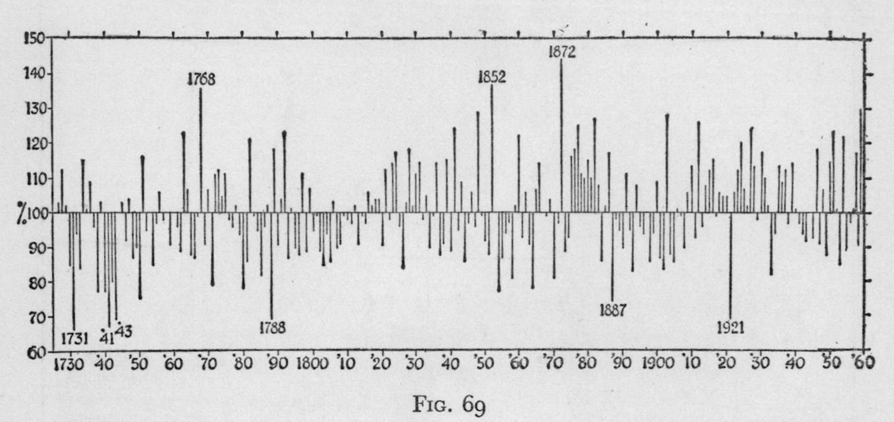

FIG. 69

General rainfall, England and Wales; percentage of the average for 1881-1915, for the years 1727-1960 (from data by J. Glasspoole, brought to date)

Results of studies of intensity, amount and duration of rainfall and of some historic floods have been summarised by Dr. Brooks and Dr. Glasspoole of the Meteorological Office and some additional figures are given by E. L. Hawke (1942). Statistical investigations for numerous individual stations have been made; in recent years, for example, Liverpool (Reynolds, 1953), covering 1867-1951 and an exhaustive study of Wrexham and district (Ashmore, 1944) covering 1880-1942. Ashmore noted that it is usual to find that the number of years in excess of the average is appreciably less than the number below the average, a fact which will give many food for thought. Over 63 years at Wrexham the range of variation has lain from 65% to 144% of the overall annual average (64 to 145, Isle of Man, Reynolds, 1954).

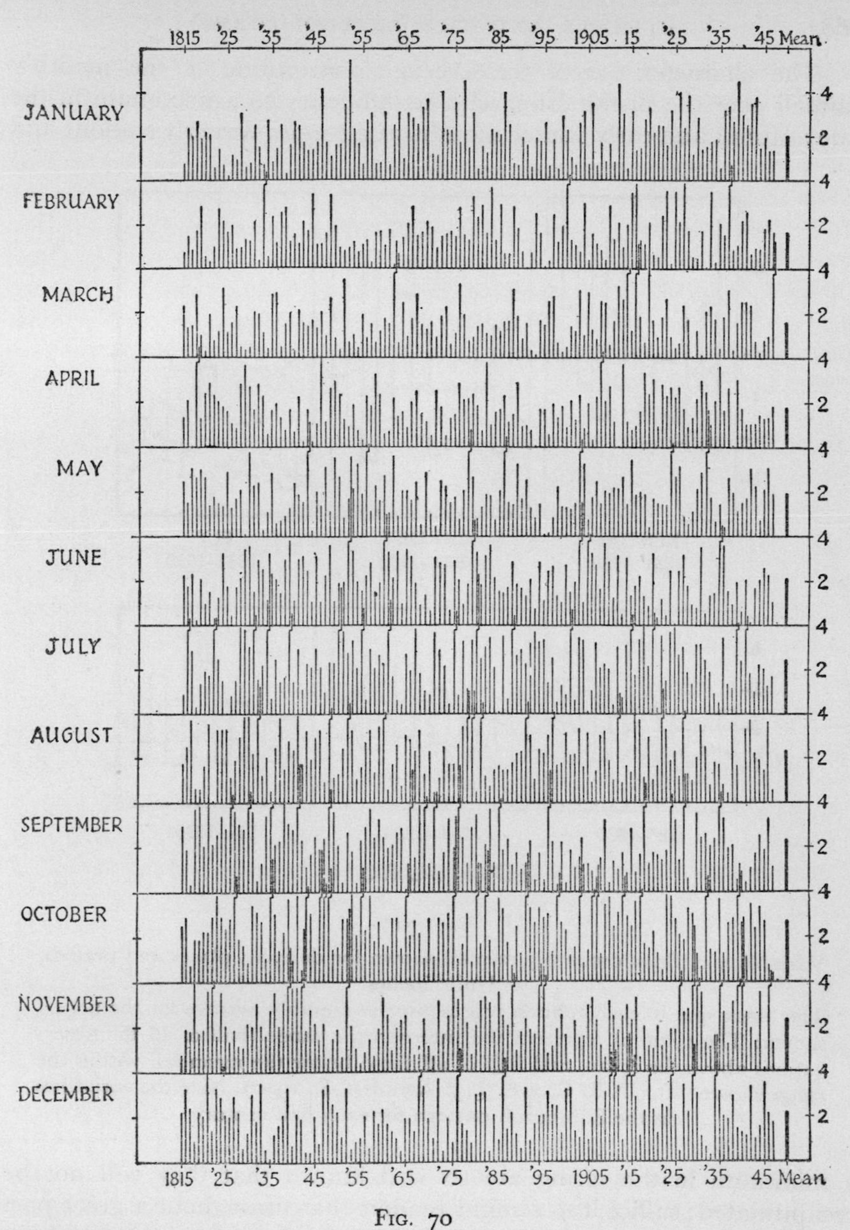

FIG. 70

Monthly rainfall in inches at Radcliffe Observatory, Oxford, from 1815 to 1947 (drawn by D. S. Brock)

The characteristics of the average distribution of the monthly rainfall over the British Isles, with its tendency to a maximum in the late autumn and early winter months at all more westerly stations and

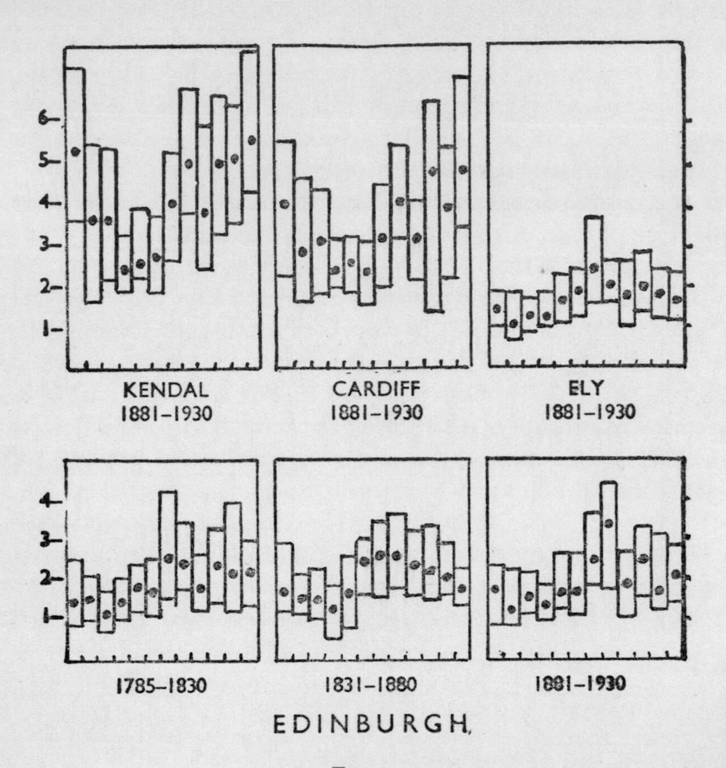

FIG. 71

Variability of British monthly rainfall shown for different places and periods. Unit: inches.

The black spot in each column represents the monthly average for the period shown: The columns represent the interquartile range (half of all the observations above the mean, and half of all those below the mean fall within the range shown (after P. R. Crowe, *Q. J. Roy. Met. S.*, 1940). Note the variations in the Edinburgh pattern for each half-century

a minimum in the spring are so well known that they will not be recapitulated; suffice it to remind readers that throughout a great part of the Midlands the rainfall of July and August commonly rivals in amount that of the autumn months. Diagrams showing the average monthly fall at representative stations can be found in the standard work by E. G. Bilham to which reference should be made by those who desire an extended discussion of the statistics and of the range of variation at individual stations in different districts; and the dispersion diagrams added to the Ordnance Survey's recent map showing the average annual rainfall over Great Britain are also to be commended. Variations in the monthly rainfall at Oxford, 1815-1947, are shown in Figure 70, p. 267, and are characteristic of a large area, over which (1916-50) November is the wettest month.

The wettest districts have already been mentioned, together with the frequency of rainfall; it may be added here that it is considered that the village of Great Wakering, near Shoeburyness in Essex is probably in the driest area, with an average annual rainfall (1881-1915) of 18·4 in., falling on about 150 days (for 1916-1950, about 19·2 in.).

The variability of the duration of sunshine from year to year is considerable, and it is not directly associated with the rainfall. The best source of information is the *Monthly Weather Report* of the Meteorological Office. A recent table of the monthly and annual totals of bright sunshine at Bognor Regis for 1924-43 was published by D. S. Hancock (1944); as it is representative of the sunniest district in Britain the averages and extremes are quoted here together with those of Durham for comparison. Both sets of figures give a useful impression of the range of variation. In 1959 Bognor recorded 2,095 hours.

BOGNOR REGIS. 1924–1943: Bright sunshine. hours. (D. S. Hancock)

	J	F	M	A	M	J	J	A	S	O	N	D	*Year*
Mean	66	83	147	166	215	241	218	209	167	117	71	61	1763
Highest	103	127	198	227	279	307	309	271	250	141	104	79	2066 (1933)
Lowest	42	29	96	125	134	203	147	172	108	75	47	43	1519 (1932)
DURHAM. 1886–1947 (E. F. Baxter, Durham University)													
Mean	48	66	107	136	162	176	159	146	124	91	57	42	1314
Highest	79	116	191	238	248	297	253	235	204	151	120	79	1606 (1901)
Lowest	13	32	48	71	87	92	74	57	65	42	29	10	982 (1912)

The Meteorological Office tabulates for a limited number of stations the frequency of temperature maxima and minima between given limits. The seasonal distribution over the British Isles of the number of days with a minimum temperature of 32° or below has been summarised by Miss L. F. Lewis (1943). This number averages from 50 to upwards of 100 at representative inland stations (Cambridge 70, Mayfield in Derbyshire 83, Eskdalemuir 110, Fort William 63, Armagh 49 are typical) diminishing to much lower figures on exposed coasts, e.g. 13 at Falmouth. From what has already been said, however, in Chapter 10 it will be evident that very large differences exist even between adjacent stations, so that figures such as those quoted above can only be taken as a general guide. Particular localities must be considered with careful attention to the characteristics of the site.

Reference may be made to the aurora borealis—the Northern Lights or the 'Merry Dancers' of Scotland. In the north of Britain the pale cold yellowish-green glow low down on the northern horizon, from which occasional faint streamers dart, momentarily, like indescribably distant searchlight beams, upward into the sky above, is quite a common sight on clear winter nights; the increase in frequency is very rapid from southern England to Shetland. In the south a keen observer far from the lights of towns may detect a faint auroral glow on perhaps five evenings of the year. Rarely in the south does it attain sufficient intensity to attract the attention even of the countrymen. But in the latitude of Aberdeen it may be observed on upwards of thirty nights annually and many more in Shetland; and correspondingly a greater number are sufficiently vivid to be widely observed. Sometimes other colours are seen, and no one who witnessed the magnificent "great aurora" in January 1938 will ever forget the wonderful deep red glow shot through by the normal greenish-yellow streamers.

Aurora is due to the impingement of a discharge from the sun, considered by many to be formed of a stream of electrically-charged

---

PLATE XXI*a*: The humid Irish summer (August). Moist maritime polar air stream with extensive strato-cumulus and cumulus over County Dublin, breaking a little as it debouches over the sea to eastward. Occasional showers inland; dark cloud base shadows indicate that the cumulus is rather deep.

b: A similar type of day in S.W. Ireland (July). Showers in the surface air stream, cumulus building up considerably over mountains. Lenticular cloud above suggests smooth flowing upper air around the margin of an anticyclone centred towards England.

PLATE XXI*a*. *Irish Tourist Association, Dublin*

b. *Irish Tourist Association, Dublin*

By courtesy of Maggs Brothers, London

PLATE XXII: The last frost fair on the Thames, January 31 to February 5, 1814. January 1814 was probably the coldest month over Britain as a whole during the past 200 years. The effect of the old bridge as a barrier impeding the movement of the ice-floes was notable.

particles, on the rarified upper atmosphere. It tends to be most frequent about the month of February but varies considerably in frequency from year to year and is closely associated with those variations of solar activity revealed by sun spots. The coldly fascinating beauty of a Highland winter night when the aurora flickers above the silent mountains offers to many of us a variation of our normal experience only to be contemplated with awe.

Mr. James Paton of the University of Edinburgh who took the photograph on Pl. VIII, p. 123, has organised the Scottish auroral observations; and it is a pleasure to acknowledge the opportunity of showing yet another aspect of the British scene.

REFERENCES

Chapter 13

Ashmore, S. E. (1944). The rainfall of the Wrexham district. *Q. J. Roy. Met. S. 70:* 241-73.

Baxter, E. F. (1948). Observations at Durham University Observatory. University Offices, Durham.

Bilham, E. G. (1938). *The Climate of the British Isles.* London, Macmillan.

Bonacina, L. C. W. (1946). The Widecombe Calamity of 1638. *Weather, 1:* 123.

Bower, S. Morris. Thunderstorm Census Organisation: Annual Reports. Published from the Organisation's Headquarters; Oakes, Huddersfield.

Brooks, C. E. P. and Glasspoole, J. (1928). *British Floods and Droughts.* London, Benn.

Defoe, D. (1927, ed. Cole). *Tour through Great Britain, vol. II,* 596-98. London, Peter Davies.

Douglas, A. C. (1947). *Gliding and Advanced Soaring.* London, John Murray.

Douglas, C. K. M. (1952). Synoptic aspects of the storm over N. Scotland on Jan. 15, 1952. *Meteor. Mag., 81:* 104-106.

Glasspoole, J. (1931). Heavy Falls of Rain in Short Periods. *Q. J. Roy. Met. S. 57:* 57-69.

and Hancock, D. S. (1936). The Distribution over the British Isles of the Average Duration of Bright Sunshine. *Q. J. Roy, Met. S. 62:* 247-59.

Gold, E. (1936). Wind in Britain. *Q. J. Roy. Met. S. 62:* 167-206.

Hancock, D. S. (1944). Sunshine at Bognor Regis. *Q. J. Roy. Met. S. 70:* 228.

Hawke, E. L. (1942). Notable Falls of Rain. *Q. J. Roy. Met. S. 68:* 279-86.

HUDLESTON, F. (1930). The Cloudbursts on Stainmore. *British Rainfall, 1930:* 287-92.

LEWIS, Lilian F. (1939). The Seasonal and Geographical Distribution of Absolute Drought in England. *Q. J. Roy. Met. S. 65:* 367-83.
(1943). Seasonal Distribution over the British Isles of the number of days with screen minimum 32° or below. *Q. J. Roy. Met. S. 69:* 155-60.

MARSHALL, W. A. L. (1934). Mean frequency of thunder over British Isles. *Q. J. Roy. Met. S. 60:* 413-24. Gives a useful map.

MILLER, A. A. (1951). Cause and effect in a Welsh cloudburst. *Weather, 6:* 172-79.

PATON, James (1946). Aurora Borealis. *Weather, 1:* 6-11.

REYNOLDS, G. (1953). Rainfall at Bidston. *Q. J. Roy. Met. S. 79:* 137-49.
(1954). Rainfall in the Isle of Man. *Q. J. Roy. Met. S. 80:* 78-88.

TINN, A. B. (1940). Local Distribution of Thundery Rains round Nottingham. *Q. J. Roy. Met. S. 66:* 47-65.

WALKER, Sir Gilbert (1930). On the mechanism of Tornadoes. *Q. J. Roy. Met. S. 56:* 59-66.

Note on Extremes of Wind and Rainfall in 1952, 1953, 1955 and 1960

Reference to the Orkney gale of January 1952 will be found in the paper by C. K. M. Douglas above. This was surpassed by the tremendous northerly gale in N.E. Scotland on January 31, 1953. On this occasion a gust of 125 m.p.h. was recorded at the wind-generator station on Costa Head, Orkney; and the average speed of the wind reached 80 m.p.h. at Lerwick in Shetland. Widespread damage to the Eastern Scottish forests resulted; and the consequent "surge" in the North Sea with its catastrophic coastal flooding is discussed in many journals (notably *Weather; Meteorological Magazine; Geographical Journal,* for 1953.

The "Lynmouth disaster" of August 15, 1952, resulted from a very exceptional localised rainfall, of thundery type but lasting for many hours, over Exmoor, which gave rise to torrential flooding in the steeply-descending rivers. At one gauge on Exmoor nine inches fell in this storm; there is, however, reason to believe that locally the fall may have approached eleven inches within the 24 hours. This storm was discussed extensively (*Meteor. Mag.; Weather,* 1952).

In another exceptional outbreak of thunderstorm rains near Weymouth, eleven inches of rain fell at Martinstown on July 18, 1955.

Since this book first appeared, several further papers have been published; notably H. H. Lamb on "Our Changing Climate" (*Weather, 14,* Oct. 1959). The present writer has now taken monthly means of temperature back to 1680 (G. Manley, *Meterological Magazine, 90,* Nov. 1961). The winter of 1684, based on the mean temperature of the three months December-February, surpassed that of 1740 and thus ranks as the coldest in the past 300 years. The winter of 1695 was notable for frequent and persistent snowfall, and was followed by a cold spring and wet summer. Indeed, cool springs and cool disturbed summers were characteristic of the decade 1692-1701 and gave rise to much distress.

CHAPTER 14

CLIMATE AND MAN

I am but mad north-north-west. When the wind is south I can tell a hawk from a handsaw.

SHAKESPEARE: *Hamlet*

ALMOST everywhere man and his works form an integral element of our British scene. The observant traveller may comprise within his vision such a view as was apostrophised by Wordsworth from the summit of Black Combe; from which "the amplest range of unobstructed prospect may be seen that British ground commands." From the hills above Ampleforth he can reflect upon the pattern and the crops of the great Yorkshire fields; he can almost espy Vanburgh's Castle Howard in addition to the villages of Anglians, Danes, and Norsemen; he can contemplate the glory of York Minister and the sense of values of that great Yorkshireman Fairfax who saved its glass; he can consider whether the adjacent splendid curve of York railway station, so finely bedecked with advertisements of diverse size and colour, is a stimulus or a deterrent to the Northern mind. Yet on another occasion the same traveller's vision will be bounded by seven sombrely-dressed individuals with colds in a gaslit railway compartment dating from 1886, while his sensations are delicately enhanced by a recognisable stream of maritime-polar air penetrating the ill-fitting window.

Beneath all these manifestations of British culture lies a climate which by virtue of its gentle extremes does not enforce the rapid condemnation of outmoded institutions and equipment. It permits the survival, and even the cultivation, of many exotics introduced from the neighbouring continent; a statement which may be applied to human institutions and ideas as well as plants. But frequently if the exotics are not given protection a slow process of adaptation begins. Vanbrugh's French-inspired palace became a school. Marxist views spread with difficulty away from the shelter of city walls. The Roman tide surged but feebly into northern England; when the protection of the Wall was removed, the northern peoples quickly swarmed into the land of the ash-tree. Megalithic man and Mediterranean shrubberies show a not dissimilar distribution.

The influence of climate on man, direct or indirect, defies measurement. Yet in a country in which man and his works are so prominent it is at least desirable to essay an opinion with regard to the extent to which climate should be taken into account, in the evolution of this integral element of the scenery. Opinions with regard to the ultimate effect of climate largely beg the question in a country whose development has been so complex as our own; historians, geographers, biologists and anthropologists of every kind will join issue with many of the inferences and analogies with which a chapter such as this must abound.

The physical appearance of the inhabitants of Britain is very varied, and if breeding could be applied to them it is probable that within this island even more varieties would be produced than there are of sheep.

As far as we can see climate places no impediment in the way of survival of many slightly differing European stocks, although it has long ago been suggested that there is a tendency for the fairer types to thrive better in the country, so that as a people we are slowly tending to become darker with increased urbanisation. Hair and eye colouring are often intermediate and only one physical attribute is sufficiently widespread to be, according to some writers, attributable to climate, namely, the relatively fair skin with more or less degree of freckling, through which the blood-vessels of the face show (*ceteris paribus*) and form what we call a complexion. Such characteristics appertain to regions of mild, cloudy and rainy climate in which no marked necessity arises for protection of the blood vessels and nerves against cold or dry cutting winds; neither is there need for a greater amount of skin pigment as a protection against the shorter wave-length radiation from sun and sky. They are widespread throughout maritime north-west Europe, from Bergen to Brest, and may well have given backing to that Venetian Ambassador who reported that England is a country where the women have the finest complexions in the world, and the wind is always blowing. Yet it becomes questionable whether urbanisation with its concomitant, the dominance of fashions derived from cities elsewhere, will not lead in this and other respects to a monotonous uniformity of style through the deliberate eschewing of that regional differentiation which hitherto has lent so much interest to all the means of expression adopted by mankind.

Some might be tempted to see the slow workings of a more favourable climate in the greater refinement of feature characteristic of those fair immigrant stocks which have found their home in Southern England, by comparison with the heavier bone so often to be seen in the north. But no invariable rule can be made; a walk along Princes Street or an inspection of the Raeburn portraits in Edinburgh will quickly reveal that delicately-chiselled lineaments are no monopoly of the men of Kent, or even Dorset. In the fact that our climate permits many types to survive with the aid of a little exertion we may see an advantage; cross-breeding of the varied stocks is undoubtedly a factor in the production of original genius, as Havelock Ellis and others have pointed out.

Mental traits and accomplishments however show considerable regional differentiation. Man's brain is the principal tool by virtue of which he survives; and as a tool it may be sharpened or blunted by environmental factors. Keenness of perception over a whole people can scarcely be measured but it may be observed in the frequency with which poetic imagery, artistic accomplishment, and scientific advances develop. That the imagery of English and Scottish poets owes an immense debt to the country in which they have been bred has been evident throughout the fifteen centuries of the use of the English language. Such imagery becomes less noticeable as we might expect in poets of the town such as Jonson and Pope, but those are relatively few. Elsewhere, passage after passage may be found in which the keen perception of, and delight in, the sights and sounds of the country appears. Throughout our literature associations between the mood of the landscape and season, that of the weather, and that of the writer can be sought and as quickly found. Can we not allow that these are indications of the mental stimulus which testify to one of the greatest of the advantages we derive from our variable climate?

The overwhelming delight in the English spring from Chaucer through the Elizabethans, Andrew Marvell's pleasure in his garden, the vivid descriptive passages of Dorothy Wordsworth, the upright determination of Kingsley's ode to the north-easter are each typical of their age. The inspiration born of the mood of the weather has set alight the imagination and creative intelligence not only of poets. It has captured the regard of the artists who for two centuries at least have struggled like the mathematicians with the problems of statics and dynamics. There are those who have wrestled with the task of capturing and stabilising the fugitive delicacy of the moment in the manner of Wordsworth's poem "composed upon an evening of extraordinary splendour and beauty."

Other men, like Turner, have striven with the representation of movement; whether in the hurrying clouds, the roaring sea, the bending trees, or in the less menacing liveliness of the autumn sowing, the pastoral scene, the hunting morning. Can we see in the clearer light of the eastern counties not merely the great focus of English and Scottish artists, but a place in which, from Cotes and Newton to Napier, thoughtful clear-sighted men have logically turned to mathematics?

Scientists too have derived at least some part of their stimulus from the weather. No better example can be quoted than Dalton, the famous proponent of the atomic theory in 1802. Born near Cockermouth on the margin of the Lake District where water in all its forms abounds, the attention he gave for several years in his youth to meteorological observations at Kendal bore fruit in the correct exposition of the behaviour of water-vapour in the atmosphere. No doubt one might add to the inspiration of the changeful clouds that of the mountain landscape, where a capacity for taking bold strides from footholds already tested is developed to the full. Can we wonder that so many of our most eminent scientists descend from those Northern, Western and Scottish stocks bred among the cloudy hills? Especially in the realm of physics it becomes quite fascinating to trace the repeated associations with environmental factors. It is entirely right to find that the need for accuracy of measurement in dealing with the manifestations of a turbulent Nature is symbolised among the Pennine moorland valleys not only by the magnificently controlled rhythm and pitch of the great choral societies. The same necessity has been the making of a great line of experimental physicists, inheritors of the same sense that lay behind early surveyors and instrument-makers, early clockmakers and those great predecessors whose instinct for accurate knowledge led to the demand for a Bible men could read: for it has already been recalled that Wycliffe and Coverdale were of the same stock. Somerset as we saw provides a smaller region in which somewhat similar attitudes of mind appear.

The flow of the air, the changeable weather and the rubato embroidery upon the fundamental rhythm of the seasons has been one of the greatest of the influences moulding the British mind. The charm and attractiveness of irregular variations in every sphere of life are the greater inasmuch as they are all underlain by the fundamental facts arising from our position between ocean and continent. Moreover, in a country in which irregularities of pattern of every sort are not only experienced but esteemed, tolerance of varieties of opinion has continually grown. We have already seen that one of the greatest assets of our island, which we can ultimately derive from the climate, lies in the diversity of local varieties of plants and animals. Inasmuch as the mental attitude in childhood owes much to the results of comparative observation of surrounding objects and patterns of behaviour we can see reasons why local varieties of men tend to differ and why in different parts of our island there should arise a predominant "flavour of mind". Otherwise it is not easy to see why, through all the mixed strains that have populated this island there should appear a persistent interest in Quakerism and in geological investigation not only in Yorkshire but again in Somerset. Likewise nonconformity and the biological sciences appear not only in East Anglia but again along the Welsh border, associated not infrequently with that wizardry that charms men and makes for political ability—however distorted when tinged with the emotions arising from the *ergastulum* rather than the pastures where the wind blows.

Prudent opportunism too is an immemorial asset; readiness to change the plan, to deviate from the policy, to refrain from putting all the eggs into one basket may well be attributed to our variable climate. Many have long ago pointed the moral—deriving from the addendum to many a village notice—"if the weather is wet, the meeting will be held in the parish hall instead." But there is also another side to the continued beneficent influence of our climate in moulding characteristics. Over many years town-dwellers in a climate of little extremes cease to be affected to any serious degree by the weather. Water continues to flow, expensive fruits and vegetables appear from overseas though our own rot, wages are regularly paid; the vicissitudes of our climate are scarcely more than an inconvenience that upsets the football match or the Sunday picnic. Life becomes too easy; and perhaps we shall see a changed attitude developing rather rapidly when it becomes clear that the food produced in our island is of more importance to our welfare than either the seductive surfeit or grudging recompense from countries overseas.

Climate and Livelihood

In the British Isles, structure as we have seen goes far to mould the diversities of our climate. Yet in its turn climate goes far to set limits to the use that can be made of the physical features of the country in which we live. There are large areas in which the normal agricultural yield is thoroughly adequate for the maintenance and accumulation of energy, a fact well shown not only by the doubling of our population in the eighteenth century, but also by the evidence of energy to spare for the graces of life whether in the form of meteorological recording, tours to the Lake District, walnut furniture or epistolary accomplishment.

But there are also large areas in which practically no material yield is obtainable at all. Thirty-five per cent of Great Britain is mountain and moorland. Practically all the land above 1,500 feet falls into this category and a great deal of it is so poorly drained that only the sourest peaty soils are found. Elsewhere on the steeper slopes leaching removes enough mineral content from the thin poor soils to render them infertile and incapable of supporting more than a poor vegetation carrying very little stock.

In all this it is evident that the amount and frequency of rainfall is very significant; local variations in the degree of slope and the mineral constitution of the soil also play their part. In the Lake District valleys which drain relatively quickly hay is sometimes cut in places where there are a hundred inches of rain; it would be tempting to discuss how far this is due to the somewhat exceptional history of settlement. In general one begins to find in the Pennines the tell-tale abandonment of old reclaimed intakes wherever the annual rainfall approaches 55″, and it is highly probable that as many of the enclosures were made late in the seventeenth or early in the eighteenth century they date from a period when the rainfall may for a decade or two have been lower than at present. Farther north this annual average tends to decrease. In Galloway unreclaimed moorland predominates above the 500-foot contour by reason of the rapid increase of rainfall inland as well as distance from markets. Around Loch Laggan and on towards Newtonmore and Speyside the limits of improvement accord well with the average rainfall; the like is true in Caithness, allowing for the higher latitude. The high altitude reached by settlement and cultivation on the Braes of Glenlivet is in agreement not only with the greater exposure to light, already mentioned, but also the annual average rainfall of 36″ at Tomintoul (1,150 feet, in the lee of the Cairngorms). Rainfall indeed is more important than temperature in regard to the use we can make of our land under present conditions.

Granted this general principle we may also observe that there is a considerable area in Eastern England on which the average annual rainfall is under 25 inches. The average annual evaporation from open water surfaces, based on existing data, is often quoted as about 16 inches; but with clear air and unsheltered sites at sea level this figure is probably rather greater. A recently published map is reproduced (based on calculations given by J. Wadsworth, 1948); these data refer to tanks in open situations. Tank measurements do not however give a complete picture; more recently Dr. H. L. Penman has given figures for actual evaporation from natural land surfaces which tend to be slightly less than these tank figures, although the principles of distribution are broadly similar. It will readily be seen that in a dry year the net loss of moisture from the ground will approach closely and occasionally even exceed the total rainfall, even in less exposed situations, so that irrigation is well worthy of consideration. Percolation is also important where the rocks are permeable. The allowance to be made for percolation has been estimated by W. V. Lewis (1943); he based his studies on the behaviour of the Breckland meres in S.W. Norfolk, some of which dried up completely during the dry years

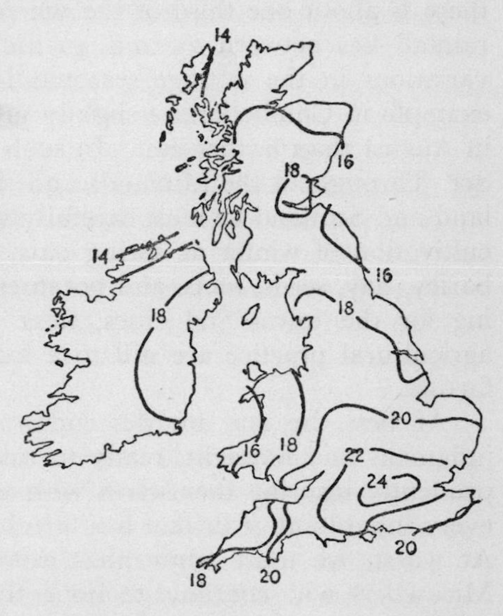

FIG. 72
Evaporation from tanks in open situation; average annual amount in inches (after Wadsworth)

1933–34. Further, as the evaporation during the summer months almost invariably exceeds the rainfall it is broadly true that in much of East Anglia the retention of meadows for hay is unprofitable in the majority of years. Hence while on the one hand a very large part of Britain is too wet for profitable agriculture, there is a considerable part in which arable farming for grain must be practised unless the farmer is to run the risk of heavy losses through drought. Between, there is about one third of the whole surface of Britain in which the rainfall lies between 25 and 40 inches. Slight but significant local variations in the average seasonal incidence can also be found; for example in Cumberland a slightly greater proportion of the total falls in August than farther east. In such a climate, what is the farmer to do? Throughout the Midlands and the lower parts of Wales, N. England and Scotland he must carefully weigh the chances attached to the cultivation of winter or spring oats, winter wheat, sometimes spring barley, hay, seeds, roots, and potatoes against stock rearing and dairying for the towns and cities, near or far. Regional differences in agricultural practice are still to a fair extent attributable to climatic factors.

At best, he and his descendants have become men of shrewd judgment and foresight, ready to take a risk on the one hand while prudently insuring themselves with some rival crop; and practically every inhabitant of Britain has farming blood within two generations. At worst, we must admit that among our people there are many Micawbers who continue to hope that "something will turn up" to offset losses through carelessness or undue optimism. The British climate is rarely so extreme as to lead to catastrophic loss of all crops and stock. Suffice it to remember February 1947, when in place of the expected reversion to the soft mild intervals of the normal winter the snowy cold continued. Those who are prone to believe that the British climate is the best in the world should always bear in mind that although its extremes may be mildly damaging, they are rarely lethal. Western Ireland for long found it possible to develop charm and to drift through life in the draughty, smoky, Celtic cottage; a sharp contrast with the cleanly and weatherproof Norwegian dwelling, demanded by a climate of rather greater extremes. It will for long remain disputable which sets of tendencies in our make-up are from time to time called out, even in a single life's span, by the varying incidence of climatic and economic factors.

Enough has been said to show that the effect of the British climate as an element in the human environment is more than a little paradoxical; shrewd opportunism may indeed be plausibly attributable to climatic factors, but so may languid carelessness. Becky Sharp, Vernon Whitford and Mr. Micawber were nearly contemporary. Moreover, he who would advocate climatic determinism in such a realm as housing cannot afford to overlook the admirable constructive ability of languid Malta, or the heedless squalor of the bracing Argentine pampa. Many factors beside climate enter into such arguments.

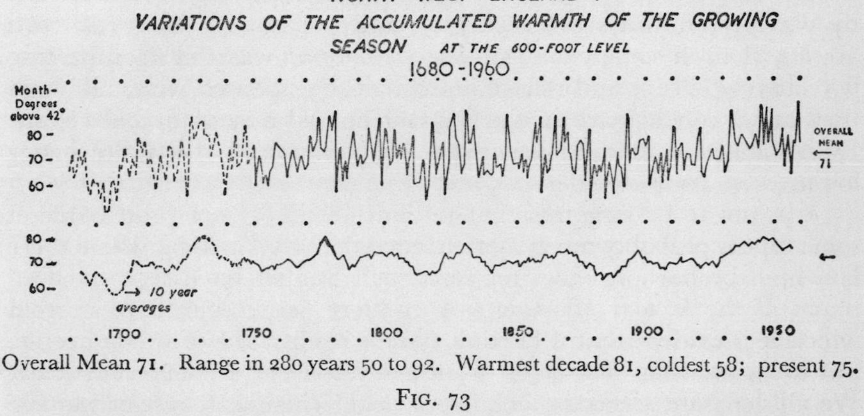

Overall Mean 71. Range in 280 years 50 to 92. Warmest decade 81, coldest 58; present 75.

FIG. 73

Accumulated temperatures during the growing season, in 'month' degrees, 1747-1950, for 600 ft. above sea level on the Western Pennines

In regard to agricultural production it is probable that no countries owe more to the efforts of man, through drainage, shelter, and careful choice of seed and stock, than the three realms of Great Britain. On the credit side we may safely say that the British climate is one in which such improvements can be counted on to give their reward; this was recognised in the sixteenth century, as the quotation at the end of this chapter will show. Rarely do calamities attain such magnitude that there is widespread ruin.

The influence of climate on the primary farming activities of man in this country is by no means simple. The manifest charms, the mental stimulus and the consequent economic advantages appear very evident, and have all been mentioned. But some may well ask whether a little more extreme behaviour (such as that of the winter of 1947) would not be a benefit to the country in the long run. At least we should pay more attention to the problems of conserving and transporting our food and raw materials, as well as those of domestic construction and heating.

With regard to manufacturing industry our climate is probably an asset of some value, although it does not demand technical achievements in such realms as air-conditioning. The variations of output arising from great heat or cold are probably less than in many other countries; and many smaller workshops need but little winter heating or shelter. This as in France is an advantage to the small scale producer. Ship-building and other outdoor industries are rarely impeded by weather; railways are relatively easily operated. Domestic costs are less than in some countries; in spite of our wasteful fires, heating is relatively cheap, and refrigerators can be dispensed with, although they are a convenience in the days of limited access to cool cellars. Even Charles I enjoyed ice-cream, and may have been the better for it.

It is not to be forgotten that radiant heat from a bright source of some type is probably much more needed than in America, where there is a high proportion of winter sunshine. Studies have recently been made of the factors affecting our comfort within our homes; from which it is evident that a heating system is not used to warm people, but to ensure that heat is lost from one's body at a comfortable rate. We all generate a certain amount of heat, whether at rest or moving about; if we are to remain comfortable heat must be lost at the same rate as it is being generated. Loss of heat may arise by evaporation, conduction to the surrounding air, or radiation.

In a very cold climate the objects which surround the body such as floors, walls and ceilings are at a lower temperature than the body itself, hence there is a considerable loss by radiation to them. If on the other hand the *air* is kept warm, the conduction loss of heat is less. Part of the reason for the greater warmth of the air in winter, characteristic of American offices and houses—or for that matter modern Russian buildings—no doubt arises from this; it helps to balance the radiation loss. Much advantage is also to be gained by careful insulation of walls and ceilings; the doubling of windows is also helpful as the air space between is also a poor conductor.

In the four months November–February many of the great American cities enjoy from three to five times as much bright sunshine as we do during the same period, and in the lower latitude it is much stronger. It is considered by many, therefore, that we have a physiological need for a high-temperature source of radiation in the form of an open fire or its equivalent.

In the realm of horticulture and gardening we find that with a very little shelter by means of walls, as in many a kitchen garden of the kind attached to country houses great and small, much can be done to mitigate the incidence of severe frost and to decrease the damaging effects of strong wind. Choice of site, combined with the planting of trees, has again been the foundation of many a successful nursery. Such precautions again lie well within the compass of the individual; more so than in many countries where greater precautions are necessary. By way of illustration of these advantages we may recall how often the English gardener can adequately protect his delicate plants in severe weather by throwing a sack over each; in Stockholm the countrymen of Linnaeus must devise much more elaborate hoods for their carefully nurtured exotics. By contrast we may also remember the grateful shade of the many deciduous trees in the Buen Retiro at Madrid; but each is surrounded at the base by a stone-lined pit to which the water can be led at intervals during the pitiless drought of summer. The English gardener often gets his results with his own watering-can. May we not deduce from all this that by comparison with many other countries our climatic environment offers especial encouragement to small-scale individual effort and enterprise, of which the most prominent expression to the foreign visitor is the endless variety of suburban gardens relieving the universal monotony of the semi-detached villa?

The astonishing variety of temperate flowers, fruits and vegetables that we can with care cultivate has already been noted. From the time that the Crusaders brought us cauliflower, men have combined to embroider the British landscape with new introductions. Small-scale enterprise in yet another respect has thus been encouraged. But our climate has its limitations, shown very well by the fact that few Mediterranean fruits will ripen, owing to the lack of summer sunshine and warmth. The olive baulks at the Cornish Riviera. As for the vine, it remains a shadowy ghost along the slopes of the North Downs and medieval Cotswolds. Even though our summer climate may have been a shade warmer during several decades of the eleventh and thirteenth centuries the evidence is doubtful that any wine worthy of the name was ever made on a large scale from English grapes. Yet experiments in vine growing are again being tried in Surrey and the fact that even the pomegranate has been ripened in 1933 and 1949 is noteworthy.

Climate then has been one of the major factors in the evolution of our present landscape, which in England at least owes so much to the work of man. The characteristic park-like vistas of the Midlands; the varied trees, native and exotic; the hedgerows and irregular enclosures all testify to the individualism of successive generations of land improvers, following the many centuries when as Trevelyan has said, the sound of the axe in the woodlands was the ground-bass of English history. Scotland and Wales are less easily tamed. Yet even in the far north-east the stone flags fencing the fields of Caithness and the great consumption dykes outside the Aberdeenshire farms not only form an element of the scenery; they testify again to a climate in which hard individual effort has more often been rewarded rather than discouraged. The Aberdeenshire farmer will for long continue to cast the loose stones from his ploughlands on the adjacent pile, that he may save his plough-share; and when the September west wind is drying the pale oatfields of Buchan he may well take pride in his accomplishment together with the resultant progeny of divines and dominies.

Climate too has indirectly played its part in the widespread and early growth of small-scale industry; the ever-open sea and variable but lively air were rather a stimulus than a deterrent to navigation. Toynbee's environmental challenge was always present in one form or another. Yet we must not forget that it is a corollary of our climatic environment that round the corner out of the wind there are numerous sheltered places in which the grasses flourish, the animals feed themselves and life for many can appear to be dangerously comfortable and secure until the local resources begin to fail. East of Dartmoor, South Devonshire has a long tradition of conservatism.

Climate, Disease and the Death-rate

Crude statistics of the death-rate do not at first suggest that the British climate is extravagantly healthy. But the summer and winter temperatures are such that protection against disease is relatively easy. The multiplication of fungi and bacteria, the spread of insect-borne and water-borne diseases are in large measure affected by air temperature. The American visitor is often horrified to see the meat exposed for several hours on the butcher's slab open to the street; he forgets that many risks can be taken with a July mean temperature of 58° which are quite unpardonable when the mean reaches 75°.

Moreover the overwhelming majority of our population dwell in towns and cities; if we take account of this it becomes evident that our climate is one of the healthiest in the world, largely because the sanitary precautions necessary for the continued existence of urban communities can be less elaborate, both in summer and winter. Severe winter cold in the more primitive communities of Eastern Europe leads to overcrowded and uncleanly dwellings among which typhus and other louse-borne diseases are endemic.

Sanitary measures were relatively easy to apply; but before they became effectively organised in the middle of last century the death-rate of London had become truly appalling by our present standards. Not only were eighteenth-century writers fond of commenting on the health and stature of many country folk; the development of the public schools owes an enormous amount to the instinctive recognition that even in our temperate climate children throve better away from the town.

Eighteenth-century physicians were passionately fond of discussing the relationship between diseases and the weather; some of these habits of mind persisted, in lengthy Victorian publications regarding the merits of British watering-places. In early Victorian days we find a remarkably widespread interest in meteorological records at our south-western resorts, probably deriving from Dr. Forbes' work on the climate of Penzance (1807–1828). Careful digests of observations can be found for Helston, Truro, Falmouth, Plymouth, Torquay and Sidmouth among other places dating from the age when the many delightful little houses were being built from naval prizemoney, and from the profits made in tin, copper, East Indiamen and Brunel's grand conceptions of what a railway should be. To this day the tradition survives among the many retired naval officers and others who maintain a rain-gauge for the British Rainfall Organisation in those parts.

With its limited range of temperature and the normal absence of persistent steady heat or cold our climate is easier to live in, and so imposes less of a strain on our organism. The daily individual precautions against disease, which may become an irritation in Egypt and a nerve-racking anxiety in Nigeria, are scarcely necessary thanks to public sanitary measures of relatively simple and inexpensive application.

Yet other ailments beset us besides those devastating epidemics against which fences have been erected. There appears to be an undoubted association between dampness and some forms of rheumatism, while the drag on our energies imposed by the common cold is significant. How much this is due to excessive urbanisation, and how much to the initial chill induced by damp clothes, internal or external, combined with a penetrating gusty wind in the lower forties, it would be hard to say. With all their central heating and consequent dry warmth the Americans are similarly troubled. It is by no means improbable that the changeability of our weather from day to day tends to build up the resistance to such minor infections. At least one commentator on tropical hygiene has pointed out that monotony tends to lower the resistance, so that occasional departures from the normal are often marked by an outbreak of colds.

Longevity is not quite so great as in the Scandinavian countries, and the infant death-rate is by no means the lowest in the world; but obviously many factors besides climate must be taken into account. It is significant that the general well-being, as measured by vital statistics, of the excessively rainy, cloudy and cool Faeroe Islands is higher than that of Western Ireland or the Hebrides; this illustration will serve as an adequate reminder of the dangers of climatic determinism. Recently the fact that unemployment was greater in Lancashire than near London in the 1930's was advanced as an argument in favour of the climate of the south-east; it would be interesting to see whether Macaulay would not have advanced exactly the opposite view a hundred years ago. We had better content ourselves by saying that at least the British climate as a whole is less inimical than others to physical and mental liveliness; and that within it we can perceive differences whose explanation may now be attempted.

Bracing and Relaxing Climates

So conscious are we of the two opposed effects resulting from our maritime climate that we are prone to refer to many places as bracing or

---

Plate XXIII*a*: Breezy August afternoon on the Norfolk Broads; westerly wind with fair weather cumulus and cumulo-stratus.

b: Harvest in East Sutherlandshire. Frontal cirrus and cirro-stratus above heralding the approach of more rain; mid-September.

PLATE XXIII*a*. *Eastern Daily Press, Norwich*

b. *G. Douglas Bolton*

PLATE XXIV*a*. *C. L. Curwen*

b. *G. Douglas Bolton*

relaxing in a manner which may well puzzle an American visitor who compares the meteorological statistics. Falmouth and Blackpool have almost the same July mean temperatures and a very similar rainfall. Harrogate and Abernethy record very similar temperatures in summer, yet again the Scottish townlet at the head of the Tay estuary is regarded as relaxing while Harrogate has an almost universal appeal as one of the most bracing inland resorts to be found. Buxton lies in an upland valley with a considerably higher relative humidity in the summer months than Cambridge or Oxford. But no one who knows the sleepy languor of an August afternoon with a light south-west breeze at Cambridge will accept the view that owing to the greater dampness of Buxton that resort is more relaxing; and any comparison of opinion regarding the local climatic merits of British resorts reveals no simple explanation.

The bracing or relaxing qualities of the air are not yet measurable, and opinion regarding them varies immensely among people of different age and temperament. Yet after all psychological factors have been discounted, there is on many occasions a stimulus to be perceived in the atmosphere of some localities which elsewhere is more or less lacking. The like indeed may be said of times as well as places; "*Afflavit Deus, et dissipati sunt*" on the Armada medal epitomised an age, as well as the joyous turbulence of the fresh west wind of the English summer.

Bracing qualities are commonly attributed to places where the air comes free from pollution over a wide stretch of open sea, moorland or mountain; and it is possible that many town-dwellers sense a real if intangible benefit from a purer air. In 1687 we find Celia Fiennes complaining of the climate of Bath. In 1750 it became a custom to take a Sunday walk on Westminster Bridge to escape from the smell of the town. Further, there is little doubt that small daily changes in the local derivation of the air are associated with changes in quality, notably as regards temperature and humidity and possibly in other respects of which we are as yet ignorant. Such changes are appreciated

---

PLATE XXIV*a*: Late summer cloud-cap on Glamaig, Skye; illustrating the results of humid air flowing fairly freely over summits.

b: Midsummer midnight on the north coast of Scotland. Daytime cloud has subsided into long streaks of stratus; air nearly calm over the land. Distant remnants of cumulus over the sea.

as a stimulus to the surface of the skin and presumably therefore to the nervous system. Variations in wind speed also play their part. It is noteworthy, as we saw in previous chapters, that it is particularly among mountains or in the neighbourhood of the sea that local breezes are often superimposed on the generally prevailing wind system; hence no doubt the so-called bracing qualities of many seaside and mountain locations. But this is not the whole story. The incidence of the sea-breeze in summer varies considerably with differences in the aspect of the coast; and we have already seen (page 158) that its scouring effects in the streets of Eastbourne and Bournemouth are likely to be less felt than at Brighton. On the North Sea coast the sea tends to be markedly cooler than the land in summer, more so than in the Channel; hence in quiet weather the thermal component of the coastal breezes is better developed (compare pp. 156–60 and Fig. 52). The North Wales coast resorts are generally considered to be less bracing than those of Lancashire and it will be evident that if the gradient wind is on the whole W.S.W. or W., the sea breezes reinforce the wind on the Lancashire coast, and incidentally more markedly at Blackpool than at Southport; whereas in N. Wales the thermal component is at right angles, or even slightly opposed to the gradient wind. Moreover there is some evidence that when the wind descends warm and dry from the mountains a sense of relaxation prevails; the warm and gusty south wind, at places such as Forres or Nairn, from time to time in summer represents the small-scale British modification of the Swiss föhn. (Cf. p. 256). Whether Scotsmen are then given to displaying even a small measure of Viennese temperament has apparently not been ascertained.

Windiness alone does not make a bracing climate. Too much wind even if it is merely audible provides for many a degree of nervous irritation which may prove tiring; this appears to be especially true if the air is extremely dry, as is the föhn of the Swiss valleys. Strong wind with a low humidity is however extremely rare in the British Isles. More typical are the endless days when the mild and nearly saturated south-wester roars unimpeded over the flatter Outer Hebrides and Donegal coast; and it must be admitted that many who know these windswept fringes of Britain do not regard them as bracing either physically or mentally. Speyside is a different matter. Here and elsewhere we must also remind ourselves that the mental stimulus derived from sojourn in a given locality is not merely a product of the climate; it owes far more to the abundance of personal historical, literary and scientific associations which the landscape calls to mind. Hence an ornithologist such as James Fisher might be tempted to ascribe more bracing qualities to South Uist than would a literary historian. Discounting such varied opinions as far as we can it still remains true that under British climatic conditions the majority regard too much damp wind as a deterrent rather than a stimulus to activity. Plants, animals and man tend to save energy by adopting the prone position. On the long winter nights there is little to do but "sing me a song of the lad that is gone", and the beauty of calm days in the Hebrides is the more idealised on account of their rarity.

Bracing and relaxing qualities may well owe a good deal to the frequency with which the layer of air within a few feet of the ground is changed, for one differing slightly in quality. The rate of change with height of the moisture content of the atmosphere is probably important and requires further investigation. Under British climatic conditions there is generally plenty of moisture in the ground and when convection is active, evaporation into the surface layer and the rising air currents implies that if there is little wind a man near the ground finds himself for many hours in air of much the same quality. The lack of stimulus in the humid tropics may thus be explained. But if for example on a sunny day in May a dry north-wester is blowing, the surface air is turbulent and rising bubbles of humid air from ground level alternate frequently and rapidly with descending packets of the dry, and physiologically cooler air above. Even in the streets of London warm and cool puffs alternate; perhaps indeed they are more frequent on account of the different degree of heating on the sunny and shady sides of streets of varying width. At least one woman meteorologist has verbally expressed her strictly unofficial opinion that London is more bracing than the Thames valley above the city. But on days when a more humid, tropical air stream prevails, the humidity gradient from the surface upward is small. The wind may be quite strong and yet the descending bubbles of humid air scarcely allow of any more evaporation from the skin than those which are rising. Little stimulus is experienced; and as in summer the temperature is often high, exertion itself quickly leads to overheating.

The notably relaxing qualities of a valley-bottom resort such as Bath probably derive from the fact that even on a 'bracing' sunny day with polar air prevailing the proportion of moist packets of air rising from the bottom and slopes of such a sheltered valley is, by comparison with an open plain, greater than that of the descending packets. Indeed replacement of an ascending packet may be effected by a similar humid packet rolling up the valley. It appears that it is the combination of moisture with inland warmth and restricted air movement that makes for relaxing qualities, and this is borne out by the general opinion regarding the Breckland of West Norfolk, around Thetford and Brandon. Here the sandy heaths and pine forests cover many square miles; the soil drains very quickly and with a fairly low rainfall the area is one of the driest in England. Although it lies a mere 50–100 feet above sea level many people find the air decidedly more bracing than that of Cambridge.

This hypothesis, that relaxing qualities result from ground moisture combined with a high humidity in the surface layers and a lack of air movement favouring the interchange from above, agrees well with experience in what many deem to be the most overwhelmingly sleepy localities in the British Isles. At the heads of the western Scottish lochs, and even more those of South-west Ireland, the conditions of a place such as Bath or the lower parts of Exeter are exaggerated. The slopes are almost always moist and readily steam in the sunshine; if there is any wind to scour away the rising humidity, it must always blow along the length of the water so that the surface layers acquire to a considerable height very much the same damp quality as the air adjacent to the slopes. The majority of visitors arrive in the summer or towards autumn when gentle south-west to westerly winds prevail. There are many days when a gentle drizzle falls at intervals from the soft grey stratus begotten of maritime tropical air, or even maritime polar air with a long Atlantic travel; when even that spur to utilitarian activity, the Nonconformist conscience, is lulled into a helpless uneasy lassitude. Harassed barristers and nerve-racked business men find no better cure than fishing under such conditions.

Another type of day which almost everyone finds particularly enervating occurs when in the summer months a 'warm grey sky' prevails. Occasionally, with tropical air in particular, much of the country lies under a continuous sheet of low stratus cloud above which decidedly warmer air is found; in such an event the surface temperature is in the neighbourhood of 70°, the humidity is high, and the radiation from the warm air above through the cloud-sheet has a peculiarly oppressive quality, very different from the wider assortment of wave-lengths which we experience on cloudless days of dry air from a brilliant sun. It is noteworthy that the enervating qualities of such days increase rapidly with temperature as the diagram on p. 292 will show; and that such days tend to be more numerous towards the west, notably in Southern Ireland.

The diagram is taken from the Presidential address by Sir David Brunt to the Physical Society in 1947, in which he brought together the results of several recent investigations on the reactions of the human body to its physical environment. It will be observed that with high relative humidities the critical temperature which the body finds oppressive lies between 67° and 80°; above such a temperature when we exert ourselves our cooling arrangements are insufficient to prevent the body temperature from rising above normal. Even near that value conditions are far removed from comfort and a considerable sense of oppression will be felt by anyone attempting vigorous exertion. It must however, be remembered that the cooling power, that is the rate of removal of heat per second by air in movement increases rapidly unless the temperature is above 80°; for example at 70° a wind of 20 m.p.h. has broadly speaking about double the cooling power of a light breeze of 5 m.p.h. and four times the cooling power of calm air. But, as Brunt remarks, "with strong winds there will be a tendency for the wind to blow through the material of the clothing, and to blow through the openings at neck, wrist and legs; it is not possible to assess these effects quantitatively." He has also produced a tentative classification of climates, shown on the diagram. It will be observed that with British temperatures between 30° and 80° and humidities between 20% and saturation, every one of the descriptive adjectives can from time to time be applied with the possible exception of "becoming irritating" when high temperatures are combined with very low humidity. For most British stations it appears very doubtful whether the relative humidity ever really falls below 10%. Suffice it to recall the many months of raw cold and incessant wind below 40° which our mountains provide, shown in Appendix, Table III.

Exposure to the wind at a moderate elevation on a well-drained slope is likely to offer the most bracing conditions to the town-dwellers inland, a fact confirmed by the siting of many favoured residential areas. Hampstead; Shotover Hill near Oxford; and Alderley Edge near Manchester afford good examples. Sheltered valleys debouching on to the south and south-west coasts are often remarkably free from

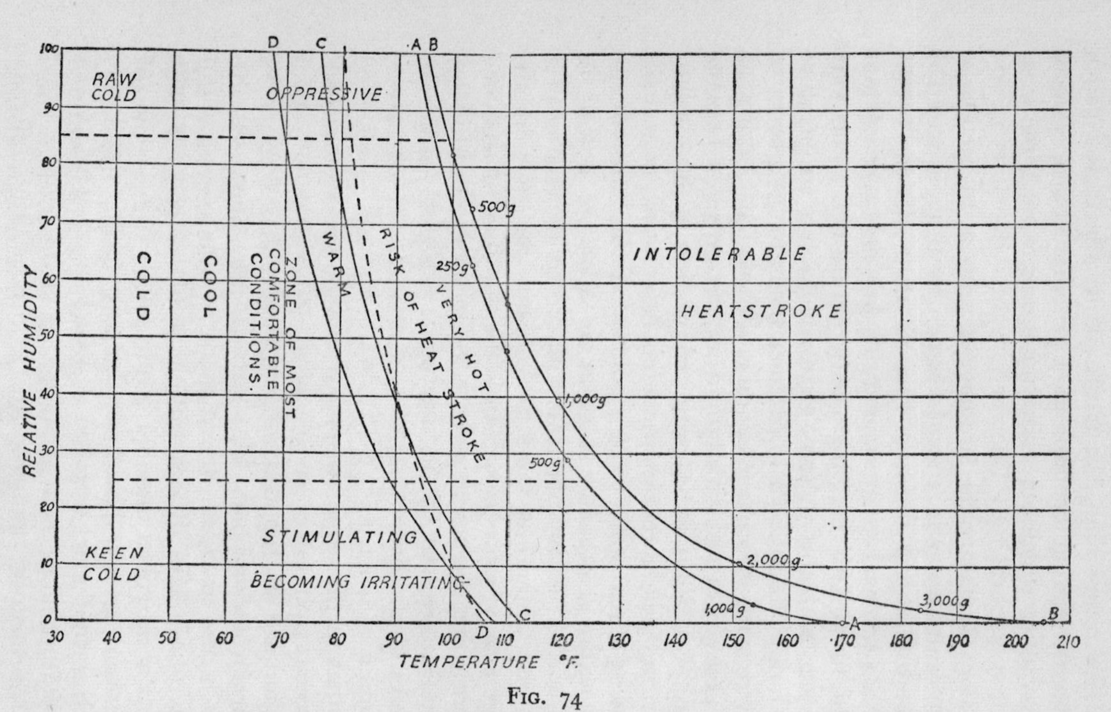

FIG. 74

AA. Heat-stroke limits for nude man resting in still air. BB. Heat-stroke limits for nude man resting in air moving 200 ft. per minute. CC. Limiting conditions for clothed man resting in sunshine with about one-third of skin wetted with sweat. DD. Limiting conditions for clothed man walking 3 m.p.h. with about one-third of skin wetted with sweat. The broken line represents equivalent temperature 80°F. The figures 500 g. etc., indicate rate of evaporation of sweat in grammes per hour for men of average size in order to maintain heat balance of the body. (Reproduced by permission of Sir David Brunt and the Physical Society from Proc. Phys. S., 1947)

frost, but are considered to be relaxing as we have seen especially when the hills on either side are high. In North Wales the Clwyd valley round Denbigh has a similar reputation. Open flat coasts adjacent to our cooler seas, over which sea and land breezes can play freely, are often regarded as bracing. Elsewhere, much depends on soil moisture; the clayey Chiltern ridges are not generally considered to be as bracing as the sandy uplands of south-west Surrey. In spite of the relative lack of sunshine and the high rainfall many of the Pennine industrial towns at higher levels are remarkably bracing, and here it is probable that the small daily exchanges between hill and valley play some part. Coupled with this we may recognise another factor. The arrival of warm humid Atlantic air in summer is commonly associated with a peculiarly relaxing effect at many inland towns. At Cambridge, an afternoon maximum of 75° with humid air, a dew-point between 60° and 65°, patchy cumulus and vigorous convection in the sunshine is characteristic. But nearer the approaching Atlantic depression the south-westerly wind often blows more strongly on the northern uplands; and even with a dew-point as high as 60° a force 5 wind in the suburbs of Huddersfield gives appreciably more cooling effect to the body than the force 3 breeze of the East Midland afternoon. Moreover, there is generally more cloud and less sunshine under such conditions, so that when the temperature is 75° in the South Midlands it is probably not more than 67° in the streets of Bradford. Under conditions of high humidity every degree makes an appreciable difference to human comfort, as the diagram above will again show. Hence we find that in the same air-mass a mere six hundred feet of elevation combined with rather more cloud and wind make for a considerable difference in the comfort of those engaged in active exertion, whether work or play. This goes far to explain the perceptible sense of increased vigour which many visitors readily admit.

In all this we have said little about the stimulating or relaxing qualities of the air in the cooler months. Strong winds even at a moderately low temperature remove heat from the body more rapidly than any other agency normally met with. The cooling power of a 30 m.p.h. wind at a temperature of 40° is as great as that of calm air at a temperature of 15°. At lower temperatures it makes little difference whether the air is dry or damp. Over the range from 35° to 50° or so dry air cools us more rapidly than moist air; hence the cutting qualities of the dry keen north-easter of spring even though its temperature may approach 50°. Damp air in the thirties owes its peculiar raw chill to the fact that evaporation from the surface of the skin beneath the clothes soon leads to saturation of the adjacent air and hence a moist layer in the garment next to the skin. The moisture forms a good conductor of heat from the body to the cool air outside and unless further thicknesses of clothing are worn a sensation of chill quickly develops. Wool in a damp climate has the advantage that it is absorbent and that the addition of moisture does not increase the conductivity so rapidly as would be the case where the fabric does not enclose so much air between the strands, for example cotton and its derivatives. Wool moreover is not airtight, a manifest advantage to those who exert themselves under humid conditions unless the temperature is far lower than we normally experience. Oats and sheep are an essential accompaniment of the north-west British climate; porridge for breakfast and wool next to the skin however unfashionable should remain the portion of those who would enjoy the British winter.

Clothing and diet, gardens and literature, parklands and pleasure resorts all form part of the British scene derived from the work of man, yet it is evident that the characteristic development of all these elements in our landscape has repeatedly been affected by climate. Indeed we have essayed the view that these emblems of British culture would not have taken the form they have were it not for our climate, which continually shows us that at one place or time its effects are stimulating and persuasive, while elsewhere on another occasion they are relaxing and permissive. We may ask ourselves whether the varying predominance of the associated mental traits at different times in our history cannot also be recognised; the cheerful and rather casual acceptance of Nell Gwynne, "let sleeping dogs lie", and the era of appeasement contrasted with the decision implicit in Cromwell's Navigation Act, the defeat of the Armada and the emancipation of the slaves.

Twenty years ago Professor Ellsworth Huntington of Yale expressed the view that the climate of South-east England might even be more advantageous than that of Yale with regard to the development and progress of civilisation. We now cast doubt upon his somewhat crude basis of assessment, yet the fact remains that a civilisation based on human effort is likely to thrive better in a region where both effort and forethought are steadily demanded and rewarded. In many lands they are discouraged, whether by natural calamity, prolonged uniformity of heat or cold, the rivalry of other forms of life, or by the more subtle results of soil exhaustion which again is largely a matter of climate. Such territories can only maintain a civilisation with the aid of large-scale technique, applied by some form of collective effort under a management with all that it implies. The battle for maintenance of lively modern communities in the Missouri valley, or in that of the Yenesei, is only beginning; a century hence we may be able to estimate the permanence of such extensions of the civilisations derived from the Western European nursery of small-scale individual effort in which all our technical accomplishment has been initiated.

Under the alternation of irritating blandishment and kindly asperity which the British climate provides we are at least secured from soil erosion; and every feature of our native environment is conducive to diversity, deviation and individual differentiation in plants, animals and men. Combined with an agreement to differ we find co-operation for common ends, in legal matters, in defence and occasionally in development. It is only in the man-made environment of our cities that a degree of monotony has been developed which is potentially similar in its effects to the Russian plains or the American prairie. It remains to be seen whether this group of islands, in which diversities of climate and structure have played so large a part in moulding the attitude of mind, can possibly be administered by methods and systems begotten of cogitation in more uniform lands, however greatly their logic may appeal to that large proportion of our population which has for generations been removed from the sublimely irregular complexities and subtle adjustments so characteristic of the country we know.

Whatever be the result, oxymoron as a figure of speech will continue to be appropriate in descriptions of that paradoxical British climate which defies definition and in which squalid culture, orderly disintegration, and untidy neatness have repeatedly been found by the bewildered observer from abroad; sufficient use of the figure of speech has been made already to justify the assertion that the critic of British institutions may praise or blame the climate as he wishes. No precise quantitative evaluation of its effects capable of satisfying a scientific inquiry can yet be made. No precise equivalent can be found elsewhere in which we can study the reactions of an immigrant community.

New Zealand is free from the effects of a continental air supply in winter and has more powerful sunshine. Vancouver Island at its southern end has similar temperatures but a drier and more settled summer. Tierra del Fuego is altogether cooler, though generally drier apart from the mountainous westward fringe.

The fundamental advantages of our climate lie firstly, in the encouragement it gives to the development of local differentiation; the consequent variety of objects composing our scenery in a short distance cannot fail to act as a mental stimulus. Moreover, such contrasts lie well within the reach of the poorest individual; they are not a matter of expensive travel. Secondly, our climate imposes a relatively small tax on human energy. Since glass became common after the Reformation, such protection as we require has lain for the most part well within the compass of individual enterprise and conscience. Drainage of the farmlands and improvement of tillage followed; with the result that over the last three centuries there has been a widespread release of energy for other purposes. But such advantages were also enjoyed in Holland and the other countries of north-west Europe; and it must not be forgotten that the potential demerits of the British climate are exemplified in those regions of Western Ireland where nascent enthusiasm is too easily damped down, from which the younger generation continues to migrate while the older generation recounts the stories of a long-past heroic age associated not only with invasion but also with diminished rainfall and a keener air. Britain may well be but one stage removed from a like fate. Further, who can fail to recognise the immense drag upon our productive energies imposed by the maintenance of our obese capital city, whose size has so far outgrown our resources—limited again as they ultimately are by our climate?

It must rest with future scientists to devise methods of measurement of qualities such as climatic stimulus, productivity of original accomplishment in arts or engineering, loss of energy due to friction from overcrowding. Eventually a balance might be struck between the expenditure of human energy necessary for continued existence and the gain due to natural causes. When that has been done we shall be in a position to declare whether the British climate is ultimately an asset. At present many who judge by the historical and geographical results, of which an indication has been attempted in this and earlier chapters will be tempted to the opinion that the astringent mildness of our dominant maritime-polar air is one of the few unalterable comforts we possess, however much we may personally deplore its harassing benevolence. Truly the air that reaches us, in the words of Shakespeare's Ariel

> "......... suffers a sea-change
> Into something rich and strange."

Yet with Harrison the Elizabethan topographer we can also agree:—

> "We have if need be sufficient help to cherish our ground withall, and to make it more fruitful ...... neither is there anything found in the aire of our region, that is not usually seen amongst other nations living beyond the seas".

REFERENCES

Chapter 14

BRUNT, D. (1947). Some physical aspects of the heat balance of the human body. Presidential Address: *Proc. Phys. S. 59:* p. 713–726.

BROOKS, C. E. P. (1950). *Climate in Everyday Life.* London: Benn.

ELLIS, Havelock (1927). A Study of British Genius. New edition, revised and enlarged. London, Constable; chapter II, p. 36, 40 *et seq.*

FLEURE, H. J. (1951). *The Natural History of Man in Britain.* London: Collins' New Naturalist.

HANDISYDE, C. C., (1947). The Climate of the Home. *Weather*, **2**: *82–88:* gives further references to recent work.

HOOKER, R. H. (1922). The Weather and the Crops in S. England, 1885-1921. *Q. J. Roy. Met. S. 48:* 115-38.

HUNTINGTON, Ellsworth (1924). *Civilisation and Climate.* 224-26, 229. Third edition, New Haven: Yale Univ. Press.

(1945). *Mainsprings of Civilisation.* 384. New York, John Wiley; London, Chapman & Hall.

LEWIS, W. V. (1943). Some aspects of percolation in S.E. England. *Proc. Geol. Ass. 54:* 171-84.

MANLEY, G. (1957). Climate fluctuations and full requirements. *Scot. Geogr. Mag., 73:*, No. 1, 19-28.

NORMAND, C. W. B. (1920). The effect of high temperatures, humidity and wind on the human body. *Q. J. Roy. Met. S. 46:* 1-14.

STONE, R. G. (1941). Health in Tropical Climates: *in* Climate and Man. *Yearb. U.S. Dep. Agri.* (1941)*:* 246-61.

WADSWORTH, J. (1948). Evaporation from tanks in the British Isles. *Weather, 3:* 322-24.

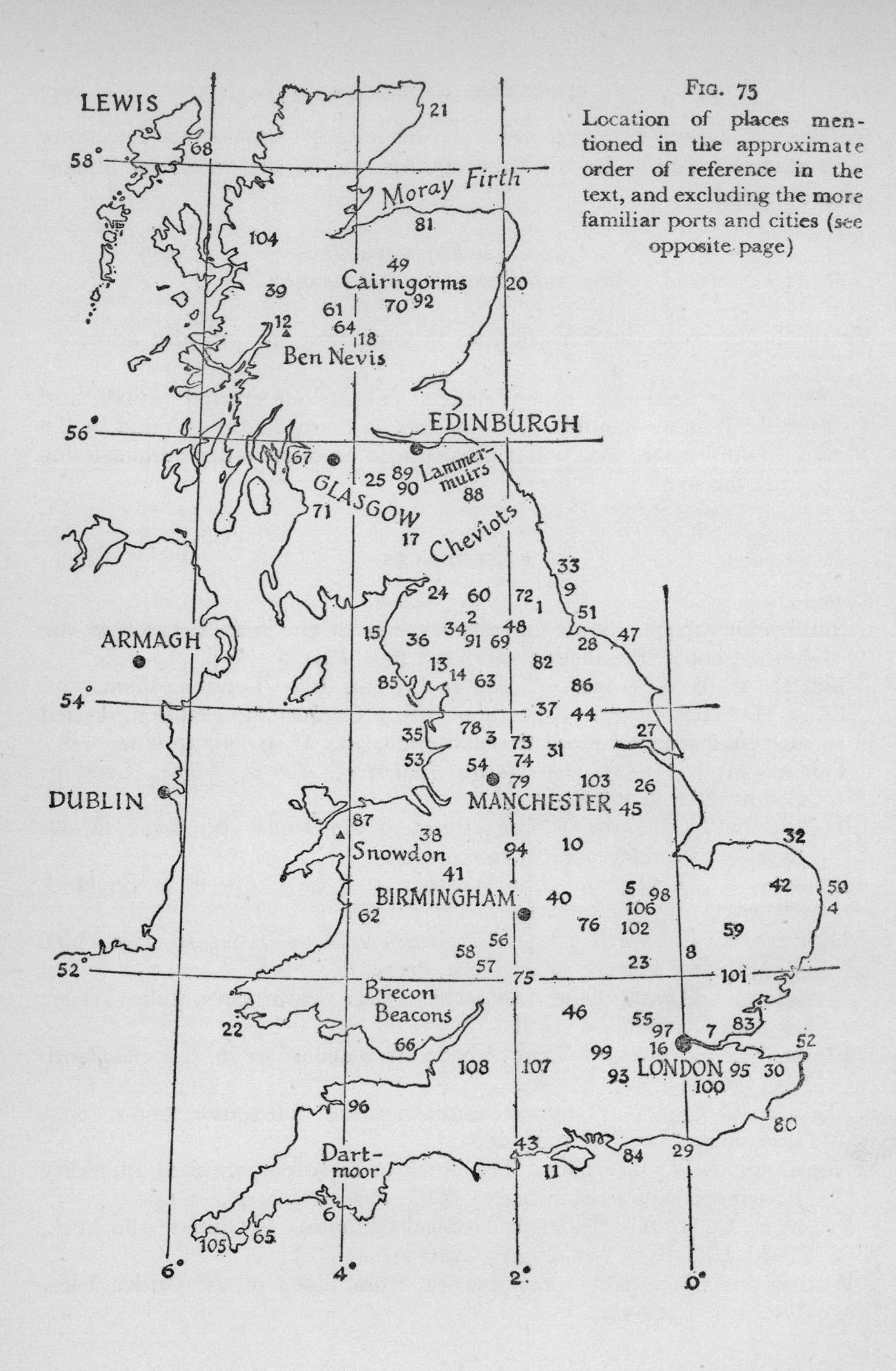

Fig. 75

Location of places mentioned in the approximate order of reference in the text, and excluding the more familiar ports and cities (see opposite page)

List of places referred to in the text—with reference numbers as given on the map

1. Durham: Ushaw, Houghall
2. Crossfell: Moorhouse
3. Burnley
4. Lowestoft
5. Lyndon: Stamford
6. Plymouth
7. Upminster
8. Cambridge
9. Sunderland
10. Belpter
11. Totland Bay
12. Ben Nevis: Fort William
13. Windermere
14. Kendal: Leven-Kent estuary
15. Whitehaven
16. Kew
17. Eskdalemuir
18. Blair Atholl
19. Lunds: Aisgill, Garsdale
20. Aberdeen
21. Wick

Leven-Kent: see 14

Houghall: see 1

22. St. Ann's Head
23. Cardington-Bedford
24. Carlisle: R. Eden
25. Carstairs
26. Lindsey
27. Holderness
28. Cleveland
29. Brighton
30. Canterbury
31. Wakefield
32. Cromer
33. Tynemouth
34. Penrith
35. Blackpool
36. Buttermere: Keswick, Derwentwater
37. Harrogate
38. Wrexham
39. Glengarry-Glenquoich
40. Atherstone
41. Shrewsbury
42. Norwich
43. Bournemouth
44. York
45. Lincoln
46. Oxford
47. Whitby
48. Teesdale: Weardale
49. Braes of Glenlivet
50. Yarmouth
51. Hartlepool
52. Margate
53. Southport
54. Bolton

Keswick-Derwentwater, see 36

55. Rickmansworth

Ushaw, see 1

56. Perdiswell-Droitwich
57. Malvern
58. Bromyard
59. Breckland: Mildenhall, Lynford
60. Alston

Moorhouse, see 2

61. Dalwhinnie
62. Plynlimmon
63. Ingleborough
64. Dalnaspidal
65. Falmouth
66. Cardiff
67. Greenock

Fort William, see 12

68. Stornoway
69. Stainmore
70. Braemar
71. Prestwick
72. Consett
73. Queensbury

Stamford, see 5

74. Huddersfield
75. Snowshill
76. Cold Ashby
77. Maiden Castle
78. Fairsnape Fell: Stonyhurst
79. Kinder Scout
80. Dungeness
81. Stonyhurst, see 78
81. Gordon Castle
82. Catterick
83. Great Wakering
84. Bognor
85. Black Combe
86. Ampleforth
87. Carnedd Llewelyn
88. Kelso
89. West Linton

Walton-on-Thames, see 16

90. Peebles
91. Appleby
92. Balmoral
93. South Farnborough
94. Cheadle
95. Maidstone
96. Barnstaple
97. Wealdstone
98. Peterborough

Mildenhall, see 59

99. Reading
100. Tonbridge
101. Halstead
102. Raunds
103. Bawtry
104. Achnashellach
105. Penzance
106. Oundle
107. Marlborough
108. Bath

TABLE I: MONTHLY AVERAGE TEMPERATURES AND AVERAGE DAILY RANGE (°F.)
INHABITED ENGLAND: EXTREME VALUES FOR SHELTERED CITY AND BLEAK UPLAND

	Alt.	Period	J	F	M	A	M	J	J	A	S	O	N	D
(*a*) London (Westminster, St. James's Pk.)	27	1916/45	41·1	41·3	44·7	49·1	55·9	60·9	64·1	63·8	59·5	52·5	45·4	41·8
			9·4	10·2	13·1	14·3	16·3	16·1	15·2	15·2	14·1	12·3	10·2	9·0
(*b*) Moorhouse, Upper Teesdale (approximate limit of cultivation)	1840	Reduced to same period	32·5	32·4	34·5	38·5	45·0	49·5	52·9	52·5	48·5	42·3	37·0	34·0
			7·2	8·0	9·6	12·2	14·2	15·2	13·6	12·6	12·0	10·2	8·0	6·6

Average annual rainfall (a) 23″ (b) 70″. Annual average, days with snow-cover at 9h. (a) 3 (b) 80 (approx.)
Extremes on record (a) 93, 15, (b) 80, -2 (latter 15 years only). 97° reached elsewhere in London, 1932.

TABLE II: AVERAGE DURATION OF BRIGHT SUNSHINE FOR EACH MONTH: HOURS AND % OF POSSIBLE.

			J	F	M	A	M	J	J	A	S	O	N	D	*Year*
Cambridge	1911–40	*hours*	50	67	114	141	194	207	185	180	145	108	60	40	1491
"Agricultural E. England"		%	19	24	31	34	40	41	37	40	38	33	23	17	33
Perth	1911–40	*hours*	41	64	106	143	171	203	162	146	128	88	53	34	1339
"Agricultural C. Scotland"		%	18	24	29	34	34	39	31	31	34	27	22	16	30
Fowey	1911–40	*hours*	57	75	130	172	197	224	200	187	155	115	73	56	1641
S. Cornwall, coast		%	22	26	36	42	41	46	41	42	41	35	27	23	37
Eskdalemuir (Dumfriesshire)	1911–40	*hours*	36	55	95	129	162	177	140	125	109	83	53	33	1197
(N. Uplands, away from smoke)		%	15	20	26	30	33	34	27	27	29	26	21	15	27
Eastbourne (Sussex)	1911–40	*hours*	59	84	139	173	238	242	233	221	175	121	71	53	1809
"Bright south coast"		%	23	30	38	42	50	49	47	49	46	37	26	22	41
Baltasound (N. Shetland)	1911–40	*hours*	18	38	82	129	164	158	126	126	97	64	27	10	1039
Cloudy hilly islands, 60° 44′ N.		%	9	15	23	30	30	28	23	26	25	20	12	5	23

Calculated by the author from Meteorological Office published data. Since this was done, averages of temperature and sunshine have now been published for 1921-50 (M.O. 571, and 572, H.M.S.O.).

TABLE III

Averages of Temperature for each Month (°F.), Lowland and Coastal Stations

		J	F	M	A	M	J	J	A	S	O	N	D	Daily Range Jan.	Daily Range July	Recorded extremes	
Baltasound (Shetland)	1917–47	38·9	38·9	40·0	42·5	46·7	50·3	54·1	54·3	51·0	46·5	42·5	40·5	7·9	9·4	77	14
Nairn	1911–20 31–49	37·9	38·1	41·0	45·0	49·5	54·4	57·7	57·3	53·7	48·0	42·5	39·6	10·1	13·2	87	8
Tiree	1927–49	41·5	41·4	43·1	45·9	49·9	53·7	56·5	56·9	54·6	50·4	45·9	43·4	6·1	8·4	78	20
Keswick	1911–49	39·6	39·7	42·0	45·6	51·5	56·1	59·2	58·5	54·6	49·2	43·4	41·0	10·6	13·5	91	0
Durham (*extremes since 1847)	1911–49	37·6	38·4	40·7	44·8	49·9	55·3	59·0	58·4	54·4	48·2	42·3	39·2	9·7	16·5	89*	-1*
Wakefield	1916–49	38·7	39·1	42·7	46·9	52·5	57·7	61·4	60·7	56·6	49·8	43·1	39·9	10·0	17·6	90	5
Cromer	1911–20 1926–49	39·0	39·7	42·6	46·8	51·9	57·6	61·3	61·4	58·5	51·3	44·9	40·9	9·3	14·5	94	14
Hereford	1911–49	39·1	39·4	42·4	47·1	52·8	57·8	61·0	60·3	55·9	49·1	43·1	40·2	10·7	18·2	92	-4
Aberystwyth	1911–20 1926–47	41·3	41·3	43·4	47·2	52·8	57·0	59·9	60·7	57·5	51·8	46·4	43·0	7·7	9·8	91	13
Reading (University)	1911–47	39·4	40·1	43·3	47·8	54·5	60·3	62·5	62·0	57·2	50·7	43·5	40·6	10·7	18·2	95	9
Brighton	1911–40	40·9	41·1	43·7	47·7	54·4	58·8	61·8	62·7	59·1	52·7	46·5	42·3	7·8	11·5	90	14
Cullompton (Exeter)	1916–49	41·1	41·2	43·9	48·0	54·0	59·2	62·0	61·3	57·5	50·9	44·4	41·5	11·5	18·8	92	2
Penzance	1911–20 1926–49	45·1	45·1	46·7	49·8	54·3	58·2	61·4	62·1	59·4	54·2	49·3	46·3	7·9	11·4	85	17
Armagh (N. Ireland)	1911–49	40·2	40·9	42·7	46·2	51·3	56·1	58·9	58·5	54·7	49·2	43·5	41·0	9·5	14·6	88	10
Ballinacurra (Cork)	1911–40	42·5	43·0	44·1	46·8	52·0	56·6	59·5	59·3	56·1	50·9	45·3	43·0	10·6	13·7	86	19

Calculated by the author from Meteorological Office published data, to 1949 where practicable. Extremes revised to 1960. Averages of temperatures for 1921-50 have since been published (M.O. 571, H.M.S.O.).

TABLE III (CONTINUED)

UPLAND AND MOUNTAIN STATIONS

Station	Height	Period	J	F	M	A	M	J	J	A	S	O	N	D				
Dalwhinnie (Grampians)	1176	1931–49	33·0	34·1	36·9	40·1	46·3	50·2	54·7	54·3	50·1	44·1	38·7	35·6	8·2	14·9	86	−5
West Linton (Peebles)	820	1911–47	35·2	36·0	38·1	42·0	47·5	52·7	55·8	55·1	50·7	45·1	39·3	36·8	11·7	16·2	87	−6
Bellingham (Northumb.)	848	1926–47	35·0	35·6	38·4	42·7	48·6	53·5	56·8	56·1	51·8	45·7	40·0	36·6	10·2	15·5	86	−3
Rhayader (Radnor)	757	1919–49	38·3	37·8	40·6	44·6	49·9	54·8	58·0	57·9	54·0	47·9	42·0	39·5	10·4	15·1	87	−10
Princetown (Dartmoor)	1359	1916–20, 1931–49	37·1	36·7	39·9	43·8	49·4	54·3	56·6	57·4	53·9	47·6	42·6	38·8	8·5	12·6	86	8
Ben Nevis	4406	Estimated for 1911-49; based on old or shorter records	25·0	24·0	25·4	28·5	33·2	39·3	41·6	41·2	38·3	32·8	29·0	27.0	(Yr., 32·1)		66 (1884–1903)	1
Dun Fell (Westmorland)	2735		28·8	29·0	31·0	34·8	40·5	45·8	49·0	48·8	45·0	39·0	33·8	31·0	(Yr., 38·6)		(76) (1937–1940)	(7)

Calculated by the author from M.O. published data. For snow, see p. 207 and Table VII. Averages of temperature for 1921-50 have since been published (M.O. 571, H.M.S.O.).

TABLE IV

AVERAGE ANNUAL NUMBER OF DAYS WITH THUNDER: PERIOD 1931–50

Station	Days	Station	Days
Inverness	3·4	Wakefield	20·4
Glasgow (Renfrew)	9·3	Birmingham	14·7
Bolton (Lancs.)	10·9	London (Kew)	16·1

Other representative data are given by W. A. L. Marshall, *Q. J. Roy. Met. S.*: 1934; and maps in the *Climatological Atlas* (M.O. 488; see p. 54).

TABLE V

AVERAGES OF TEMPERATURE WITH DAILY, MONTHLY AND EXTREME RANGES, °F.

	J	F	M	A	M	J	J	A	S	O	N	D
Cambridge 1911–47 (Botanic Gardens)	38·8	39·5	42·6	47·0	53·6	58·5	62·1	61·7	57·2	50·1	43·3	39·8
Average daily range	10·5	12·2	15·9	17·6	19·7	19·5	18·8	18·9	18·0	15·7	12·1	10·3
Average monthly	54/20	55/21	62/24	69/27	78/31	82/38	83/43	83/42	78/35	68/29	59/25	55/21
Recorded extremes (1876–1949)	58/4	67/1	73/11	84/21	88/24	93/31	95/36	96/38	93/28	80/21	70/8	60/0
Perth 1911–47	37·1	38·4	40·9	45·4	50·7	56·2	59·5	58·2	53·8	47·6	41·1	38·3
Average daily range	10·8	11·7	13·6	16·1	17·5	18·3	16·9	15·8	16·0	13·8	12·2	10·8
Average monthly	53/18	53/21	59/23	65/26	72/30	77/37	78/41	77/39	71/32	64/27	57/22	54/19
Recorded extremes	57/0	58/3	68/4	75/19	80/25	89/31	90/35	84/31	78/25	77/19	64/-7	59/7
Fowey 1911–47	43·9	43·9	45·5	48·9	53·9	58·2	61·2	61·5	58·6	53·5	47·6	44·9
Average daily range	9·6	10·3	12·3	13·6	13·7	14·1	13·1	13·6	13·5	12·0	11·1	10·0
Average monthly	54/28	55/29	59/29	65/33	71/38	74/43	75/47	76/47	73/42	67/36	59/31	55/29
Recorded extremes	56/13	59/18	66/22	75/28	79/32	83/38	84/43	85/40	83/35	73/30	64/25	60/22
Wick 1921–47	38·9	39·1	40·5	42·9	46·6	50·9	54·7	54·4	52·1	47·6	43·0	40·7
Average daily range	6·8	7·6	8·8	9·3	9·1	9·7	9·2	9·2	9·5	8·7	7·2	6·5
Average monthly (1911–47)	50/25	51/24	54/25	57/28	61/31	65/37	68/41	67/40	64/36	60/32	55/28	51/25
Recorded extremes (1911–47)	55/9	58/9	61/14	67/19	69/25	80/30	75/34	78/35	71/31	66/26	62/18	57/15

Calculated by the author from Meteorological Office published data. The new averages of temperature for 1921-50 are in general slightly higher than those for 1911-47 given above.

TABLE V (CONTINUED)

	J	F	M	A	M	J	J	A	S	O	N	D
Braemar 1912–47, 1120 feet	34·0	34·6	36·6	40·9	46·5	51·8	55·4	54·2	49·7	43·8	37·8	35·3
Average daily range	10·9	11·7	13·9	15·2	17·8	18·6	16·9	16·6	15·7	13·6	11·6	10·5
Average monthly extremes	49/11	49/13	55/13	61/21	69/25	74/32	75/36	74/34	68/29	62/22	54/15	50/12
Recorded extremes	56/–7	56/–4	69/–6	70/8	77/16	83/26	85/30	82/28	75/21	75/13	60/–10	54/–7
Oakes (Huddersfield) 762 feet 1925–47	37·1	37·2	40·7	45·0	50·1	55·9	59·6	59·2	54·9	48·3	42·4	38·6
Average daily range	8·9	9·4	12·2	14·1	16·2	16·5	15·3	14·9	13·3	11·2	9·1	8·1
Average monthly extremes	51/22	51/23	59/25	64/29	71/32	78/39	804/5	78/44	73/39	64/33	56/29	52/25
Recorded extremes	56/11	58/13	72/16	74/26	83/29	87/35	88/38	87/40	81/34	71/26	67/22	56/18

Notes. From the average daily range the average daily maxima and minima can be obtained and these in turn can be used to compare the normal chances of frost or great heat. Before 1912, Braemar recorded –17° in February 1895 but for the most part the above extremes have been but slightly surpassed, if at all, during the previous century as the Durham and Oxford records indicate. For Westminster in Table I it should be noted that other city stations have recorded up to 97° in the same period.

Oakes is included to illustrate the decreased severity of extremes, especially with regard to frost on an upland ridge by comparison with valley sites such as Braemar, Perth and Cambridge.

TABLE VI

AVERAGE MONTHLY RAINFALL, INCHES; FOR REPRESENTATIVE STATIONS

		Alt.	Period	J	F	M	A	M	J	J	A	S	O	N	D	Year
Baltasound	(Shetland)	31	1910–49	4·8	3·9	3·7	3·0	2·4	2·4	2·5	2·8	3·9	4·5	5·2	5·2	44·3
Wick	(Caithness)	115	1910–49	2·8	2·0	1·9	2·0	1·9	2·0	2·5	2·6	2·7	3·0	3·3	3·0	29·6
Nairn	(Nairn)	20	1910–49	2·0	1·5	1·4	1·6	2·0	1·9	2·9	2·7	2·3	2·6	2·4	2·0	25·4
Braemar	(Aberdeen)	114	1910–49	4·1	2·9	2·4	2·2	2·5	1·9	2·8	3·1	2·7	4·1	3·9	4·1	36·7
Perth	(Perth)	115	1910–49	3·0	2·3	2·0	1·7	2·2	1·9	3·0	3·0	2·4	3·2	2·8	3·0	30·4
Greenock	(Renfrew)	200	1914–49	7·8	5·3	4·1	3·6	3·6	3·3	4·0	4·5	5·0	7·2	6·6	7·2	62·2
Keswick	(Cumberland)	254	1910–49	6·4	4·5	3·6	3·3	3·1	3·4	4·3	5·2	4·9	6·4	6·1	6·7	57·9
Harrogate	(Yorks.)	478	1910–49	3·3	2·4	2·0	2·0	2·3	2·2	2·9	2·9	2·3	3·0	3·0	3·0	31·3
Manchester *W.P.*	(Lancs.)	125	1910–49	3·3	2·3	2·0	1·9	2·5	2·4	3·2	3·4	2·6	3·3	3·2	3·2	33·4
Northrepps (Cromer)	(Norfolk)	225	1910–49	2·6	1·8	1·7	1·8	1·8	1·6	2·6	2·3	2·3	2·7	2·9	2·7	26·8
Cambridge	(Cambs.)	41	1910–49	2·0	1·3	1·4	1·7	1·7	1·6	2·2	2·0	1·9	2·0	2·0	1·9	21·7
Aberystwyth *P.B.S.*	(Card.)	452	1925–49	4·2	2·8	2·2	2·1	2·6	3·0	3·9	3·7	3·7	5·0	4·7	3·9	41·8
Rhayader	(Radnor)	757	1917–49	5·8	3·8	2·9	3·0	3·0	2·5	3·5	3·6	3·7	5·0	5·1	5·4	47·3
Hereford	(Here.)	255	1910–49	2·6	1·8	1·8	1·7	2·1	1·7	2·3	2·5	2·1	2·8	2·6	2·6	26·6
Reading	(Berks.)	152	1910–49	2·4	1·8	1·7	1·8	1·9	1·7	2·4	2·2	2·0	2·7	2·6	2·6	25·8
Brighton	(Sussex)	140	1910–49	3·3	2·4	2·2	2·1	2·0	1·8	2·6	2·8	2·5	3·7	4·2	3·8	33·5
Canterbury	(Kent)	27	1910–49	2·5	1·8	1·9	1·9	1·9	1·6	2·3	2·2	2·1	3·3	3·3	2·9	27·7
Cullompton (Exeter)	(Devon)	202	1910–49	4·2	2·9	2·7	2·4	2·4	2·1	2·6	3·1	2·7	3·8	4·1	4·5	37·5
Penzance	(Cornwall)	55	1910–49	5·0	3·5	3·3	2·5	2·4	2·1	2·7	3·0	2·9	4·4	4·8	5·2	41·8
Armagh	(N. Ireland)	205	1901–40	3·0	2·4	2·2	2·1	2·4	2·4	3·1	3·4	2·5	3·2	2·8	3·2	32·7
Cork	(Eire)	100	1901–40	4·5	3·8	3·4	2·6	2·6	2·2	2·9	3·0	2·8	3·8	4·1	4·7	40·4

The newly published "Averages of Rainfall, for 1916-50" (M.O. 635, H.M.S.O.) came out in 1958.

TABLE VII: Number of Days with Precipitation 0·01″ or more and Months with Greatest and Least Average Number of Days. Period 1881–1915.

Station	Days	Greatest	Least
Shetland (Lerwick)	260	Dec. Jan. 27	June 15
Leith	182	Several : 17	June 13
Stonyhurst (Lancs.)	206	Dec. Jan. 20	June 14
Cambridge	163	Dec. 16	Sept. 11
Falmouth	207	Dec. 23	June 13
Armagh	215	Dec. 21	May, June, Sept. 16
Cities: Glasgow 202, Manchester 194, London 167, Cardiff 196			

Based on M.O. Book of Normals, H.M.S.O. Part IV, 1923.

TABLE VIII

Average Annual Number of Days with (a) Snow or Sleet Observed to Fall, (b) Snow Lying at 9h.

	Alt.	*Period*	J	F	M	A	M	J	S	O	N	D	*Year*	*Range of Variation: (Snow-Cover)*
Braemar	1120	1913–49	10·5	8·6	7·7	5·6	2·0	<0·1	0·1	1·7	4·4	7·2	47·8	
E. Highlands			17·3	13·8	12·3	4·0	0·7	—	0·1	1·6	5·1	11·5	66·4	32 to 122+
West Linton	770	1912–49	7·9	6·8	7·4	4·6	1·5	<0·1	<0·1	0·9	3·3	5·8	38·2	
S. Uplands			10·9	9·0	7·0	1·3	0·1	—	—	0·6	2·8	6·1	37·8	13 to 88
Harrogate	478	1912–49	6·5	5·1	4·6	1·7	0·2	—	0·1	0·4	1·5	3·6	23·6	
Pennine slopes, E.			7·3	5·8	4·3	0·4	<0·1	—	—	0·2	0·8	4·2	23·1	3 to 69
Stonyhurst	377	1921–49	5·3	4·9	1·3	1·6	0·3	—	—	0·2	1·0	3·3	20·9	
Pennine slopes, W.			3·6	3·2	2·0	<0·1	<0·1	—	—	0·1	0·2	2·0	11·2	1 to 56
Cambridge	41	1912–49	4·1	3·7	3·0	0·9	0·1	—	—	0·1	0·7	1·9	14·5	
Eastern lowland			3·5	3·2	1·4	0·2	—	—	—	—	0·2	1·4	9·9	0 to 51
Hampstead, London	450	1912–49	5·6	5·3	4·5	2·1	0·2	—	trace	0·1	1·1	3·4	22·3	
South-east, hill			4·4	4·0	2·3	0·4	—	—	—	—	0·4	2·1	13·6	1 to 61
Rhayader	757	1918–49	5·2	4·0	3·6	1·8	0·4	—	—	0·3	1·1	3·2	19·6	
C. Wales, upland			4·6	4·0	3·0	0·9	<0·1	—	—	—	0·5	2·3	15·4	2 to 60
Cullompton, Exeter	202	1912–49	1·9	1·6	1·4	0·6	<0·1	—	—	<0·1	0·3	0·9	6.8	
Inland, Devon		1914–49	1·5	1·9	0·6	<0·1			—	—	<0·1	0·4	4·5	0 to 25

Other annual averages include: London (Kew) 15 and 4; Lerwick (Shetland) 38 and 9; Falmouth 5 and 0·5; Durham 23 and 18; Norwich 19 and 14; Armagh 15 and 8. Figures computed by the author from M.O. published observations.

TABLE IX

AVERAGE ANNUAL NUMBER OF DAYS WITH SNOW OR SLEET OBSERVED, FOR PAST PERIODS AND PLACES

Place	Period	Days	(Present-day)	Place	Period	Days	(Present-day)
Westminster	1669–1700	20	(13)	London (S.W.)	1770–1820	14	(13)
Plymouth	1728–1752	7	(7)	London (S.E.)	1820–1860	15	(14)
Liverpool	1768–1793	16	(13)	Sunderland	1859–1912	29	(23)
Stroud	1771–1813	15	(13)	Stonyhurst	1886–1915	26	(21)

Present-day average by similar standards estimated in brackets

TABLE X

AVERAGE NUMBER OF "DAYS WITH GALE"

	J	F	M	A	M	J	J	A	S	O	N	D	*Year*
Lerwick (Shetland)	9	6	2	2	0.9	0·5	0·2	0·5	1	5	4	7	38
Stornoway (Hebrides)	10	6	4	3	1	1	0·5	0·7	3	7	5	7	48
Blacksod Pt. (W. Ireland)	6	3	2	2	0·7	0·9	0·5	0·7	1	4	4	6	31
Bidston (Liverpool)	2	1	0·6	0·4	0·1	0·3	0·1	0·2	0·5	1	0·8	2	9
Tynemouth	0·8	1	0·4	0·1	0·1	0·1	0	0	0·3	0·4	0·4	0·5	4
Dungeness (E. Channel)	2	2	1	1	0·1	0·3	0·5	0·7	1	2	2	3	16
Scilly	5	4	2	2	0·3	0·4	0·4	0·5	0·7	2	3	5	25
Renfrew (inland airfield)	2	1	0·7	0·3	0·1	0·1	0·1	0·1	0·3	0·9	0·5	1	7
Kew, London	none recorded												0

Data from M.O. "Weather in Home Waters," published H.M.S.O. 1940, which gives numerous other stations.
A gale is recorded when at any time during the day the hourly wind exceeds 38 m.p.h.
Great variations in frequency result from differences in exposure; the above are broadly representative of the seasonal incidence over periods of 10–14 years.

INDEX

Figures in heavy type are pages opposite which illustrations are to be found.

THE NEW NATURALIST

1. BUTTERFLIES—*E. B. Ford*
2. BRITISH GAME —*Brian Vesey-FitzGerald*
3. LONDON'S NATURAL HISTORY —*R. S. R. Fitter*
4. BRITAIN'S STRUCTURE AND SCENERY—*L. Dudley Stamp*
5. WILD FLOWERS—*John Gilmour & Max Walters*
6. NATURAL HISTORY IN THE HIGHLANDS AND ISLANDS —*F. Frazer Darling*
7. MUSHROOMS AND TOADSTOOLS —*John Ramsbottom*
8. INSECT NATURAL HISTORY —*A. D. Imms*
10. BRITISH PLANT LIFE —*W. B. Turrill*
11. MOUNTAINS AND MOORLANDS —*W. H. Pearsall*
12. THE SEA SHORE—*C. M. Yonge*
14. THE ART OF BOTANICAL ILLUSTRATION—*Wilfred Blunt*
15. LIFE IN LAKES AND RIVERS —*T. T. Macan & E. B. Worthington*
16. WILD FLOWERS OF CHALK AND LIMESTONE—*J. E. Lousley*
17. BIRDS AND MEN—*E. M. Nicholson*
18. A NATURAL HISTORY OF MAN IN BRITAIN—*H. J. Fleure*
19. WILD ORCHIDS OF BRITAIN —*V. S. Summerhayes*
20. THE BRITISH AMPHIBIANS AND REPTILES—*Malcolm Smith*
21. BRITISH MAMMALS —*L. Harrison Matthews*
22. CLIMATE AND THE BRITISH SCENE —*Gordon Manley*
23. AN ANGLER'S ENTOMOLOGY —*J. R. Harris*
24. FLOWERS OF THE COAST —*Ian Hepburn*
25. THE SEA COAST—*J. A. Steers*
26. THE WEALD—*S. W. Wooldridge & Frederick Goldring*
27. DARTMOOR—*L. A. Harvey & D. St. Leger-Gordon*
28. SEA-BIRDS—*James Fisher & R. M. Lockley*
29. THE WORLD OF THE HONEYBEE —*Colin G. Butler*
30. MOTHS—*E. B. Ford*
31. MAN AND THE LAND —*L. Dudley Stamp*
32. TREES, WOODS AND MAN —*H. L. Edlin*
33. MOUNTAIN FLOWERS —*John Raven & Max Walters*
34. THE OPEN SEA: The World of Plankton—*Sir Alister Hardy*
35. THE WORLD OF THE SOIL —*Sir E. John Russell*
36. INSECT MIGRATION —*C. B. Williams*
37. THE OPEN SEA: Fish and Fisheries —*Sir Alister Hardy*
38. THE WORLD OF SPIDERS —*W. S. Bristowe*
39. THE FOLKLORE OF BIRDS —*Edward A. Armstrong*
40. BUMBLEBEES—*John B. Free & Colin G. Butler*
41. DRAGONFLIES—*Philip S. Corbet, Cynthia Longfield & N. W. Moore*
42. FOSSILS—*H. H. Swinnerton*
43. WEEDS AND ALIENS —*Sir Edward Salisbury*
44. THE PEAK DISTRICT —*K. C. Edwards*
45. THE COMMON LANDS OF ENGLAND AND WALES—*W. G. Hoskins & L. Dudley Stamp*

SPECIAL VOLUMES

THE BADGER — *Ernest Neal*

THE REDSTART — *John Buxton*

THE WREN — *Edward A. Armstrong*

THE YELLOW WAGTAIL — *Stuart Smith*

THE GREENSHANK — *D. Nethersole-Thompson*

FLEAS, FLUKES AND CUCKOOS — *M. Rothschild and T. Clay*

THE HERRING GULL'S WORLD — *Niko Tinbergen*

THE HERON — *Frank A. Lowe*

SQUIRRELS — *Monica Shorten*

THE RABBIT — *H. V. Thompson & A. N. Worden*

THE HAWFINCH — *Guy Mountfort*

THE SALMON — *J. W. Jones*

LORDS AND LADIES — *C. T. Prime*

OYSTERS — *C. M. Yonge*

THE HOUSE SPARROW — *J. D. Summers-Smith*

BUTTERFLIES

E. B. FORD

"Dr. Ford combines the ardour of the naturalist with the rigorous discipline of the experienced research worker. These qualities are reflected in the book, so that while one outstanding feature is the beautiful colour photographs of live butterflies in the wild, another is the frequent reference to gaps in our knowledge, which naturalists might fill by careful observation. The book is thus not just another butterfly book, but is an introduction to the scientific study of butterflies. Evolution is taken as the key-note, as we might expect from Dr. Ford, who is himself a geneticist of distinction. This is certainly a book that every naturalist, budding or fully fledged, will want to have."

BIRMINGHAM POST